Theory of Approximation

With Applications

ACADEMIC PRESS RAPID MANUSCRIPT REPRODUCTION

Proceedings of a conference conducted by The University of Calgary and The University of Regina, at The University of Calgary, Alberta, Canada, August 11-13, 1975.

Theory of Approximation

With Applications

Edited by

Alan G. Law
Department of Mathematics
University of Regina
Regina, Canada

Badri N. Sahney
Department of Mathematics
University of Calgary
Calgary, Canada

ACADEMIC PRESS, INC.

New York San Francisco London 1976

A Subsidiary of Harcourt Brace Jovanovich, Publishers

ACADEMIC PRESS, INC.
111 Fifth Avenue, New York, New York 10003

United Kingdom Edition published by
ACADEMIC PRESS, INC. (LONDON) LTD.
24/28 Oval Road, London NW1

Library of Congress Cataloging in Publication Data

Conference on the Theory of Approximation, with Applications, University of Calgary, 1975.
Theory of approximation, with applications.

"Proceedings of a conference conducted by the University of Calgary and the University of Regina at the University of Calgary, Alberta, Canada, August 11-13, 1975."
Includes bibliographies and indexes.
1. Approximation theory–Congresses. I. Law, Alan G. II. Sahney, Badri N. III. Calgary, Alta. University. IV. University of Regina. V. Title
QA297.5.C68 1975 511'.4 76-1839
ISBN 0–12–438950–3

PRINTED IN THE UNITED STATES OF AMERICA

To the Memory of
Eckard Schmidt
1938-1975

Contents

List of Contributors and Participants *xi*
Preface *xvii*

APPROXIMATION IN ABSTRACT SPACES

Applications of Fixed-point Theorems to Approximation Theory 1
E.W. Cheney

Weighted Polynomial Approximation and K–Functionals 9
Géza Freud

Continuity Theorems for the Product Approximation Operator 24
M.S. Henry and D. Schmidt

Some Remarks on the Estimation of Quadratic Functionals 43
F.M. Larkin

On Alternation in the Restricted Range Problem 64
William H. Ling

Some Results on the Dual of an Approximation Problem 76
S.J. Poreda

Best Least Square Approximation by Minimum Property of Fourier Expansions 86
Chung-Lie Wang

INTERPOLATION, SPLINE, AND GENERAL APPROXIMATION

Splines in Orlicz Spaces 101
J. Baumeister

Geometric Convergence of Rational Functions to the Reciprocal of Exponential Sums on $[0, \infty]$ 111
Hans-Peter Blatt

On Local Linear Functionals Which Vanish at All B-Splines but One 120
Carl de Boor

Spline Approximation in Banach Function Spaces 146
Stephen Demko

Chebyshev Nodes for Interpolation on a Class of Ellipses 155
K.O. Geddes

RRAS Approximation of Differentiable Functions 171
Darell J. Johnson

On the Expansion of Exponential Type Integrals in Series of Chebyshev Polynomials 180
Yudell L. Luke

Neville-Aitken Algorithms for Interpolation by Functions of Cebysev-Systems in the Sense of Newton and in a Generalized Sense of Hermite 200
G. Mühlbach

Best Simultaneous Approximation in the L_1 and L_2 Norms 213
G.M. Phillips and B.N. Sahney

Comonotone Approximation and Piecewise Monotone Interpolation 220
Louis Raymon

Angular Overconvergence for Rational Functions Converging Geometrically on $(0, +\infty)$ 238
E.B. Saff and R.S. Varga

A Generalization of Monosplines and Perfect Splines 257
A. Sharma and J. Tzimbalario

Zero Properties of Splines
J. Tzimbalario 268

APPLICATIONS

A Review of the Remez Exchange Algorithm for Approximation with Linear Restrictions 278
Bruce L. Chalmers

An Algorithm for Constrained, Non-linear, Tchebycheff Approximations 298
Gary A. Gislason

Transfinite Interpolation and Applications to Engineering Problems 308
Charles A. Hall

Some Remarks on Best Simultaneous Approximation 332
A.S.B. Holland and B.N. Sahney

The Numerical Solution of the Hilbert Problem 338
Y. Ikebe, T. Y. Li, and F. Stenger

On Adaptive Piecewise Polynomial Approximation 359
John R. Rice

Extremals and Zeros in Markov Systems Are Monotone Functions of One Endpoint 387
Roy L. Streit

Author Index 402
Subject Index 404

List of Contributors and Participants

An asterisk denotes a contributor to this volume.

K. Andersen
Department of Mathematics
University of Alberta
Edmonton, Alberta T6G 2G1

T.H. Andres
Department of Computer Science
University of Manitoba
Winnipeg, Manitoba R3T 2N2

*J. Baumeister
Freie Universität Berlin
Fachbereich Mathematik I
D-1 Berlin 33
Arnimallee 2-6, Germany

J.A.R. Blais
Energy, Mines & Resources Canada
Surveys and Mapping Branch
615 Booth Street
Ottawa, Canada

*Hans-Peter Blatt
Universität Erlangen-Nürnberg
Department of Mathematics
852 Erlangen
Martenstrasse 1, Federal Republic of Germany

*Carl de Boor
Mathematics Research Center
University of Wisconsin
610 Walnut Street
Madison, Wisconsin 53706

*Bruce L. Chalmers
Department of Mathematics
University of California
Riverside, California 92507

*E.W. Cheney
Department of Mathematics
University of Texas at Austin
Austin, Texas 78712

D.L. Crawford
3603 Arkansas Drive
Anchorage, Alaska 99503

*Stephen Demko
School of Mathematics
Georgia Institute of Technology
Atlanta, Georgia 30332

Z. Ditzian
Department of Mathematics
The University of Alberta
Edmonton, Alberta T6G 2G1

M.F. Estabrooks
Department of Mathematics
Red Deer Coolege, P.O. Box 5005
Red Deer, Alberta T4N 5H5

D.L. Evans
5 Roseview Drive
Calgary, Alberta, T2K 1N8

J. L. Fields
Department of Mathematics
Evelyn Frank
P.O. Box 361
Evanston, Illinois 60204

W. Fraser
Department of Mathematics
University of Guelph
Guelph, Ontario

*Géza Freud
Department of Mathematics
The Ohio State University
231 West 18th Avenue
Columbus, Ohio 43210

*K.O. Geddes
Computer Science Department
University of Waterloo
Waterloo N2L 3G1, Ontario

*Gary A. Gislason
Department of Mathematics
University of Alaska
College, Alaska 99701

W.J. Gordon
Department of Mathematics
GM Research Laboratories
Warren, Michigan 48090

*Charles A. Hall
Department of Mathematics
University of Pittsburgh
Schenley Hall
Pittsburgh, Pennsylvania 15213

G.A. Hamoud
Department of Mathematics
University of Saskatchewan
Saskatoon, Saskatchewan

P.C. Hein
3436 Wascana Street
Regina, Saskatchewan S4S 2H3

W. Hengartner
Department of Mathematics
Laval University
Quebec, Quebec G1K 7P4

J. N. Henry
Department of Mathematics
Montana State University
Bozeman, Montana 59715

*M.S. Henry
Department of Mathematics
Montana State University
Bozeman, Montana 59715

*A.S.B. Holland
Department of Mathematics
The University of Calgary
2920 24th Avenue N.W.
Calgary, Alberta T2N 1N4

W.D. Hoskins
Computer Science Department
The University of Manitoba
Winnipeg, Manitoba R3T 2N2

L.R. Huff
Department of Mathematics
Pennsylvania State University
University Park, Pennsylvania 16802

Mourad E.H. Ismail
Mathematics Research Center
University of Wisconsin
Madison, Wisconsin

*Darell J. Johnson
Department of Mathematics
Massachusetts Institute of Technology
Cambridge, Massachusetts 02139

Z.V. Kovarik
Department of Mathematics
McMaster University
Hamilton, Ontario L8S 4K1

R.P. Kurshan
Bell Laboratories (Rm. 2C-362)
Murray Hill
New Jersey 07974

*F.M. Larkin
Department of Mathematics
Queen's University
Kingston, Ontario K7L 3N6

A.G. Law
Department of Mathematics
University of Regina
Regina, Saskatchewan S4S 0A2

S.J. Lwon
Department of Mathematics
Weber State College
3750 Harrison Boulevard
Ogden, Utah 84408

D. Leviatan
Department of Mathematics
Tel-Aviv University
Ramat-Aviv
Tel-Aviv, Israel

*William H. Ling
Department of Mathematics
Union College
Schenectady, New York 12308

*Yudell L. Luke
Department of Mathematics
University of Missouri
Kansas City, Missouri 64110

A. Magnus
708 Dartmouth Trail
Fort Collins, Colorado 80521

J.H. McCabe
Mathematical Institute
University of St. Andrews
St. Andrews, Fife
Scotland

I. McDonald
1204 Shannon Road
Regina, Saskatchewan S4S 5L1

W. Weston Meyer
General Motors Research
GM Technical Center
Warren, Michigan 48090

P. Milman
222 Major Street
Toronto, Ontario

*G. Mülbach
Department of Mathematics
Technische Universität Hannover
3000 Hannover
Welfengarten 1, Federal Republic of Germany

P.P. Narayanaswami
Department of Mathematics
Memorial University of Newfoundland
St. John's, Newfoundland A1C 5S7

C. Nasim
Department of Mathematics
The University of Calgary
2920 24th Avenue N.W.
Calgary, Alberta T2N 1N4

R. Panda
Department of Mathematics
University of Victoria
P.O. Box 1700
Victoria, B.C. V8W 2Y2

*G.M. Phillips
The Mathematical Institute
University of St. Andrews
St. Andrews, Fife, Scotland

*S.J. Poreda
Department of Mathematics
Clark University
Worcester, Massachusetts 01610

J.A. Pulsifer
Department of Mathematics
Acadia University
Wolfville, Nova Scotia

*Louis Raymon
Department of Mathematics
Temple University
Philadelphia, Pennsylvania 19122

*John R. Rice
Division of Mathematical Sciences
Purdue University
Lafayette, Indiana 47907

S.D. Riemenschneider
Department of Mathematics
The University of Alberta
Edmonton, Alberta T6G 2G1

*B.N. Sahney
Department of Mathematics
The University of Calgary
2920 24th Avenue N.W.
Calgary, Alberta T2N 1N4

K. Salkauskas
Department of Mathematics
The University of Calgary
2920 24th Avenue N.W.
Calgary, Alberta T2N 1N4

*D. Schmidt
217 W. Koch #310
Bozeman, Montana 59715

*A. Sharma
Department of Mathematics
University of Alberta
Edmonton, Alberta T6G 2G1

P.N. Shivakumar
Department of Mathematics
University of Manitoba
Winnipeg, Manitoba R3T 2N2

H.M. Srivastava
Department of Mathematics
University of Victoria
P.O. Box 1700
Victoria, B.C. V8W 2Y2

*F. Stenger
Department of Mathematics
University of Utah
Salt Lake City, Utah 84112

*Roy L. Streit
New London Laboratory
U.S. Navy Underwater Systems Center
New London, Connecticut 06320

*J. Tzimbalario
Department of Mathematics
The University of Alberta
Edmonton, Alberta T6G 2G1

*R.S. Varga
Department of Mathematics
Kent State University
Kent, Ohio 44242

A.K. Varma
Department of Mathematics
University of Florida
Gainesville, Florida 32601

*Chung-Lie Wang
Department of Mathematics
University of Regina
Regina, Saskatchewan S4S 0A2

K.L. Wiggins
Box 12945
Charleston, South Carolina 29412

H. Wozniakowski
Department of Computer Science
Carnegie-Mellon University
Schenley Park
Pittsburgh, Pennsylvania 15213

D. Wulbert
Department of Mathematics
University of California
P.O. Box 109
La Jolla, California 92037

Z. Ziegler
Department of Mathematics
Technion
Israel Institute of Technology
Haifa, Israel

Preface

This book contains invited and refereed papers that were presented at the Conference on the Theory of Approximation, with Applications, which was held at the University of Calgary, August 11–13, 1975. The conference was organized both by the University of Calgary and the University of Regina—that is, by two institutions, in neighboring provinces, which are some 500 miles apart. The possibility of joint sponsoring for such a conference had not been explored previously; the high quality of the mathematical research recorded in this book, and the fact that the 70 conference delegates represented a number of different countries (including Australia, Canada, Germany, Israel, Scotland, and the United States) attest to the success of the collaboration.

The contents of this volume include the five major conference addresses by E. W. Cheney, C. de Boor, G. Freud, J. R. Rice, and R. S. Varga, plus 22 other papers on approximation theory and practice. Collectively, they reflect the diversity of approximation theory: Research and survey articles about current developments in general function spaces are augmented by papers covering recent progress in such areas as spline theories and applications, Tchebycheff approximation, polynomial approximation and interpolation, convergence and algorithms for rational approximations, estimation of functionals, Fourier expansions, and engineering or numerical applications. The 27 articles have been grouped under three major headings for these proceedings: Approximation in Abstract Spaces, Interpolation, Spline, and General Approximation; and Applications.

Initial planning and conference program organization was greatly assisted by efforts of K. W. Chang, A. S. B. Holland, C. Nasim, G. M. Phillips, K. Salkauskas, J. Tzimbalario, D. R. Westbrook, and, until his untimely passing, friend and colleague, E. F. Schmidt. It was a priviledge to associate with such a capable team. We also wish to express our thanks to the many mathematicians who served as referees for submissions; they met severe time constraints, which became unyielding as the conference deadline approached. A strict external refereeing system was followed for this conference, and only papers that met all the criteria are included in these proceedings. There were a number of substantial submissions that were rejected for one reason or another at various stages of the assesment procedure.

For help during the conference, and for much editorial aid afterward, the editors are indebted to Professors Holland and Salkauskas—it is a pleasure to acknowledge their unfailing assistance.

Support for this conference was derived from several sources: The National Research Council of Canada, the presidents of the two universities, the Dean of Arts and Science at the University of Calgary, the Dean of Graduate Studies and Research at the University of Regina, the two departments, and the Royal Bank of Canada. We would like to extend official recognition for this assistance. Finally, we would like to express our appreciation to Mrs. J. Watkins for her many long days at final preparation and compilation of the manuscripts submitted by the various authors.

Theory of Approximation

With Applications

APPLICATIONS OF FIXED-POINT THEOREMS TO APPROXIMATION THEORY

E. W. Cheney*

1. Introduction. My own researches in approximation theory have recently seemed to utilize more and more frequently the fixed-point theorems of analysis and their closely related theorems from nonlinear functional analysis. I therefore thought that it might be interesting and useful to summarize for this conference a number of typical instances in which fixed-point theorems were of use in this subject.

Between the fixed-point theorems and their ultimate application in approximation theory there is often another level of abstract theorems, usually stating that under suitable conditions, a nonlinear mapping has a root. Some examples of theorems on this intermediate level will be given, drawn in some cases from unpublished work of T. A. Kilgore and myself.

2. Invariance Theorems. In the subject of best approximation one often wishes to know whether some useful property of the function being approximated is inherited by the approximating function. For example, a function whose average value on an interval is 0 will have a best L^2-approximation with the same property, provided only that the approximating subspace contains constants.

Meinardus [1963] seems to have been the first to observe a general principle that could be applied here and to have employed a fixed-point theorem to establish it. See also his works [1966] and [1967]. Brosowski [1969] extended and simplified the theorem of Meinardus. Here is Brosowski's Theorem:

THEOREM. Let A be a contractive linear operator on a normed space X. Let Y be an A-invariant subset of X, and x an A-invariant point. If the set of best Y-approximants to x is nonvoid, convex, and compact, then it contains an A-invariant point.

Proof. The set K of best Y-approximants to x is A-invariant, because if $y \in K$ then $Ay \in K$, as shown by the inequality $\|x-Ay\| = \|A(x-y)\| \leq \|x-y\|$. Thus by the Schauder Fixed-point Theorem, K contains a fixed point of A. $\square$

Analogues of the preceding theorem, for generalized rational approximation in a space C(T), have been given by Meinardus [1966] and Brosowski [1969]. The one due to Brosowski follows.

*Research sponsored in part by NSF Grant MPS75-07939.

<u>THEOREM</u>. Let U and V be finite-dimensional subspaces of C(T) and let Y be the set of functions u/v where $u \in U$, $v \in V$, and $v \geq \epsilon$, ϵ being fixed and positive. Let A be a contractive linear operator on C(T). If Y is A-invariant and if x is A-invariant, then Y contains an A-invariant best approximation to x.

It seems likely that a fruitful area for further research is that of invariance theorems for other nonlinear approximating families.

3. <u>Suns and Tchebycheff Sets</u>. An ingenious use of a fixed-point theorem occurs in a theorem of Vlasov [1961] concerning Tchebycheff sets which are suns. Recall that a set M in a normed space is a <u>Tchebycheff</u> set if each x has a unique closest point x' in M. The Tchebycheff set M is termed a <u>sun</u> if for every x, x' is the closest point in M for every point on the ray x'x.

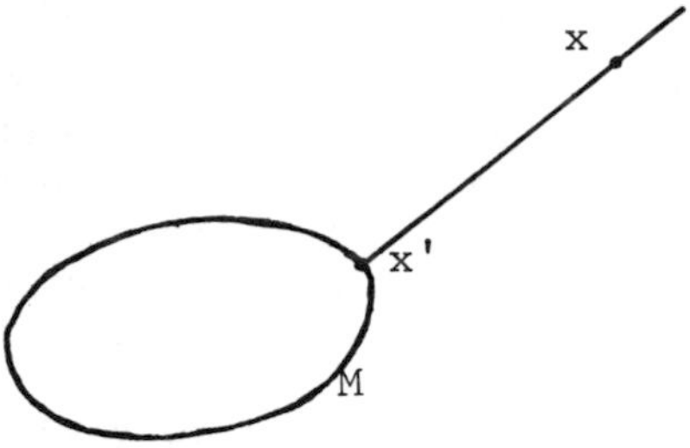

M is a sun

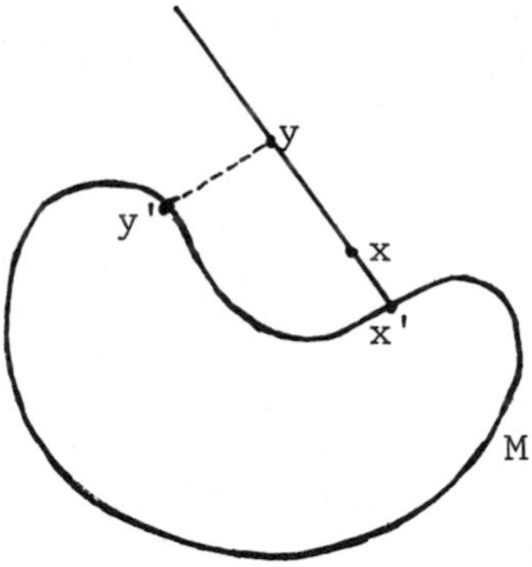

M is not a sun

In attempting to illustrate these notions, I have drawn a set which is not a sun, and the same set appears to lack the Tchebycheff property as well. This is more or less inevitable because of Vlasov's result:

<u>THEOREM</u>. Every boundedly compact Tchebycheff set is a sun.

<u>Proof</u>. Recall that a set is boundedly compact if it intersects each closed ball in a compact set. Now take a boundedly compact Tchebycheff set M which is not a sun. A short argument establishes the existence of a point z such that no point beyond z on the ray z'z has z' for its closest point. Define now the "Vlasov map" (so named by Brosowski) $f(x) = \frac{4}{3}z - \frac{1}{3}x'$. Let K denote the convex closure of $\{f(x):$

$\|x-z\| \leq \|z-z'\|\}$. One shows easily that K is compact, using the fact that M is boundedly compact. Furthermore, $f(K) \subset K$, and f is continuous. By the Schauder Fixed-point theorem, f has a fixed point y in K. Thus $y = (4/3)z - y'$, or $z = (3/4)y + (1/4)y'$. Since z is on the line segment joining y to y' we conclude that $z' = y'$. But this contradicts our choice of z. $\square$

A convenient source for this theorem and related matters is the treatise of Singer [1970]. Other investigations about suns and Tchebycheff sets occur in Blatter-Morris-Wulbert [1968] and Brosowski [1968], [1969].

4. Extremal Interpolation. In 1956, Chandler Davis posed the following problem: to give necessary and sufficient conditions on a set of real numbers $\lambda_1,\ldots,\lambda_{n-1}$ in order that a polynomial p of degree at most n exist which assumes the values $\lambda_1,\ldots,\lambda_{n-1}$ at n-1 successive points of the interval (-1,1) and whose derivative vanishes precisely at the same points. Davis' theorem [1956] is this:

THEOREM. If numbers $\lambda_0,\ldots,\lambda_n$ are prescribed and satisfy $(-1)^i(\lambda_i-\lambda_{i-1}) > 0$ for $i = 1,\ldots,n$, then there exist nodes $t_0,\ldots,t_n$ and a polynomial p of degree n such that

$$p(t_i) = \lambda_i \qquad (0 \leq i \leq n)$$

$$p'(t_i) = 0 \qquad (0 < i < n)$$

$$-1 = t_0 < t_1 < \ldots < t_{n-1} < t_n = 1$$

Proof. Denote by K the simplex in $\mathbb{R}^{n-1}$ consisting of all points $T = (t_1,\ldots,t_{n-1})$ which satisfy the inequalities

$$-1 = t_0 < t_1 < t_2 < \ldots < t_{n-1} < t_n = 1.$$

We call such a point a node-vector. For any node vector there is a polynomial $p \in \prod_n$ such that $p(t_i) = \lambda_i$ for $0 \leq i \leq n$. By the hypotheses on λ_i, p' must have exactly one root in each of the intervals (t_0,t_2), (t_1,t_3), ..., (t_{n-2},t_n). We label these roots $s_1,\ldots,s_{n-1}$ and contemplate the mapping $(t_1,\ldots,t_{n-1}) \to (s_1,\ldots,s_{n-1})$. This mapping carries K continuously into itself, and we would like to conclude that it has a fixed point. This cannot be inferred directly from the Brouwer Theorem because K is not closed. However, the map does have a fixed point in K, and it can be computed by simple iteration. For the details of the proof see Kammerer [1961]. $\square$

Similar theorems concerning polynomials which oscillate between two given functions have been given by Karlin [1963], and we refer the reader also to Karlin and Studden [1966], especially Section 6 of Chapter 6. The theorem of Davis was proved by the use of differential equations by Fitzgerald and Schumaker [1969]. See also Videnskii [1965], Davis [1967].

5. Optimization of Interpolation Operators. A problem in interpolation theory quite different from those in the preceding section is also amenable to the use of a fixed-point theorem. I refer to the question of whether interpolation nodes can be so chosen that the resulting Lebesgue function exhibits n+1 equal maxima. If the interpolation process is defined by Lagrange's formula,

$$Lf = \sum_{i=1}^{n} f(t_i)\ell_i \qquad \ell_i(t) = \prod_{\substack{j=1\\ j\neq i}}^{n} \frac{t-t_j}{t_i-t_j},$$

then the "Lebesgue Function" is $\Lambda = \sum_{i=1}^{n} |\ell_i|$. Kilgore and I [1974] proved the existence of nodes $t_1,\ldots,t_n$ in $[-1,1]$ such that Λ reaches its maximum value at n+1 points. From our proof it is possible to exhume the following abstract theorem governing systems of nonlinear equations (Kilgore and Cheney [1975]).

THEOREM. Let K be the open simplex in $\mathbb{R}^n$ consisting of points $T = (t_1,\ldots,t_n)$ satisfying $-1 = t_0 < t_1 < \cdots < t_n < t_{n+1} = 1$. Assume that the i-th component, F_i, of F is a strictly increasing function of t_i when all other components of T are held fixed. Assume also the existence of a nondecreasing function ϕ such that $\lim_{\delta\to 0} \phi(\delta) = 0$, $F_i(T) < 0$ when $\phi(t_i - t_{i-1}) < t_{i+1} - t_i$, and $F_i(T) > 0$ when $\phi(t_{i+1}-t_i) < t_i - t_{i-1}$. Then $0 \in F(K)$.

Theorems of this type, which infer the existence of solutions of equations from easily verified hypotheses, seem to be quite useful in this subject and elsewhere. Here is another example which we recently proved.

THEOREM. Let K be a polyhedron in $\mathbb{R}^n$ defined by a system of linear inequalities $(A^i,x) \leq b_i$ $(1 \leq i \leq m)$. Let $F\colon K \to \mathbb{R}^n$ be a continuous map such that $(A^i,F(x)) \leq 0$ whenever $(A^i,x) = b_i$. Then $0 \in F(K)$.

An interesting special case of this, first shown to me by Kilgore, is the following. Suppose that F is a continuous map of the cube $[-1,1]^n$ into $\mathbb{R}^n$ such that $F_i(x)F_i(y) \leq 0$ whenever x and y are points of the cube satisfying $x_iy_i = -1$. Then F has a root in the cube.

These theorems, and other related ones, will be published in full elsewhere.

6. The Hobby-Rice Theorem on L^1-Approximation. This theorem is not an example of the application of a fixed-point theorem but rather its cousin, the Antipodal-mapping Theorem. According to that result, the n-ball cannot be continuously mapped onto its surface in such a way that pairs of antipodal points map into pairs of antipodal points. The theorem proved by Hobby and Rice [1965] is as follows.

THEOREM. Let μ be a finite Borel measure on [0,1] such that all finite sets are assigned 0 measure. Let W be an n-dimensional subspace of $L^1(\mu)$. Then there exist points $0 = t_0 \leq t_1 \leq \cdots \leq t_{n+1} = 1$ such that

$$\sum_{i=0}^{n} (-1)^i \int_{t_i}^{t_{i+1}} w(t)\,d\mu(t) = 0 \text{ for all } w \in W.$$

This powerful theorem is of basic importance in the theory of best approximation in the L^1-norm. An outline of the proof is as follows. Let K denote the simplex of points $(t_1,\ldots,t_n)$ such that $0 = t_0 \leq t_1 \leq \cdots \leq t_n \leq t_{n+1} = 1$. Define $F\colon K \to \mathbb{R}^n$ by putting

$$F_i(t_1,\ldots,t_n) = \sum_{j=0}^{n} (-1)^j \int_{t_j}^{t_{j+1}} w_i\,d\mu$$

where $\{w_1,\ldots,w_n\}$ is any basis for W. The proof will be complete if we can show that F has a root in K. Now the mapping F has two special properties:

(A) $F(t_1,\ldots,t_{n-1},1) + F(0,t_1,\ldots,t_{n-1}) = 0$

(B) For each i, the expression $F(t_1,\ldots,t_{i-1},s,s,t_{i+2},\ldots,t_n)$ is independent of the variable s.

The authors complete their proof by establishing, in effect, the following abstract theorem:

THEOREM. Any continuous map F of K into $\mathbb{R}^n$ which has properties (A) and (B) above has a root.

Since the authors' proof of the abstract theorem is rather complicated, it would be worthwhile to find a simpler proof of it, or alternatively a simpler proof of the first theorem on L^1-approximation.

7. Proximity Maps. If K is a closed convex set in Hilbert space, the proximity map for K is the map P such that Px is the point of K closest to x. If K_1, K_2 are two such sets and P_1, P_2 their respective proximity maps then every fixed point of $Q \equiv P_1P_2$ is a point of K_1 closest to K_2, as Goldstein and I proved [1959]. We proved further the following result.

THEOREM. Convergence of the iterates $Q^n x$ to a fixed point of Q occurs if one of K_1,K_2 is compact. It also occurs if both sets are finite-dimensional polytopes.

This can be interpreted as a result in approximation theory inasmuch as best approximations are nearest points in appropriate function spaces. The theorem also gives a means of computing best simultaneous approximations to sets of functions.

8. Conclusion. Further examples of the use of fixed-point theorems can be found in the list of references, for which no claim of completeness is made. Also, as was pointed out to me by Dr. Ziegler, some algorithms such as the 2nd Algorithm of Remez can be interpreted as iteration processes for fixed-points.

References

J. Blatter, P. Morris, D. E. Wulbert, "Continuity of the set-valued metric projection," Math. Ann. 178 (1968), 12-24.

C. de Boor, "A remark concerning perfect splines," Bull. Amer. Math. Soc. 80 (1974), 724-727.

B. Brosowski, "Fixed-point theorems in approximation theory," in "Theory and Applications of Monotone Operators," NATO Institute, Venice, 1968.

______, "Fixpunktsätze in der Approximationstheorie," Mathematica (Cluj) 11 (1969), 195-220.

______, "Über eine Fixpunkteigenschaft der Metrischen Projektion," Computing 5 (1970), 295-302.

______, K. H. Hoffmann, E. Schäfer, H. Weber, "Stetigkeitssätze für metrische Projektionen. III. Eine Fixpunkteigenschaft der metrischen Projektion," Max Planck Institute, Munich, Report Astro 12, March 1969.

E. W. Cheney and A. A. Goldstein, "Proximity maps for convex sets, Proc. Amer. Math. Soc. 10 (1959), 448-450.

C. Davis, "Problem 4714," Amer. Math. Monthly 63 (1956), 729. Solution, ibid 64 (1957), 679-680.

______, "Mapping properties of some Čebyšev systems," Dokl. Akad. Nauk SSSR 175 (1967) = Soviet Math. Dokl. 8 (1967), 840-843. MR 1968.

C. H. Fitzgerald and L. L. Schumaker, "A differential equation approach to interpolation at extremal points," J. d'Analyse Math. 22 (1969), 117-134.

M. S. Henry and K. Wiggins, "Applications of approximation theory to the initial-value problem," to appear, J. Approximation Theory.

C. R. Hobby and J. R. Rice, "A moment problem in L_1 approximation," Proc. Amer. Math. Soc. (1965), 665-670.

R. S. Johnson, "On monosplines of least deviation," Trans. Amer. Math. Soc. 96 (1960), 458-477.

W. J. Kammerer, "Polynomial approximations to finitely oscillating functions," Math. Comp. 15 (1961), 115-119. MR 23 #A1187.

S. Karlin, "Representation theorems for positive functions," J. Math. Mech 12 (1963), 599-618.

______, "Some variational problems on certain Sobolev spaces and perfect splines," Bull. Amer. Math. Soc. 79 (1963), 124-128.

______ and L. S. Shapley, "Geometry of Moment Spaces," Mem. Amer. Math. Soc. No. 12 (1953).

S. Karlin and W. J. Studden, "Tchebycheff Systems: with Applications in Analysis and Statistics," Interscience Publishers, New York, 1966.

T. A. Kilgore and E. W. Cheney, "A theorem on interpolation in Haar subspaces," to appear, Aequationes Mathematicae. Preprint, Center for Numerical Analysis, Report No. 80, University of Texas at Austin, January 1974.

____________, "An existence theorem for solutions of systems of nonlinear equations," Center for Numerical Analysis, Report No. 97, University of Texas at Austin, January 1975.

G. Meinardus, "Invarianz bei linearen Approximation," Arch. Rational Mech. Anal. 14 (1963), 301-303.

______, "Invarianz bei rationalen Approximation," Computing 2 (1966), 115-118.

______, "Approximation of Functions: Theory and Numerical Methods," Springer-Verlag, New York, 1967.

I. Singer, "Best Approximation in Normed Linear Spaces by Elements of Linear Subspaces," Springer-Verlag, Berlin, 1970.

V. S. Videnskii, "A class of interpolation polynomials with nonfixed nodes," Soviet Math. Doklady 6 (1965), 637-640. = Doklady Akad. Nauk SSSR 162 (1965), 251.

L. P. Vlasov, "Čebyšev Sets in Banach Spaces," Soviet Math. Doklady 2 (1961), 1373-1374. = Dokl. Akad. Nauk 141 (1961), 19-20.

Weighted polynomial approximation and K-functionals.

Géza Freud

1. Introduction.

We refer to the following result of J. Peetre [8]: Let A_0 be the space of bounded uniformly continuous functions on $(-\infty,\infty)$ with the supremum norm $\|\cdot\|$, and let A_1 the subspace of those elements of A_0 which have a derivative belonging to A_0. We define the K-functional as

$$K(f:\delta) = \inf_{f=f_0+f_1} (\|f_0\|+\delta\|f_1'\|) \tag{1.1}$$

where the inf is taken for all decompositions of f into a sum of an $f_0 \in A_0$ and an $f_1 \in A_1$. Peetre proved that

$$K(f;\delta) = \frac{1}{2}\,\omega^*(f;\delta) \tag{1.2}$$

where ω^* is the least concave majorant of the modulus of continuity $\omega(f;\delta)$ of f.
Note that

$$\omega(f;\delta) \leqq \omega^*(f;\delta) \leqq 2\omega(f;\delta) \tag{1.3}$$

(see G. G. Lorentz [7] p. 45) hence

$$K(f;\delta) \sim \omega(f;\delta) \tag{1.4}$$

This relation seems to be the deeper reason why the modulus of continuity of a function is the quantity entering into Jackson's and Bernstein's theorems for trigonometric approximation. In recent investigations we extended the Jackson and Bernstein theorems to weighted polynomial approximation on the real line; in this case we needed a modified expression of the continuity modulus. In our present lecture we establish the analogue of (1.4) for our modified modulus of continuity. We are also giving a brief account of the related Jackson and Bernstein type theorems.

By $c_k(k=1,2,...)$ we denote positive numbers depending only on Q(see below).

2. The weighted polynomial approximation problem

Let $Q(x)$ be an even continuously differentiable convex function which is twice continuously differentiable in $(0,\infty)$ and which satisfies

$$\lim_{x\to\infty} Q'(x) = \infty , \tag{2.1}$$

and let

$$w_Q(x) = \exp\{-Q(x)\} \tag{2.2}$$

We characterized the polynomial approximation with respect to the weight w_Q on the whole real line by the following generalized modulus of continuity:

$$\omega(\mathcal{L}p,w_Q;\phi,\delta) \overset{\text{def}}{=} \sup_{|h|\leq\delta} \|T_h(w_Q\phi)-w_Q\phi\|_p + \delta\|Q_\delta' w_Q\phi\|_p . \tag{2.3}$$

Here $\|\cdot\|$ is the usual norm of the space $\mathcal{L}_p=\mathcal{L}_p(-\infty,\infty)$ $(1\leq p\leq\infty)$, T_h is the translation operator defined by

$$T_h f(x) = f(x+h) \tag{2.4}$$

and

$$Q_\delta'(x) = \min\{|Q'(x)|,\delta^{-1}\} \tag{2.5}$$

Our expression (2.3) is well defined and tends to zero for $\delta\to+0$ whenever $1 \leq p < \infty$ and $w_Q\phi\in\mathcal{L}_p$ resp. if ϕ is continuous on the whole real line $w_Q\phi$ is bounded and it tends to zero for $|x|\to\infty$. Note that these conditions describe (under the conditions imposed on w_Q) precisely the elements of the weighted $\mathcal{L}_p$-closure of the polynomials.

Let us denote by $\mathcal{P}_n$ the set of polynomials the degree of which is not greater than n and let

$$\epsilon_n^{(p)}(w_Q;f) \overset{\text{def}}{=} \inf_{P \in \mathcal{P}_n} \|(f-P)w_Q\|_p \tag{2.6}$$

Let q_n be the positive solution of the equation

$$q_n Q'(q_n) = n \tag{2.7}$$

As a consequence of $Q'(x) \to \infty$ there exists for every sufficiently great n a unique $q_n > 0$ satisfying (2.7).

We proved under suitable conditions satisfied by Q the two following results:

Jackson type theorem. If $w_Q \phi^{(r)} \in \mathcal{L}_p$ then

$$\epsilon_{n+r}^{(p)}(w_Q;\phi) \leqq c_1 e^{c_2 r} \left(\frac{q_n}{n}\right)^r \omega(\mathcal{L}_p, w_Q; \phi^{(r)}, \frac{q_n}{n}) \tag{2.8}$$

Here as well as in the next statement we apply the usual convention $\phi^{(0)}=\phi$.

Bernstein type theorem. Let r be a natural number and $0 < \rho < 1$ then

$$\epsilon_n^{(p)}(\mathcal{L}_p, w_Q; \phi, \frac{q_n}{n}) = O\{(\frac{q_n}{n})^{r+\rho}\} \tag{2.9}$$

implies that ϕ has an r-th order derivative satisfying $\phi^r w_Q \in \mathcal{L}_p$ and we have

$$\omega(\mathcal{L}_p, w_Q; \phi^{(r)}, \delta) = O(\delta^\rho) \tag{2.10}$$

We proved the Jackson type theorem in [4] for Q which satisfy the following conditions (i),(ii),(iii) and (iv):

(i) Q is even, convex, twice continuously differentiable in $(0,\infty)$ and $Q'(\infty)=\infty$.

(ii) $0 < Q''(t) \leqq (1+c_3)Q''(x) \quad (c_4 \leqq t < x)$ (2.11)

(iii) $t\,\dfrac{Q''(t)}{Q'(t)} \leq c_5 \qquad (t \geqq c_6)$ (2.12)

(iv) $Q''(2t) \geqq (1 + c_7)Q''(t) \qquad (t \geqq c_8)$ (2.13)

We proved in [1] that the Jackson type theorem is valid also for the weights $(1+x^2)^{\beta/2}\exp\{-x^2/2\}$ which do not satisfy (2.13).

For special classes of weights the Bernstein type theorem was proved in our papers [2] and [3]. We know that the extension of both Jackson's and Bernstein's theorem is valid in every $\mathcal{L}_p$ $(1 < p \leq \infty)$ if the weight satisfies (i), (ii),(iii) and

$$1 < \varliminf_{x\to\infty} \frac{Q'(2x)}{Q'(x)} < \varlimsup_{x=\infty} \frac{Q'(2x)}{Q'(x)} < \infty \tag{2.14}$$

(Lectures of the author in 1973/74 at the University of Szeged, his seminar in winter 1975 at the Université de Montréal and his course in spring 1975 at The Ohio State University, Columbus). Note that the second half of (2.14) is implied by (2.12).

Moreover, for the case p=1 the Jackson theorem holds even if we assume only (i) and (2.14); we are going to prove this in part 6 of this paper.

3. The K-functional of the weighted polynomial approximation problem.

We define the K-functionals with the weights w_Q as

$$K(\mathcal{L}_p, w_Q; \phi, \delta) = \inf_{\phi = f_1 + f_2} \{\|w_Q f_1\|_p + \delta \|w_Q f_2'\|_p\} \tag{3.1}$$

where the infimum is taken over all representations of ϕ as a sum of functions f_1, f_2 satisfying $w_Q f \in \mathcal{L}_p$ and $w_Q f_2' \in \mathcal{L}_p$.

Before establishing the link between the K-functional (3.1) and the continuity modulus (2.3) we make an observation of principal interest. Replacing ϕ by $\phi - k$, where k is constant, the value of $\epsilon_n^{(p)}(w_Q;\phi)$ remains unchanged, i.e.

$$\epsilon_n^{(p)}(w_Q;\phi-k) = \epsilon_n^{(p)}(w_Q;\phi) \tag{3.2}$$

but our continuity modulus is altered. To avoid this complication we introduce

$$\Omega(\mathcal{L}_p, w_Q; \phi, \delta) \overset{\text{def}}{=} \inf_{-\infty<k<\infty} \omega(\mathcal{L}_p, w_Q; \phi - k; \delta) \tag{3.3}$$

Clearly our Jackson-type and Bernstein-type theorems stay valid after replacing ω by Ω in (2.8) resp. (2.10). Our main result is now as follows:

Theorem 3.1. Let $Q(x)$ satisfy conditions (i) and let

$$\overline{\lim_{t\to\infty}}\, Q''(t)/[Q'(t)]^2 = \theta < 1 \tag{3.4}$$

then we have for every $1 \leq p \leq \infty$ and every ϕ with $w_Q\phi \in \mathcal{L}_p$

$$c_9 \leqq K(\mathcal{L}_p, w_Q; \phi, \delta)/\Omega(\mathcal{L}_p, w_Q; \phi, \delta) \leqq c_{10} (0 < \delta \leqq 1) \tag{3.5}$$

We obtain Theorem 3.1 by combining the lower estimate (Theorem 4.1) to be proven in part 4 and the upper estimate (Theorem 5.1 and Lemma 5.2) to be proven in part 5 of this paper. Observe that (3.4) is implied by (i) and (iii).

Theorem 3.2.

Let us assume that for the weight w_Q and for some $1 \leqq p \leq \infty$ the second half of (3.5) is true and that the Jackson inequality

$$\epsilon_n^{(p)}(w_Q; \phi) \leqq c_{11} \frac{q_n}{n} \|w_Q \phi'\|_p \tag{3.6}$$

is valid for every differentiable ϕ, then

$$\epsilon_{n+r}^{(p)}(w_Q; \phi) \leq c_{12} e^{c_{13} r} \left(\frac{q_n}{n}\right)^r \Omega(\mathcal{L}_p, w_Q; \phi^{(r)}, \frac{q_n}{n}) \tag{3.7}$$

Note. Observe that ω is a majorant of Ω thus (3.7) implies (2.8).

Proof of Theorem 3.2. (In three steps.)

a., By (3.1) and (3.5) there exists a differentiable φ such that

$$\|w_Q(\phi - \varphi)\|_p + \frac{q_n}{n} \|w_Q \varphi'\|_p \leqq 2K(\mathcal{L}_p, w_Q; \phi, \frac{q_n}{n}) \leqq$$

$$\leqq 2\, c_{10}\, \Omega\, (\mathcal{L}_p, w_Q; \phi, \frac{q_n}{n})$$

In virtue of (3.6) there exists $\tau \in \mathcal{P}_n$ for which

$$\|w_Q(\varphi-\tau)\|_p \leqq 2\ c_{11}\ \frac{q_n}{n}\ \|w_Q\varphi'\|_p$$

hence

$$\|w_Q(\phi-\tau)\|_p \leqq \|w_Q(\phi-\varphi)\|_p + \|w_Q(\varphi-\tau)\|_p \leqq$$

$$\leqq \|w_Q(\phi-\varphi)\|_p + 2\ c_{11}\ \frac{q_n}{n}\ \|w_Q\varphi'\|_p \leqq$$

$$\leqq (1 + 2c_{11})[\|w_Q(\phi-\varphi)\|_p + \frac{q_n}{n}\ \|w_Q\varphi'\|_p] \leqq$$

$$\leqq 2\ c_{10}(1 + 2\ c_{11})\ \Omega(\mathcal{L}_p, w_Q; \phi, \frac{q_n}{n})$$

i.e.

$$\epsilon_n^{(p)}(w_Q;\phi) \leqq c_{12}\ \Omega\ (\mathcal{L}_p, w_Q; \phi, \frac{q_n}{n}) \tag{3.8}$$

b., Let $\rho \in \mathcal{P}_n$ be such that

$$\|w_Q(\phi'-\rho)\|_p \leqq 2\ \epsilon_n^{(p)}(w_Q;\phi')\ .$$

Let $R \in \mathcal{P}_{n+1}$ be an integral of ρ. In consequence of (3.6)

$$\epsilon_{n+1}^{(p)}(w_Q;\phi) = \epsilon_{n+1}^{(p)}(w_Q;\phi-R) \leqq c_{11}\ \frac{q_n}{n}\ \|w_Q(\phi'-\rho)\| \leqq$$

$$\leqq 2\ c_{11}\ \frac{q_n}{n}\ \epsilon_n^{(p)}(w_Q;\phi')$$

i.e.

$$\epsilon_{n+1}^{(p)}(w_Q;\phi) \leqq e^{c_{13}}\ \frac{q_n}{n}\ \epsilon_n^{(p)}(w_Q;\phi'). \tag{3.9}$$

c., We apply r-times (3.9) and subsequently we apply (3.8):

$$\epsilon_{n+r}^{(p)}(w_Q;\phi) \leqq e^{c_{13}r}\ \frac{q_{n+r-1}}{n+r-1}\cdots\frac{q_n}{n}\ \epsilon_n^{(p)}(w_Q;\phi^{(r)}) \leqq$$

$$\leqq c_{12}e^{c_{13}r}\ (\frac{q_n}{n})^r\ \Omega(\mathcal{L}_p, w_Q; \phi^{(r)}, \frac{q_n}{n})\ . \qquad \text{Q.E.D.}$$

4., The lower estimate of the K-functional.

Lemma 4.1. The conditions (i) and (3.4) imply that

$$Q'(x)e^{-Q(x)} \int_0^x e^{Q(t)}dt < c_{18} \tag{4.1}$$

Proof. Let

$$Q''(t)/[Q(t)]^2 < \theta_1 < 1 \text{ for } t > a\ .$$

Hence

$$\int_a^x e^{Q(t)}dt = \int_a^x [Q'(t)]^{-1} e^{Q(t)} Q'(t)dt =$$

$$= [Q'(t)]^{-1} e^{Q(t)} \Big|_a^x + \int_a^x \frac{Q''(t)}{[Q'(t)]^2} e^{Q(t)}dt <$$

$$< [Q'(x)]^{-1} e^{Q(x)} + \theta_1 \int_a^x e^{Q(t)}dt$$

thus

$$Q'(x)e^{-Q(x)} \int_a^x e^{Q(t)}dt < (1 - \theta_1)^{-1}. \tag{4.2}$$

We see from (3.4) that the differential quotient

$$\frac{d}{dx} [Q'(x)e^{-Q(x)}] = \{Q''(x) - [Q'(x)]^2\}e^{-Q(x)}$$

is negative for large x. Consequently $Q'(x)e^{-Q(x)}$ is positive and decreasing for large x hence

$$Q'(x)e^{-Q(x)} \leqq c_{19} \tag{4.3}$$

By (4.2) and (4.3)

$$Q'(x)e^{-Q(x)}\int_0^x e^{Q(t)}dt < c_{19}\int_0^a e^{Q(t)}dt+(1-\theta_1)^{-1}$$

Q.E.D.

Lemma 4.2. Let Q satisfies (i) and (3.4) and let f be locally absulutely continuous, f(0) = 0 then we have for every $1 \leqq p \leqq \infty$

$$\|Q'w_Q f\|_p \leqq c_{20} \|w_Q f'\|_p. \tag{4.4}$$

For p = 1 (4.4) is valid even without the assumption (3.4).

Proof. Let $\phi = w_Q f$ and $\psi = w_Q f'$ thus

$$\phi(x) = e^{-Q(x)} \int_0^x e^{Q(t)}\psi(t)dt \tag{4.5}$$

We obtain

$$\int_0^\infty Q'(x)e^{-Q(x)}\int_0^x e^{Q(t)}[|\psi(t)|+|\psi(-t)|]dt =$$

$$= \int_0^\infty e^{Q(t)}[|\psi(t)|+|\psi(-t)|]\int_t^\infty Q'(x)e^{-Q(x)}dx$$

$$= \int_0^\infty [|\psi(t)|+|\psi(-t)|]dt = \|w_Q f'\|_1$$

This proves (4.4) for $p = 1$. (Without having assumed (3.4)). Now let $p = \infty$. Then we have by (4.5) and Lemma 1

$$\|Q'w_Q f\|_\infty = \|Q'\phi\|_\infty \leq \|\psi\|_\infty \|Q'(x)e^{-Q(x)}\int_0^x e^{Q(t)}dt\|_\infty \leq$$

$$\leq c_{18} \|\psi\|_\infty = c_{18} \|w_Q f'\|_\infty$$

so that (4.4) is valid also for $p = \infty$.

Since the transformation $w_Q f' \to Q'w_Q f$ is linear it follows from the Riesz-Thorin interpolation theorem that (4.4) holds for every $1 \leq p \leq \infty$, Q.E.D.

<u>Theorem 4.1</u>. We have for every $1 \leq p \leq \infty$ and ϕ satisfying $w_Q\phi \in \mathcal{L}_p$

$$K(\mathcal{L}_p, w_Q;\phi,\delta) \geq c_9\Omega(\mathcal{L}_p, w_Q;\phi,\delta). \tag{4.6}$$

<u>Proof</u>. Let $\phi = f_1+f_2$, $w_Q f_1 \in \mathcal{L}_p$ and $w_Q f_2' \in \mathcal{L}_p$ and let us put $\phi^* = \phi - f_2(0)$, $f_2^* = f_2 - f_2(0)$, $\phi^* = f_1 + f_2^*$. By (2.3) we have

$$\omega(\mathcal{L}_p, w_Q;\phi^*,\delta) \leq \sup_{|h|\leq\delta} \|T_h(w_Q f_1) - w_Q f_1\|_p + \\ + \sup_{|h|\leq\delta} \|T_h(w_Q f_2^*) - w_Q f_2\| + \delta\|Q_\delta' w_Q f_1\|_p + \delta\|Q_\delta' w_Q f_2^*\|_p . \tag{4.7}$$

Clearly

$$\|T_h(w_Q f) - w_Q f\|_p \leq 2\|w_Q f_1\|_p . \tag{4.8}$$

By (2.5) we have $\|\delta Q_\delta'\|_\infty = 1$ and consequently

$$\delta\|Q_\delta' w_Q f_1\|_p \leq \|w_Q f_1\|_p . \tag{4.9}$$

By Minkovski's inequality and a subsequent application of Lemma 4.2

$$\|T_h(w_Q f_2^*) - w_Q f_2^*\| \leq |h| \|(w_Q f_2^*)'\|_p \leq$$

$$\leq |h| [\|w_Q f_2^{*\prime}\|_p + \|Q' w_Q f_2^*\|] \leq (1+c_{20}) |h| \|w_Q f_2^{*\prime}\|_p =$$

$$= (1+c_{20}) |h| \|w_Q f_2'\|_p .$$

By (2.5) $0 \leq Q_\delta'(x) \leq |Q'(x)|$ so that in view of Lemma 4.2

$$\|\delta Q_\delta' w_Q f_2^*\|_p \leq \delta \|Q' w_Q f_2^*\|_p \leq c_{20} \delta \|w_Q f_2^{*\prime}\|_p = c_{20} \delta \|w_Q f_2'\|_p \tag{4.11}$$

Inserting the estimates (4.8) to (4.11) in (4.7)

$$\Omega(\mathcal{L}_p, w_Q; \phi, \delta) \leq \omega(\mathcal{L}_p, w_Q; \phi^*, \delta) \leq$$

$$\leq 3\|w_Q f_1\|_p + (1+2c_{20}) \delta \|w_Q f_2'\|_p \leq \tag{4.11}$$

$$\leq (3+2c_{20}) [\|w_Q f_1\|_p + \delta \|w_Q f_2'\|_p]$$

This estimate is valid for every decomposition $\phi = f_1 + f_2$ hence $\Omega(\mathcal{L}_p, w_Q; \phi, \delta) \leq (3+2c_{20}) K(\mathcal{L}_p, w_Q; \phi, \delta)$. Q.E.D.

Note that for $p = 1$ the condition (3.4) can be dropped.

5., The Upper Estimate of The K-Functional

Our reasoning in this chapter is, apart from a different notation, the same as in our paper [4].

Let $Q(x)$ satisfy (i) and

$$\begin{aligned} [Q'(x)] &\leq \delta^{-1} \text{ for } |x| \leq x_\delta \\ [Q'(x)] &\geq \delta^{-1} \text{ for } |x| > x_\delta . \end{aligned} \tag{5.1}$$

We set

$$\varphi_\delta(x) = \begin{cases} w_Q(x)\phi(x) & \text{for } |x| \leq x_\delta \\ 0 & \text{for } |x| > x_\delta \end{cases} \tag{5.2}$$

and

$$\phi_\delta(x) = \delta^{-1} w_Q^{-1}(x) \int_x^{x+\delta} \varphi_\delta(t)\,dt \tag{5.3}$$

Lemma 5.1. Let (i) hold and let us assume that $\delta Q'(x_\delta+\delta)$ is bounded thus

$$\|w_Q(\phi-\phi_\delta)\|_p \leq \omega(\mathcal{L}_p;w_Q;\phi,\delta) \tag{5.4}$$

and

$$\|w_Q\phi'_\delta\|_p \leq c_{21}\delta^{-1}(\mathcal{L}_p;w_Q;\phi,\delta) \ . \tag{5.5}$$

Proof. By repeated use of Minkovski's inequality we obtain

$$\begin{aligned}\|w_Q(\phi-\phi_\delta)\|_p &= \|\delta^{-1}\int_0^\delta [w_Q(x)\phi(x)-\varphi_\delta(x+t)]dt\|_p \leq \\ &\|\delta^{-1}\int_0^\delta [w_Q(x)\phi(x)-w_Q(x+t)\phi(x+t)]dt\|_p \qquad (5.6)\\ &+ \|\delta^{-1}\int_0^\delta [w_Q(x+t)\phi(x+t)-\varphi_\delta(x+t)]dt\|_p\end{aligned}$$

Clearly

$$\|\delta^{-1}\int_0^\delta [w_Q(x)\phi(x)-w_Q(x+t)\phi(x+t)]\|_p \leq \sup_{0\leq t\leq\delta} \|T_t(w_Qf)-w_Qf\|_p. \tag{5.7}$$

As to the last term in (5.6), let us observe that $(w_Qf-\phi_\delta)(x)$ vanishes for $|x| \leq x_\delta$ by (5.2), (5.1) and (2.5) and $(w_Q\phi\ -\varphi_\delta)(x) = (w_Q\phi)(x) = \delta(Q'_\delta w_Q\phi)(x)$ otherwise.

Consequently

$$\|\delta^{-1}\int_0^\delta [w_Q(x+t)\phi(x)-\varphi_\delta(x+t)]dt\|_p \leq \|w_Q\phi-\varphi_\delta\|_p \leq \delta\|Q'_\delta w_Q\phi\| \tag{5.8}$$

Inequalities (5.6)-(5.8) prove our first assertion (5.4). To prove (5.5) let us first observe that by (5.3) $\phi_\delta(x)$ is differentiable for every $x \neq 0$ and

$$w_Q\phi'_\delta = \delta^{-1}[\varphi_\delta(x+\delta)-\varphi_\delta(x)] + \delta^{-1}Q'(x)\int_x^{x+\delta} \varphi_\delta(t)dt. \tag{5.9}$$

By (5.8) we have for every $|t| \leq \delta$

$$\|\varphi_\delta(x+t)-\varphi_\delta(x)\|_p \leq \|w_Q(x+t)\phi(x+t)-w_Q(x)\phi(x)\|_p + $$
$$+ 2\|w_Q(x)\phi(x)-\varphi_\delta(x)\|_p \leq \|T_t(w_Q\phi)-w_Q\phi\|_p + \tag{5.10}$$
$$+ 2\delta\|Q_\delta' w_Q\phi\| \leq 2\omega(\mathcal{L}_p, w_Q;\phi,\delta)$$

by the last part of (5.8).

We turn now to the second term in (5.9).

$$\delta^{-1}\|Q'(x)\int_x^{x+\delta}\varphi_\delta(t)dt\|_p \leq \|Q'\varphi_\delta\|_p + \delta^{-1}\|Q'(x)\int_0^\delta[\varphi_\delta(x+t)-$$
$$- \varphi_\delta(x)]dt\|_p \tag{5.11}$$

In view of (5.2) $\varphi_\delta(x)$ vanishes for $|x| \geq x_\delta$ thus we have

$$\|Q'\varphi_\delta\|_p = \|Q_\delta'\varphi_\delta\|_p \leq \|Q_\delta' w_Q\phi\|_p \leq \delta^{-1}\omega(\mathcal{L}_p, w_Q;\phi,\delta) \tag{5.12}$$

In the second term on the right of (5.11) the integral vanishes for $|x| > x_\delta+\delta$ hence by applying (5.10)

$$\|Q'(x)\int_0^\delta[\varphi_\delta(x+t)-\varphi_\delta(x)]dt\|_p \leq Q'(x_\delta+\delta)\delta \sup_{|t|\leq\delta}\|\varphi_\delta(x+t)-\varphi_\delta(x)\| \leq \tag{5.13}$$
$$\leq 2\,\delta Q'(x_\delta+\delta)\omega(\mathcal{L}_p, w_Q;\phi,\delta) \leq c_{22}\omega(\mathcal{L}_p, w_Q;\phi,\delta)\ .$$

By (5.9)-(5.13)

$$\|w_Q\phi_\delta'\| \leq (3 + c_{22})\delta^{-1}\omega(\mathcal{L}_p, w_Q;\phi,\delta) \tag{5.15}$$

i.e. also the second assertion of Lemma 5.1 is true. Q.E.D.

Lemma 5.2. $\delta Q'(x_\delta+\delta)$ is bounded provided one of the following to conditions holds:

a., (3.4) is valid

b., the second half of (2.14) is satisfied.

Proof. a., In virtue of (3.4) we have for sufficiently great x_δ (i.e. small δ)

$$\delta-[Q'(x_\delta+\delta)]^{-1}=[Q'(x_\delta)]^{-1}-[Q'(x_\delta+\delta)]^{-1} =$$

$$= \int_{x_\delta}^{x_\delta+\delta} Q''(t)[Q'(t)]^{-2}dt < \tfrac{1}{2}(1+\theta)\delta$$

hence

$$\delta Q'(x_\delta+\delta) < 2(1-\theta)^{-1} .$$

b., For $\delta\to 0$ $x_\delta\to\infty$ thus $x_\delta+\delta < 2x_\delta$ for great x_δ. Thus by the second half of (2.14) $\delta Q'(x_\delta+\delta) = Q'(x_\delta+\delta)/Q'(x_\delta) < Q'(2x_\delta)/Q'(x_\delta)$ is bounded. Q.E.D.

Lemma 5.3. We have for every real constant k

$$K(\mathcal{L}_p,w_Q;\phi-k,\delta) = K(\mathcal{L}_p,w_Q;\phi,\delta). \tag{5.16}$$

Proof. To every decomposition $\phi=f_1+f_2$ let correspond the decomposition $\phi-k=f_1+(f_2-k)$ and apply (3.1).

Lemma 5.4. We have for every $0 < s < t$

$$\Omega(\mathcal{L}_p,w_Q;\phi,t) \leq 2\frac{t}{s}\,\Omega(\mathcal{L}_p,w_Q;\phi,s) \tag{5.17}$$

Proof. Obviously

$$\|T_{2h}(w_Q\phi)-w_Q\phi)\|_p \leq 2\|T_h(w_Q\phi)-w_Q\phi\|_p . \tag{5.18}$$

By its definition $x_{2\delta} \leq x_\delta$ hence for every real x $0 \leq Q'_{2\delta}(x) \leq Q'_\delta(x)$ so that

$$2\delta\|Q'_{2\delta}w_Q\phi\|_p \leq 2\ \{\delta\|Q'_\delta w_Q\phi\|_p\} \tag{5.19}$$

In consequence of the two last inequalities

$$\omega(\mathcal{L}_p,w_Q;\phi,2\delta) \leq 2\omega(\mathcal{L}_p,w_Q;\phi,\delta) \tag{5.20}$$

and this implies that for $2^\nu \leq \frac{t}{s} < 2^{\nu+1}$ we have

$$\begin{aligned}\omega(\mathcal{L}_p,w_Q;\phi,t) &\leq 2^{\nu+1}\omega(\mathcal{L}_p,w_Q;\phi,2^{-\nu-1}t) \leq \\ &\leq 2^{\nu+1}\omega(\mathcal{L}_p,w_Q;\phi,s) \leq 2\frac{t}{s}\,\omega(\mathcal{L}_p,w_Q;\phi,s).\end{aligned} \tag{5.21}$$

It follows that for every real k

$$\Omega(\mathcal{L}_p,w_Q;\phi,t) \leq \omega(\mathcal{L}_p,w_Q;\phi-k,t) \leq 2\frac{t}{s}\,\omega(\mathcal{L}_p,w_Q;\phi-k,s)$$

Taking the infimum of the right side with respect to k, we obtain (5.17) Q.E.D.

Theorem 5.1. Let Q satisfy condition (i) and let $\delta Q'(x_\delta+\delta)$ be bounded than we have for every $1 \leq p \leq \infty$, every $\phi \in \mathcal{L}_p$ and every $0 < \delta \leq 1$

$$K(\mathcal{L}_p, w_Q; \phi, \delta) \leq c_{10} \Omega(\mathcal{L}_p, w_Q; \phi, \delta). \tag{5.22}$$

Proof. By Lemma 5.1 we have for $0 \leq \delta \leq c_{23}$

$$\begin{aligned} K(\mathcal{L}_p, w_Q; \phi, \delta) &\leq \|w_Q(\phi - \phi_\delta)\|_p + \delta \|w_Q \phi_\delta'\|_p \leq \\ &(1 + c_{21}) \omega(\mathcal{L}_p, w_Q; \phi, \delta) \end{aligned} \tag{5.23}$$

Thus in virtue of Lemma 5.2

$$\begin{aligned} K(\mathcal{L}_p, w_Q; \phi, \delta) &\leq (1+c_{21}) \inf_k \omega(\mathcal{L}_p, w_Q; \phi - k; \delta) = \\ &= (1+c_{21})\, \Omega(\mathcal{L}_p, w_Q; \phi, \delta). \qquad (0 < \delta \leq c_{23}) \end{aligned} \tag{5.24}$$

For $\delta > c_{23}$ we observe that in virtue of the inequality

$$K(\mathcal{L}_p, w_Q; \phi, t) \leq \max(1, \tfrac{t}{s})\, K(\mathcal{L}_p, w_Q; \phi, s)$$

(see J. Peetre [8], p. 9) we have

$$\begin{aligned} K(\mathcal{L}_p, w_Q; \phi, \delta) &\leq c_{23}^{-1}\, \delta K(\mathcal{L}_p, w_Q; \phi, c_{23}) \leq \\ &\leq c_{23}^{-1}(1+c_{21})\delta\, \Omega(\mathcal{L}_p, w_Q; \phi, c_{23}) \leq \\ &\leq c_{23}^{-1}(1+c_{21})\, \Omega(\mathcal{L}_p, w_Q; \phi, \delta) \qquad (c_{23} \leq \delta \leq 1) \end{aligned} \tag{5.25}$$

(5.24) and (5.25) together show that (5.22) is valid.

Q.E.D.

6., On weighted $\mathcal{L}_1$-approximation.

We assume in this chapter that Q satisfies (i) and (2.14)(and nothing else). Let $\{p_\nu(w_Q^2;\xi)\}$ be the sequence of orthonormal polynomials with respect to w_Q^2 and

$$\lambda_n(w_Q^2;\xi) = \{\sum_{\nu=0}^{n-1}[p_\nu(w_Q^2;\xi)]^2\}^{-1}$$

be the related Christoffel functions. We proved in [5] that (i) and (2.14) imply

$$\lambda_n(w_Q^2;\xi) \leqq c_{25}\frac{q_n}{n}\, w_Q^2(\xi) \quad (|\xi| \leqq c_{26}q_n)\ . \tag{6.1}$$

In [6] we have shown that from (6.1) follows for every ϕ having locally bounded variation

$$\varepsilon_n^{(1)}(w_Q^2;\phi) \leqq c_{27}\ \frac{q_n}{n}\int_{-\infty}^{\infty} w_Q^2(\xi)\,|d\phi(\xi)| \tag{6.2}$$

We replace Q by Q/2 i.e. w_Q^2 by w_Q. In view of (2.7) we have $q_n(Q/2) = q_{2n}(Q) \leqq q_n(Q)$. Hence taking for ϕ an absolutely continuous function we get the Jackson inequality

$$\varepsilon_n^{(1)}(w_Q;\phi) \leqq 2\ c_{27}\ \frac{q_n}{n}\ \|w_Q\phi'\|_1 \tag{6.3}$$

In virtue of Lemma 5.2 the conditions of Theorem 5.1 are satisfied so that (5.22) holds true. Consequently Theorem 3.2 with p=1 is applicable. Hence we obtain

Theorem 6.1. If Q satisfies (i) and (2.14) then the estimate (3.6)(Jackson-type theorem) is valid for p=1, every natural r and every ϕ satisfying $\phi^{(r)}w_Q \in \mathcal{L}_1$.

Bibliography

[1] Freud, G. A contribution to the problem of weighted polynomial approximation. Linear Operators and Approximations, edited by P. L. Butzer, J. P. Kahane and B. Sz-Nagy. I.S.N.M. vol. 20, Birkhauser (Basel) 1972 pp. 431-447.

[2] Freud, G. On direct and converse theorems of weighted polynomial approximation. Mathem. Zeitschrift 126 (1972) 123-134.

[3] On converse theorems of weighted polynomial approximation. Acta Math. Aced. Sci. Hungar. 24 (1973)389-397.

[4] On polynomial approximation with respect to general weights. Functional Analysis and its Appl. Edited by H. G. Garnir, K. R. Unni and J. H. Williamson. Lecture Notes vol. 399 Springer Berlin 1974 pp. 149-179.

[5] Freud, G. On the Theory of Onesided Weighted Polynomial Approximation. Approximation Theory and Functional Analysis, edited by P. L. Butzer and B. Sz-Nagy I.S.N.M. Vol. 25 Birkhauser Basel 1975, pp. 285-303.

[6] Freud, G. Extension of the Dirichlet-Jordan Convergence Criterion to a General Class of Orthogonal Polynomial Expansions. Acta. Math. Ac. Sci. Hung. 25 (1974) 109-122.

[7] Lorentz, G. G. Approximation of functions. Holt, Rinehart and Winston, New York 1966.

[8] Peetre, J. A Theory of interpolation of normed spaces. Notas de Matematica vol. 39 (1968).

[9] Peetre, J. Exact interpolation theorems for Lipschitz continuous functions. Ricerche Mathematich 18 (1969) 239-259.

Continuity Theorems for the Product Approximation Operator

M.S. Henry and D. Schmidt

1. INTRODUCTION. A number of recent papers [1,2,6,8,14] have considered various extensions of the concept of <u>product approximation</u>, first formally considered by Weinstein [13]. A brief description of this type of approximation follows.

Let D denote the rectangle $I \times J = [a,b] \times [c,d]$, and let $F \in C(D)$. For each $y \in J$, let $F_y \in C(I)$ be given by $F_y(x) = F(x,y)$. Suppose that $\{\phi_i\}_{i=1}^n$ is a Chebyshev system of real-valued functions on I. Let Φ denote the linear space spanned by $\{\phi_i\}_{i=1}^n$. For each $y \in J$, there exists a unique "polynomial"

$$T(F_y,\cdot) = \Sigma_{i=1}^n f_i(y)\phi_i,$$

$$f(y) = (f_1(y),\ldots,f_n(y)) \in E_n,$$

which is the best approximation (sup norm) to F_y on I. The resulting $f_i(y)$, $i = 1,\ldots,n$, are continuous functions (see [13]) on J. If $\{\psi_j\}_{j=1}^m$ is a Chebyshev system of real-valued functions on J, and if

$$Q(f_i,\cdot) = \Sigma_{j=1}^m f_{ij}\psi_j,$$

$$\alpha_i = (f_{i1},\ldots,f_{im}) \in E_m,$$

are the unique best polynomial approximations (sup norm) to the $f_i(y)$, $i = 1,\ldots,n$, respectively, from the linear space Ψ spanned by $\{\psi_j\}_{j=1}^m$, then the product approximation to F on D is

$$(1.1) \qquad \mathcal{P}F = \Sigma_{i=1}^n Q(f_i,\cdot)\phi_i = \Sigma_{i=1}^n \Sigma_{j=1}^m f_{ij}\psi_j\phi_i .$$

Hereafter, the operator $\mathcal{P}$ in (1.1) will be called the product approximation operator.

The term product approximation is due to the following observation: if $F(x,y) = g(x)h(y)$, then the product approximation to F on D is the product of the best approximations to g and h on I and J, respectively. If $F(x,y) = g(x) + h(y)$, then under appropriate conditions, the product approximation is the sum of the best approximations to g and h, respectively. The latter case was studied in a different setting by Newman and Shapiro [10]. Reasons for studying product approximations are noted in all the papers [1,2,6,8,13,14], and each considers existence and computations in a variety of situations. However, such classical results as the continuity of the best approximation operator and a point Lipschitz theorem have not been studied in the context of product approximation. In this paper, we consider these two classical results for product approximation.

2. CONTINUITY OF THE PRODUCT APPROXIMATION OPERATOR. In this section, we prove that the product approximation operator is a continuous mapping from C(D) into $\Phi \cdot \Psi$, the linear space spanned by

the products $\{\phi_i\psi_j\}_{i,j=1}^{n,m}$.

Let $F \in C(D)$. As in Section 1, the best approximation to F_y from Φ on I is denoted by

$$(2.1) \qquad T(F_y,\cdot) = \Sigma_{i=1}^{n} f_i(y)\phi_i,$$

where the coefficient vector

$$(2.2) \qquad f(y) = (f_1(y),\ldots,f_n(y)) \in E_n.$$

Define

$$\rho(y) = \|F_y - T(F_y,\cdot)\|_I.$$

By Theorem 2.1 of [13], ρ is continuous on J. For coefficient vectors $\alpha = (a_1,\ldots,a_n)$ and $\beta = (b_1,\ldots,b_n)$ in E_n, define

$$\sigma(\alpha,\beta) = \max_{1\leq i\leq n} |a_i - b_i|.$$

The continuity of the product approximation operator $\mathcal{P}$ follows from the next two lemmas. The proof of the first is similar to that of Theorem 2.2 of [13].

LEMMA 1. *Given* $\epsilon > 0$ *there is a* $\delta = \delta(F,\epsilon) > 0$ *such that whenever* $A(\cdot) = (\alpha_1(\cdot),\ldots,\alpha_n(\cdot))$ *is continuous on* J *and*

$$\|F_y - \Sigma_{i=1}^{n}\alpha_i(y)\phi_i\|_I \leq \rho(y) + \delta$$

for all $y \in J$, *then* $\sigma(A(y),f(y)) < \epsilon$ *for all* $y \in J$.

Proof. Assume otherwise. Then there is a sequence

$$\{(\alpha_1^k(\cdot),\ldots,\alpha_n^k(\cdot))\}_{k=1}^{\infty}$$

in $[C(J)]^n$ such that

(2.3) $$\|F_y - \alpha_1^k(y)\phi_1 - \cdots - \alpha_n^k(y)\phi_n\|_I \le \rho(y) + 1/k$$

for all $y \in J$ and such that for each k there is a $y_k \in J$ satisfying

(2.4) $$\sigma(A^k(y_k), f(y_k)) \ge \epsilon.$$

Since J is compact, we may assume by relabeling that $y_k \to y^* \in J$. Inequality (2.3) implies

$$\|\alpha_1^k(y)\phi_1 + \cdots + \alpha_n^k(y)\phi_n\|_I \le \|F\|_D + \|\rho\|_J + 1.$$

Hence, by the linear independence of $\{\phi_1,\ldots,\phi_n\}$, the sequence $\{(\alpha_1^k(y),\ldots,\alpha_n^k(y))\}_{k=1}^{\infty}$ is uniformly bounded in k and y. Thus by appropriately relabeling we can conclude that $\alpha_i^k(y_k) \to \lambda_i$, $i = 1,\ldots,n$. Since F is continuous on D and ρ is continuous on J, (2.3) implies

(2.5) $$\|F_{y^*} - \lambda_1\phi_1 - \cdots - \lambda_n\phi_n\|_I \le \rho(y^*).$$

Since the best approximation to F_{y^*} from Φ is unique, (2.5) and the definition of ρ establish that $\lambda_i = f_i(y^*)$, $i = 1,\ldots,n$. But (2.4) and the continuity of f_i, $i = 1,\ldots,n$, now imply that $\max_{1\le i\le n} |f_i(y^*) - \lambda_i| \ge \epsilon$, a contradiction.

LEMMA 2. <u>Given</u> $\epsilon > 0$ <u>there is a</u> $\delta = \delta(F,\epsilon) > 0$ <u>such that whenever</u> $G \in C(D)$ <u>and</u> $\|G-F\|_D < \delta$, <u>then</u> $\|g_i - f_i\|_J < \epsilon$, $i = 1,\ldots,n$, <u>where</u> $T(G_y,\cdot) = \Sigma_{i=1}^n g_i(y)\phi_i$ <u>and the</u> f_i <u>are as in</u> (2.1).

Proof. Let $\epsilon > 0$ be given. By Lemma 1 we may select $\delta > 0$ such that if $G \in C(D)$ and $\|F_y - T(G_y,\cdot)\|_I \leq \rho(y) + \delta$ for all $y \in J$, then $\sigma(g(y),f(y)) < \epsilon$ for all $y \in J$, where $g(y) = (g_1(y),\ldots,g_n(y))$.

Let $G \in C(D)$ with $\|G-F\|_D < \delta/2$. Denote $\mu(y) = \|G_y - T(G_y,\cdot)\|_I$. We observe that

$$\begin{aligned}\mu(y) - \rho(y) &= \|G_y - T(G_y,\cdot)\|_I \\ &\quad - \|F_y - T(F_y,\cdot)\|_I \\ &\leq \|G_y - T(F_y,\cdot)\|_I \\ &\quad - \|F_y - T(F_y,\cdot)\|_I \\ &\leq \|G_y - F_y\|_I \\ &\leq \|G - F\|_D.\end{aligned}$$

Now for all $y \in J$,

$$\begin{aligned}\|F_y - T(G_y,\cdot)\|_I &\leq \|F - G\|_D \\ &\quad + \|G_y - T(G_y,\cdot)\|_I \\ &\leq \delta/2 + \mu(y) \\ &\leq \rho(y) + \delta.\end{aligned}$$

Thus, $\sigma(g(y),f(y)) < \epsilon$. Finally, $\|g_i - f_i\|_J < \epsilon$, $i = 1,\ldots,n$.

THEOREM 1. *Suppose that Φ, Ψ and $\Phi \cdot \Psi$ are as described above. Then the product approximation operator is a continuous mapping from $C(D)$ into $\Phi \cdot \Psi$.*

Proof. Let $F \in C(D)$ and $\epsilon > 0$ be given. In the setting of Section 1, the best approximation operator of the single variable y is a continuous mapping from C(J) into Ψ(Freud's Theorem). Thus given $f(y) = (f_1(y),\ldots,f_n(y))$ and $g(y) = (g_1(y),\ldots,g_n(y))$ defined in Lemma 2 and (2.1), there is a $\tau > 0$ such that if $\|f_i - g_i\|_J \leq \tau$, then

$$\|\Sigma_{j=1}^m f_{ij}\psi_j - \Sigma_{j=1}^m g_{ij}\psi_j\|_J < \frac{\epsilon}{n\|\phi_i\|_I} .$$

Here $\Sigma_{j=1}^m f_{ij}\psi_j$ and $\Sigma_{j=1}^m g_{ij}\psi_j$ are the best approximations to f_i and g_i from Ψ, respectively. Lemma 2 implies that there exists a $\delta > 0$ such that if $\|G - F\|_D < \delta$, then $\|g_i - f_i\|_J < \tau$, $i = 1,\ldots,n$. Hence, $\|G - F\|_D < \delta$ implies that

$$\|\mathcal{P}G - \mathcal{P}F\|_D \leq \Sigma_{i=1}^n \|\phi_i\|_I \|\Sigma_{j=1}^m (f_{ij} - g_{ij})\psi_j\|_J < \epsilon,$$

and the proof is complete.

3. UNIFORM POINT LIPSCHITZ CONDITION. In this section we show that the classical best approximation operator T (see 2.1) satisfies a uniform point Lipschitz condition over certain subsets of C(I). The results of this section will be used to prove a point Lipschitz theorem for the product approximation operator $\mathcal{P}$. We first state the classical theorem of Freud [5].

THEOREM 2. <u>For</u> $f \in C(I)$, <u>there is a constant</u> $\lambda_f > 0$ <u>such that</u>

$$\|T(g,\cdot) - T(f,\cdot)\|_I \leq \lambda_f \|g - f\|_I \tag{3.1}$$

for each $g \in C(I)$.

It is known that λ_f cannot be chosen independent of f over all of C(I), (see [4, p. 164]); that is, there exists no constant $\lambda > 0$ such that for all $f,g \in C(I)$

$$\|T(g,\cdot) - T(f,\cdot)\|_I \leq \lambda \|g - f\|_I. \tag{3.2}$$

However, for appropriate subsets $\Gamma \subseteq C(I)$ and corresponding $\lambda = \lambda_\Gamma$, (3.2) may hold for all $f \in \Gamma$ and $g \in C(I)$. We propose to determine conditions on $\Gamma \subseteq C(I)$ in order to insure that λ_f can be chosen independent of f over Γ.

To do so we utilize Freud's characterization of the λ_f in (3.1). Let $a \leq x_0 < \cdots < x_n \leq b$ be a point set of alternation for $f - T(f,\cdot)$. Define

$$K(x_0,\ldots,x_n) = \sup\{\|\phi\| : \phi \in \Phi \text{ and } (-1)^i\phi(x_i) \geq -1,\ i = 0,\ldots,n\}. \tag{3.3}$$

The next lemma establishes that $K(x_0,\ldots,x_n)$ is finite. The constant of inequality (3.1) can be taken to be $\lambda_f = 2K(x_0,\ldots,x_n)$, (see [11]).

LEMMA 3. *Let* $a \leq \alpha_0 \leq \beta_0 < \alpha_1 \leq \beta_1 < \cdots < \alpha_n \leq \beta_n \leq b$. *Then*

$$\{\phi \in \Phi : \exists \xi_i \in [\alpha_i,\beta_i] \ni (-1)^i\phi(\xi_i) \geq -1,\ i = 0,\ldots,n\}$$

is $\|\cdot\|_I$*-bounded*.

Remark. This lemma generalizes an assertion of Rice [11, p. 64]. The proof parallels the proof of that assertion and is omitted.

Lemma 3 can be used to prove that λ_f is a continuous function of the point set of alternation. We now prove that for appropriate subsets Γ of $C(I)$ the λ_f may be chosen independent of $f \in \Gamma$.

THEOREM 3. *Let* Γ *be a compact subset of* $C(I)$ *such that* $\Gamma \cap \Phi = \emptyset$. *Then there exists a* $\lambda_\Gamma > 0$ *such that* (3.2) *is valid for all* $g \in C(I)$ *and* $f \in \Gamma$.

Remark. The hypothesis $\Gamma \cap \Phi = \emptyset$ cannot be excluded. An example in the next section will demonstrate the necessity of this requirement.

Proof. Assume otherwise. Then there exist sequences $\{f_k\}$ in Γ and $\{g_k\}$ in $C(I)$, $f_k \neq g_k$, such that

$$(3.4) \qquad \frac{\|T(g_k,\cdot) - T(f_k,\cdot)\|_I}{\|g_k - f_k\|_I} \to \infty .$$

Since Γ is compact we may assume (by relabeling) that $f_k \to f \in \Gamma$. Since $\Gamma \cap \Phi = \emptyset$, $f \notin \Phi$. If $R = f - T(f,\cdot)$, then $\|R\|_I > 0$.

By the alternation theorem [3, p. 75], for each k there is a point set of alternation

$$a \leq x_0^k < \dots < x_n^k \leq b$$

for $R_k = f_k - T(f_k,\cdot)$. We further extract a subsequence and relabel so that $x_i^k \to x_i$, $i = 0,\dots,n$.

Then

$$a \leq x_0 \leq \cdots \leq x_n \leq b.$$

We show that $x_0 < \cdots < x_n$. Suppose that $x_i = x_{i+1}$ for some i. Since $f_k \to f$, $x_i^k \to x_i$, and $x_{i+1}^k \to x_{i+1} = x_i$, k may be chosen large enough to insure that

$$|R(x_{i+1}^k) - R(x_i^k)| < \|R\|_I,$$

and

$$\|R - R_k\|_I < \frac{1}{4} \|R\|_I.$$

Thus

$$\begin{aligned} \|R\|_I > {} & |R(x_i^k) - R(x_{i+1}^k)| \\ & \geq |R_k(x_i^k) - R_k(x_{i+1}^k)| \\ & \quad - |R(x_i^k) - R_k(x_i^k)| \\ & \quad - |R(x_{i+1}^k) - R_k(x_{i+1}^k)| \\ & \geq 2\|R_k\|_I - 2\|R - R_k\|_I. \end{aligned}$$

But $\|R_k\|_I \geq \|R\|_I - \|R - R_k\|_I$. This implies that

$$\begin{aligned} \|R\|_I & > 2\|R\|_I - 4\|R - R_k\|_I \\ & > 2\|R\|_I - 4 \cdot \frac{1}{4} \|R\|_I = \|R\|_I, \end{aligned}$$

which is false. Hence, $x_0 < \cdots < x_n$.

Since the x_i, $i = 0,\ldots,n$, are distinct, there exist numbers α_i, β_i, $i = 0,\ldots,n$, such that $a \leq \alpha_0 < \beta_0 < \alpha_1 < \beta_1 < \cdots < \alpha_n < \beta_n \leq b$ and such that $x_i \in (\alpha_i,\beta_i)$, $i = 1,\ldots,n-1$. If $x_0 = a$, then $x_0 \in [\alpha_0,\beta_0)$ and if $x_n = b$, then $x_n \in (\alpha_n,\beta_n]$. Otherwise open intervals can be employed for either or both of these indices. By Lemma 3,

$$(3.5) \qquad \overline{K} = \sup\{\|\phi\|_I : \phi \in \Phi \text{ and } \exists \xi_i \in [\alpha_i,\beta_i] \ni (-1)^i\phi(\xi_i) \geq -1, \ i = 0,\ldots,n\}$$

is finite. Since $x_i^k \to x_i$, there exists an M such that for all $k \geq M$, $x_i^k \in [\alpha_i,\beta_i]$, $i = 0,\ldots,n$. Then Theorem 2 implies that

$$\frac{\|T(g_k,\cdot) - T(f_k,\cdot)\|_I}{\|g_k - f_k\|_I} \leq 2\, K(x_0^k,\ldots,x_n^k).$$

Now a comparison of (3.3) and (3.5) reveals that for all $k \geq M$, $K(x_0^k,\ldots,x_n^k) \leq \overline{K}$. Thus

$$\frac{\|T(g_k,\cdot) - T(f_k,\cdot)\|_I}{\|g_k - f_k\|_I} \leq 2\, \overline{K},$$

a contradiction to (3.4). This completes the proof of Theorem 3.

Related to the theorem of Freud is the strong unicity theorem [3, p. 80] which states that for each $f \in C(I)$ there is a constant $\gamma_f > 0$ such that

$\|f - \phi\|_I \geq \|f - T(f,\cdot)\|_I + \gamma_f\|\phi - T(f,\cdot)\|_I$ for each $\phi \in \Phi$. It can be shown that if $x_0 < x_1 < \cdots < x_n$ is a point set of alternation for $f - T(f,\cdot)$, then the constant γ_f can be taken to be $1/K(x_0,\ldots,x_n)$. Under the same conditions as in Theorem 3, a uniform strong unicity theorem also holds.

4. POINT LIPSCHITZ CONDITION FOR $\mathcal{P}$. In this section, we prove that for certain $F \in C(D)$ the product approximation operator $\mathcal{P}$ satisfies a point Lipschitz condition.

THEOREM 4. Let $F \in C(D)$, where $F_y \notin \Phi$ for all $y \in J$. Then there is a constant $\overline{\lambda}_F > 0$ such that $\|\mathcal{P}G - \mathcal{P}F\|_D \leq \overline{\lambda}_F\|G - F\|_D$ for each $G \in C(D)$.

Proof. We first note that $\Gamma_F = \{F_y : y \in J\}$ is a compact subset of $C(I)$ which does not meet Φ. Thus Theorem 3 implies there is a constant $\lambda_F > 0$ such that

$$\|T(G_y,\cdot) - T(F_y,\cdot)\|_I \leq \lambda_F\|G_y - F_y\|_I$$

for all $y \in J$ and $G \in C(D)$. If $f_1(y),\ldots,f_n(y)$ are as in (2.1), then from Theorem 2 there exist constants $\rho_1,\ldots,\rho_n$ depending only on $f_1,\ldots,f_n$ such that

$$\|Q(g_i,\cdot) - Q(f_i,\cdot)\|_J \leq \rho_i\|g_i - f_i\|_J$$

for $g_i \in C(J)$, $i = 1,\ldots,n$. The representation

(1.1) yields

(4.1) $$\|\mathcal{P}_G - \mathcal{P}_F\|_D \le \Sigma_{i=1}^{n} \|\phi_i\|_I \|Q(g_i,\cdot) - Q(f_i,\cdot)\|_J \le \Sigma_{i=1}^{n} \|\phi_i\|_I\, \rho_i \|g_i - f_i\|_J.$$

On the finite dimensional space Φ, the two norms $\|\cdot\|_I$ and

$$\|\alpha_1\phi_1 + \dots + \alpha_n\phi_n\|_* = \max_{1\le i\le n} |\alpha_i|$$

are equivalent. So there is a constant K^* such that $\|\cdot\|_* \le K^*\|\cdot\|_I$. For fixed $y \in J$

$$|g_i(y) - f_i(y)| \le \max_{1\le j\le n} |g_j(y) - f_j(y)| \le K^*\|T(G_y,\cdot) - T(F_y,\cdot)\|_I \le K^*\lambda_F\|G_y - F_y\|_I \le K^*\lambda_F\|G - F\|_D.$$

From (4.1) we now obtain

$$\|\mathcal{P}_G - \mathcal{P}_F\|_D \le \lambda_F K^*(\Sigma_{i=1}^{n}\|\phi_i\|_I\rho_i)\|G - F\|_D.$$

The constant $\overline{\lambda}_F$ may thus be taken as

$$\overline{\lambda}_F = \lambda_F K^*(\Sigma_{i=1}^{n}\|\phi_i\|_I\rho_i).$$

We conclude this section with an example of a function $F \in C(D)$ for which $F_y \in \Phi$ for some $y \in J$ and for which the conclusion of Theorem 4 is false. The example is motivated by Theorem 4 of Cline [4, p. 164].

Let $D = I \times J = [-1,1] \times [0,1]$, $\Phi = \text{span}[1,x]$ and $\Psi = \text{span}[1,y]$. For $y \in (0,1]$, define

$$f_y(x) = \begin{cases} 0, & -1 \le x \le -\frac{3y}{4} \\ \frac{4}{y}x + 3, & -\frac{3y}{4} \le x \le -\frac{y}{2} \\ -\frac{4}{y}x - 1, & -\frac{y}{2} \le x \le 0 \\ \frac{4}{y}x - 1, & 0 \le x \le \frac{y}{2} \\ -\frac{4}{y}x + 3, & \frac{y}{2} \le x \le \frac{3y}{4} \\ 0, & \frac{3y}{4} \le x \le 1 . \end{cases}$$

It can be shown that $f_y(x)$ is continuous for $(x,y) \in [-1,1] \times (0,1]$. Also, for each $y \in (0,1]$, $\|f_y\|_I = 1$, $f_y(-y/2) = f_y(y/2) = 1$ and $f_y(0) = -1$. Thus, $T(f_y,x) \equiv 0$ for each $y \in (0,1]$. Define

$$g_y(x) = \begin{cases} 0, & -1 \le x \le -y \\ -3x - 3y, & -y \le x \le -\frac{3y}{4} \\ x + f_y(x), & -\frac{3y}{4} \le x \le \frac{3y}{4} \\ -3x + 3y, & \frac{3y}{4} \le x \le y \\ 0, & y \le x \le 1 . \end{cases}$$

Then $g_y(x)$ is also continuous on $[-1,1] \times (0,1]$. Since $\|g_y(x) - x\|_I = 1$, $g_y(-y/2) = g_y(y/2) = 1$ and $g_y(0) = -1$, we conclude that $T(g_y,x) = x$ for each $y \in (0,1]$. It can also be seen that $\|g_y - f_y\|_I \le y$ for $y \in (0,1]$.

Define $F \in C(D)$ by

$$F(x,y) = \begin{cases} y\, f_y(x), & 0 < y \le 1 \\ 0 & , y = 0. \end{cases}$$

We note that $F_y \in \Phi$ only for $y = 0$. Also, $T(F_y,x) \equiv 0$ for all $y \in J$, and consequently, $\mathcal{P}_F \equiv 0$ on D.

Define $G^k \in C(D)$ by

$$G^k(x,y) = \begin{cases} F(x,y), \; y \in J - [\frac{1}{k+1}, \frac{1}{k}] \\ F(x,y) + (y-\frac{1}{k+1})(\frac{1}{k} - y)(g_y(x) - \\ \qquad f_y(x)), \\ \qquad y \in [\frac{1}{k+1}, \frac{1}{k}]. \end{cases}$$

Now $T(G_y^k,x) \equiv 0$ for all $y \in J - [\frac{1}{k+1}, \frac{1}{k}]$. For $y \in [\frac{1}{k+1}, \frac{1}{k}]$,

$$G_y^k(x) = [y^2 + (1 - \frac{2k+1}{k(k+1)})y + \\ + \frac{1}{k(k+1)}]f_y(x) \\ + (y - \frac{1}{k+1})(\frac{1}{k} - y)g_y(x).$$

For $y \in [\frac{1}{k+1}, \frac{1}{k}]$ and $k \geq 2$, the coefficients of f_y and g_y are non-negative. Since $f_y - T(f_y,\cdot)$ and $g_y - T(g_y,\cdot)$ have the same point set of alternation with the same sign orientation, for $y \in [\frac{1}{k+1}, \frac{1}{k}]$,

$$T(G_y^k,\cdot) = [y^2 + (1 - \frac{2k + 1}{k(k+1)})y \\ + \frac{1}{k(k+1)}]T(f_y,\cdot) \\ + (y - \frac{1}{k+1})(\frac{1}{k} - y)T(g_y,\cdot) \\ = (y - \frac{1}{k+1})(\frac{1}{k} - y)x.$$

Thus,

$$T(G_y^k,\cdot) = \begin{cases} 0, & y \in J - [\frac{1}{k+1}, \frac{1}{k}] \\ (y - \frac{1}{k+1})(\frac{1}{k} - y)x, & \\ \quad y \in [\frac{1}{k+1}, \frac{1}{k}]. & \end{cases}$$

Then

$$(\boldsymbol{P}G^k)(x,y) = \frac{1}{8k^2(k+1)^2}\, x$$

and

$$\|\boldsymbol{P}G^k - \boldsymbol{P}F\|_D = \frac{1}{8k^2(k+1)^2}\,.$$

But $\|G^k - F\|_D \leq \frac{1}{4k^3(k+1)^2}$. Consequently,

$$\frac{\|\boldsymbol{P}G^k - \boldsymbol{P}F\|_D}{\|G^k - F\|_D} \geq \frac{k}{2},$$

and $\boldsymbol{P}$ does not satisfy a point Lipschitz condition at F. Since $F_y \in \Phi$ only for $y = 0$, this example demonstrates that the hypotheses of Theorem 4 are in some sense "minimal".

In regard to the remark following Theorem 3, $\Gamma_F = \{F_y : y \in J\}$ is a compact subset of C(I) that meets Φ and over which λ_f cannot be chosen independent of f. Thus the hypotheses of Theorem 3 are also minimal.

5. CONCLUSIONS. The results of this paper establish that the product approximation operator is continuous and that if $F \in C(D)$ satisfies an additional requirement, then the product approximation operator satisfies a point Lipschitz condition at F. In addition, apparently minimal conditions are given that produce a uniform Freud's theorem for Chebyshev approximation in one variable.

Under appropriate normality conditions, these results appear to be extendable to the product rational approximation operator [7].

REFERENCES

1. J. A. Brown and M. S. Henry, Best rational product approximation of functions, J. of Approx. Theory, 9(1973), 287-294.

2. J. A. Brown and M. S. Henry, Best Chebyshev composite approximation, SIAM J. of Num. Anal., 12(1975), 336-344.

3. E. W. Cheney, "Introduction to Approximation Theory," McGraw-Hill, New York, 1966.

4. A. K. Cline, Lipschitz conditions on uniform approximation operators, J. of Approx. Theory, 8(1973), 160-172.

5. G. Freud, Eine Ungleichung für Tschebyscheffsche Approximationspolynome, Acta Sci. Math. (Szeged), 19(1958), 162-164.

6. J. Henry, Computation of rational product approximations, (submitted).

7. M. S. Henry and D. Schmidt, Continuity theorems for the rational product approximation operator, (in preparation).

8. M. S. Henry and S. E. Weinstein, Best rational product approximation of functions II, J. of Approx. Theory, 12(1974), 6-22.

9. H. Maehly and Ch. Witzgall, Tschebyscheff - Approximationen in kleinen Intervallen I: Approximation durch Polynome, Numer. Math., 2(1960), 142-150.

10. D. J. Newman and H. S. Shapiro, Some theorems of Čebyšev approximation, Duke Math. J., 30(1963), 673-682.

11. J. R. Rice, "The Approximation of Functions", Vol. 1, Addison-Wesley, Reading, MA, 1964.

12. J. R. Rice, "The Approximation of Functions", Vol. 2, Addison-Wesley, Reading, MA, 1969.

13. S. E. Weinstein, Approximation of functions of several variables: product approximations I, J. of Approx. Theory, 2(1969), 433-447.

14. S. E. Weinstein, Product approximations of functions of several variables, SIAM J. of Num. Anal., 8(1971), 178-189.

SOME REMARKS ON THE ESTIMATION OF QUADRATIC FUNCTIONALS

F. M. Larkin

Abstract

Different approaches to the problem of estimating the value of a bounded, linear functional on a Hilbert space, from given values of a number of other bounded, linear functionals, are re-examined in relation to the problem of estimating the value of a quadratic functional from the same data. It is shown that, for the non-linear problem, these approaches result in estimators which do not necessarily agree with each other, and that the assumption of a weak Gaussian distribution on the function space can be useful in correcting for bias and comparing efficiency of these estimators.

1. Introduction

If we are given the distinct abscissae $\{x_j; j=1,2,..,n\}$ and corresponding ordinate values $\{f(x_j); j=1,2,..,n\}$, and wish to estimate the value $\int_0^1 f(x).dx$, the linear quadrature problem requires us to adopt some rationale for choosing the weight coefficients $\{w_j; j=1,2,..,n\}$ so that the quantity $\sum_{j=1}^{n} w_j f(x_j)$ might be acceptable as an estimate of the required integral, for a wide variety of functions f(.). The formulation of such problems (numerical quadrature, interpolation, etc.) as problems of estimating the value of a bounded, linear functional in terms of given values of other bounded, linear functionals is now a well established technique in Numerical Analysis (e.g. Sard, 1963; Davis, 1963; Handscomb, 1966). Notice that, since the given information is generally insufficient to characterize the function f(.) completely, it is unavoidable that any numerical estimate be based also upon the assumption of extra information, and must therefore be tentative, at least to that extent.

It is interesting that, for the linear estimation problem, a number of superficially different strategies can lead to the same set of weight coefficients $\{w_j\}$. In particular, several approaches leading to weights which are said to be "optimal", or "best" in the sense of Sard, are discussed briefly in the following section.

However, when we require to estimate the value of a non-linear functional, such as

$\int_0^1 [f(x)]^2.dx$, from the foregoing information, it is necessary to re-examine the arguments leading to the choice of weight coefficients analogous to the previous $\{w_j\}$. It turns out that the points of view which, for the linear problem, ultimately lead to the same set of "best" weight coefficients, will no longer agree in the case of a non-linear problem. We are thus forced into a position, long faced by statisticians, of having a number of apparently reasonable "estimators" of the required unknown quantity. Indeed, a point of view explored in previous papers (Larkin, 1969, 1971, 1972, 1974), poses the estimation problem in the context of a Hilbert space possessing a weak Gaussian probability distribution which, within limits, permits the use of standard statistical techniques such as Maximum Likelihood and Minimum Variance estimation. This point of view naturally includes non-linear estimation problems and also suggests that we might compare the different estimators on the basis of their bias, efficiency, consistency, etc.

2. The Linear Estimation Problem

Let H be a Hilbert space of functions defined on some domain X, and let $\{L_j; j=0,1,2,...,n\}$ be bounded linear functionals on H, with corresponding Riesz representers $\{g_j \in H; j=0,1,2,...,n\}$. Suppose we are given the numerical values $\{\alpha_j; j=1,2,...,n\}$, and are told that $\exists h \in H$ such that

$$L_j h \equiv (h,g_j) = \alpha_j \quad , \quad j = 1,2,...,n \qquad (1)$$

and wish to estimate the numerical value of $L_0 h \equiv (h,g_0)$.

Notice that, if

$$L_j f = f(x_j) , \qquad \forall f \in H; \ j = 1,2,\ldots,n ,$$

and we allow the $\{x_j\}$ to be unrestricted within X, then H must possess a reproducing kernel function in order that the $\{L_j\}$ be bounded linear functionals for all $\{x_j\}$ (Aronszajn, 1950). Also, we restrict our consideration to real Hilbert spaces, in order to simplify the presentation without undue loss of generality.

As pointed out by Golomb and Weinberger (1958), relations (1) merely restrict h to a linear manifold in H, so information about some non-linear functional of h (e.g. its norm) must be given, or assumed, if we wish h to be unequivocally restricted to some bounded subset of H. The probabilistic notions, used later, provide an alternative mechanism for localizing h within H.

Suppose now that we wish to construct a linear combination $\sum_{j=1}^{n} w_j L_j h$ to use as an estimator of $L_0 h$; one argument proceeds as follows:

(a) Consider the error functional

$$Eh \overset{\text{def}}{=} L_0 h - \sum_{j=1}^{n} w_j L_j h = (h, g_0 - \sum_{j=1}^{n} w_j g_j) .$$

By the Schwarz inequality, we have

$$|Eh| \le \|h\| . \|g - \sum_{j=1}^{n} w_j g_j\| \qquad \text{——— (2)}$$

and, since we have no control over $\|h\|$, a sensible approach to limiting the magnitude of the error is to choose the weights $\{w_j\}$ so as to minimize the quantity $\|g_0 - \sum_{j=1}^{n} w_j g_j\|$. In other words, the Riesz representer $\hat{g}_0$ of the functional of h which we actually compute as an estimator of L_0h is simply the projection Pg_0 of g_0 onto the subspace G of H spanned by $\{g_j; j=1,2,\ldots,n\}$. Notice also that, since an orthogonal projection is self-adjoint,

$$(\hat{g}_0,h) = (Pg_0,h) = (g_0,Ph) \stackrel{\text{def}}{=} (g_0,\hat{h}) = L_0\hat{h} \quad ,$$

and it is well-known (e.g. Davis, 1963) that an alternative characterization of $\hat{h}$ is that element of smallest norm which satisfies the constraints (1).

If an upper bound on $\|h\|$ is known, either (2) or the Hypercircle Inequality (e.g. Davis, 1963) provides an upper bound on $|Eh|$.

(b) In order to develop quadrature rules, the early numerical analysts effectively chose to construct a simple function (usually a low order polynomial) satisfying the given constraints, and then integrate this over the required range. An analogous procedure in the present context is to construct the function $\hat{h} \in H$ with smallest norm, subject to conditions (1), and then evaluate $L_0\hat{h}$ as an estimate of L_0h.

(c) Following the point of view developed by the author in the previously mentioned papers, we might regard H as possessing a weak Gaussian probability measure with a relative likelihood functional

$$\mathscr{L}(h) = \exp(-\tfrac{\lambda}{2}.\|h\|^2) \qquad \forall h \in H \quad ,$$

where λ is an unknown, positive constant.

In that case, $\hat{h}$ is the function most likely to have resulted in the given observations on h, so the estimator $L_0\hat{h}$ is simply the Maximum Likelihood estimator of L_0h. This, of course, ultimately agrees with the procedures outlined in (a) and (b) above. However, the assumption of the distribution on H, instead of an upper bound on $\| h \|$, permits the use of standard statistical techniques to obtain further interesting results, reported in the earlier papers.

(d) Without delving too deeply into the mathematical subtleties (Kuelbs, Larkin and Williamson, 1972) we denote the expectation value of a functional F(.) by

$$\int_H F(h).\mu(dh) \quad ,$$

if it exists, where $\mu(.)$ denotes the previously introduced weak Gaussian measure. If there exists an orthogonal projection operator P_n onto an n-(finite)-dimensional subspace of H such that

$$F(h) = F(P_n h) \quad , \qquad \forall h \in H \quad ,$$

then F(.) is called a "tame" functional, and its expectation value exists in the sense of an extended-valued, n-dimensional integral. Expectation values of tame, and other simple, functionals can sometimes be evaluated analytically. In particular, if g is a fixed element of H, it may be shown (Larkin, 1972) that

$$\int_H (h,g)^{2r-1} . \mu(dh) = 0 \quad , \qquad \text{—— (3)}$$

and

$$\int_H (h,g)^{2r} . \mu(dh) = \frac{(2r)!}{r!} . \frac{\|g\|^{2r}}{(2\lambda)^r}$$

$$; \quad r = 1,2,3,.... \qquad \text{—— (4)}$$

Choosing r = 1 and

$$g = g_0 - \sum_{j=1}^{n} w_j g_j$$

for an arbitrary set of weights $\{w_h\}$, it follows (from (3)) that any homogeneous, linear estimator of $L_0 h$ is unbiased, and (from (4)) that the minimum variance, unbiased, linear estimator of $L_0 h$ is found by choosing the $\{w_j ; j=1,2,...,n\}$ so as to minimize $\|g_0 - \sum_{j=1}^{n} w_j g_j\|$, in agreement with strategies (a), (b) and (c).

We remark in passing that a useful generalization, in which a semi-norm of $g_0 - \sum_{j=1}^{n} w_j g_j$ is minimized with respect to the $\{w_j\}$, resulting in a linear estimation rule which is exact for the null space of the semi-norm and "optimal" for its orthogonal complement in H, can also be obtained by the foregoing approaches. However, in the interests of clarity of presentation we shall confine the discussion to the simple situation.

3. The Quadratic Estimation Problem

For the purpose of estimating the value of $\int_0^1 [f(x)]^2.dx$ from given values $\{f(x_j);j=1,2,...,n\}$, one might consider simply squaring the given ordinates and then using these squared values in the optimal linear estimation rule. However, on two counts this procedure should be viewed with suspicion:

(i) It is easy to find non-pathological sets of ordinate values from which this strategy leads to a negative estimate for

$$\int_0^1 [f(x)]^2.dx.$$

(ii) If the given numerical information relates to linear functionals other than those of point evaluation the procedure is quite unworkable. This can happen, for example, if the given values result from attempts to measure ordinate values with a noise-free instrument of less than perfect resolution.

Generalizing the problem somewhat, let A_0 be a bounded linear, positive mapping of H into itself and define the bounded, linear, self-adjoint operators $\{A_{jk};j,k=1,2,\ldots n\}$ by

$$(h,A_{jk}h) = (h,g_j)\ (h,g_k)\ , \quad \forall h \in H. \qquad \text{——— (5)}$$

(a) Following the approach of section 2(a) we construct a general, homogeneous, quadratic combination of the quantities given in (1) to use as an estimator of (h,A_0h). Thus, a quadratic error functional $E_1(.)$ can be defined as

$$E_1(h) = (h,A_0h) - \sum_{j=1}^{n} \sum_{k=1}^{n} w_{jk}(h,A_{jk}h) \quad ,$$

i.e.:

$$E_1(h) = (h,[A_0 - \sum_{j=1}^{n} \sum_{k=1}^{n} w_{jk}A_{jk}]h) \quad , \quad \forall h \in H, \qquad \text{——— (6)}$$

and, using the Schwarz inequality twice, we find that

$$|E_1(h)| \leq \|A_0 - \sum_{j=1}^{n} \sum_{k=1}^{n} w_{jk}A_{jk}\| . \|h\|^2 \quad . \qquad \text{——— (7)}$$

Again we have no control over the quantity $\|h\|^2$, so a reasonable choice for the weight coefficients $\{w_{jk}\}$ might be that set which minimizes

$\|A_0 - \sum_{j=1}^{n} \sum_{k=1}^{n} w_{jk} A_{jk}\|$. Although possible in principle, the determination of the $\{w_{jk}\}$ which are "optimal" in the above sense is probably quite difficult computationally, involving, as it does, a non-trivial eigenvalue problem.

(b) and (c). The approach corresponding to sections 2(b) and 2(c) is simply to find $\hat{h}$ (usually not a difficult task) and evaluate $(\hat{h}, A_0\hat{h})$ as an estimator of (h, A_0h). The associated error functional is given by

$$E_2(h) = (h, A_0h) - (Ph, A_0Ph) = (h, [A_0 - PA_0P]h), \quad \forall h \in H.$$

This Maximum Likelihood estimator at least has the advantage that it always provides a positive result if A_0 is positive, but it is generally biased in the statistical sense. If A_0 (and hence PA_0P) has finite trace it may be shown (Larkin, 1972) that

$$\int_H (h, A_0h) . \mu(dh) = \frac{\operatorname{trace}(A_0)}{\lambda}$$

and

$$\int_H (h, PA_0Ph) . \mu(dh) = \frac{\operatorname{trace}(PA_0P)}{\lambda},$$

and in general these two quantities will not be identical. However, there is nothing to prevent our using the estimator

$$\frac{\mathrm{trace}(A_0)}{\mathrm{trace}(PA_0P)} \cdot (\hat{h}, A_0\hat{h}) \equiv \frac{\mathrm{trace}(A_0)}{\mathrm{trace}(PA_0P)}(h, PA_0Ph) \qquad (8)$$

which is unbiased, as an estimator of (h,A_0h). This modification is analagous to the usual multiplicative correction to a sample variance in order to produce an unbiased estimator of the population variance. In simple problems it may be possible to evaluate the required traces analytically.

(d) A Minimum Variance estimator is found by choosing the weight coefficients to minimize the quantity

$$\int_H \{(h,A_0h) - \sum_{j=1}^{n} \sum_{k=1}^{n} w_{jk}(h,A_{jk}h)\}^2 . \mu(dh) \quad , \qquad (9)$$

which may be expressed as

$$B - 2 \sum_{j=1}^{n} \sum_{k=1}^{n} w_{jk}C_{jk} + \sum_{j=1}^{n} \sum_{k=1}^{n} \sum_{r=1}^{n} \sum_{s=1}^{n} w_{jk}w_{rs}D_{jkrs} \quad ,$$

where

$$\left.\begin{aligned} B &= \int_H (h,A_0h)^2 . \mu(dh) \quad , \\ C_{jk} &= \int_H (h,A_0h)\ (h,A_{jk}h) . \mu(dh) , \\ \text{and} \quad D_{jkrs} &= \int_H (h,A_{jk}h)\ (h,A_{rs}h) . \mu(dh) \end{aligned}\right\} . \qquad (10)$$

Thus, the $\{w_{jk}\}$ satisfy

$$\sum_{j=1}^{n} \sum_{k=1}^{n} D_{jkrs} w_{jk} = C_{rs} \quad ; \qquad r,s=1,2,\ldots,n \quad .$$

Once again, in simple cases the required expectation values over H may be evaluated analytically, and it then becomes a simple matter to find the $\{w_{jk}\}$ which are "optimal" in this sense. In general, this Minimum Variance estimator will not be unbiased, but if unbiasedness is regarded as critical the minimization could be performed subject to the extra condition

$$\int_H (h,A_0h).\mu(dh) = \sum_{j=1}^{n} \sum_{k=1}^{n} w_{jk}.\int_H (h,A_{jk}h).\mu(dh) \quad ,$$

giving a less efficient (in the statistical sense) estimator, with zero bias. In any case, this last estimator will be at least as efficient as the bias-corrected Maximum Likelihood estimator introduced in (b) and (c) above.

4. An Illustration

To illustrate the preceding theory, suppose X is a real line segment and we wish to estimate the value of $\int_a^b [h(x)]^2.q(x).dx$, for some positive, integrable q(.) given that

$$h(x_j) = \alpha_j \quad ; \quad j = 1,2,\ldots,n$$

for some set of fixed, distinct abscissae $\{x_j \in X\}$. Suppose also that H possesses the reproducing kernel function K(.,.), so that

$$(h, K(.,x)) = h(x) \quad , \quad \forall x \in X \text{ and } \forall h \in H,$$

and if $\{\phi_j ; j=1,2,3,\ldots\}$ is any ortho-normal basis for H we have

$$K(x,y) = \sum_{j=1}^{\infty} \phi_j(x).\phi_j(y)$$

We now determine the traces of the operators A_0 and PA_0P. Consider

$$(h, A_0 h) = \int_a^b (h, K(.,x))h(x).q(x).dx$$

$$= \int_a^b (h, h(x)K(.,x).q(x))dx$$

i.e.

$$(h, A_0 h) = (h, \int_a^b h(x) K(.,x).q(x).dx) \qquad \forall h \in H.$$

Hence

$$(A_0 h)(y) = \int_a^b K(y,x).h(x).q(x).dx \qquad \forall h \in H.$$

However, by definition,

$$\text{trace}(A_0) = \sum_{j=1}^{\infty} (\phi_j, A_0 \phi_j)$$

$$= \sum_{j=1}^{\infty} (\phi_j, \int_a^b \sum_{k=1}^{\infty} \phi_k(.)\phi_k(x).\phi_j(x).q(x).dx)$$

$$= \int_a^b \sum_{j=1}^{\infty} \sum_{k=1}^{\infty} \phi_j(x).\phi_k(x).(\phi_j,\phi_k).q(x).dx$$

$$= \int_a^b \sum_{j=1}^{\infty} |\phi_j(x)|^2.q(x).dx$$

i.e.

$$\text{trace}(A_0) = \int_a^b K(x,x).q(x).dx \quad . \qquad \text{——— (11)}$$

Similarly, if we now choose $\{\phi_j;j=1,2,...,n\}$ to be an ortho-normal basis for G, it is not difficult to show, by ortho-normalizing the elements $\{K(.,x_j);j=1,2,...,n\}$, that

$$\text{trace } (PA_0P) = \int_a^b \left\{\sum_{j=1}^{n} |\phi_j(x)|^2\right\}.q(x).dx$$

$$= \int_a^b \left\{\sum_{j=1}^{n} \sum_{k=1}^{n} E_{jk}K(x,x_j)K(x,x_k)\right\}.q(x).dx \ ,$$

$$\text{——— (12)}$$

where the matrix E is the inverse of the matrix whose (j,k)-th element is $K(x_j,x_k)$.

The right-hand-sides of (11) and (12) can

then be evaluated, either analytically or numerically, leading to an explicit form for the unbiased estimator (8).

We now show how to calculate the coefficients (10) for use in constructing the Minimum Variance estimator. We have

$$B = \int_H \left\{ \int_a^b [h(x)]^2 q(x).dx \right\}^2 .\mu(dh)$$

$$= \int_a^b \int_a^b \left\{ \int_H [h(x).h(y)]^2.\mu(dh) \right\} q(x).q(y).dxdy \quad ,$$

using the Fubini theorem (Kuelbs, Larkin and Williamson, 1972). For fixed x and y, $[h(x).h(y)]^2$ is a tame functional based on the subspace of H spanned by K(.,x) and K(.,y), so we obtain (Larkin, 1972)

$$\int_H [h(x)h(y)]^2 \mu(dh) = \left[\frac{\lambda}{2\pi|M|}\right]^{1/2} \times$$

$$\int_{-\infty}^{\infty} \int_{-\infty}^{\infty} (h_1h_2)^2.\exp\left(-\frac{\lambda}{2}[h_1h_2]M^{-1}\begin{bmatrix}h_1\\h_2\end{bmatrix}\right).dh_1dh_2 \quad ,$$

where

$$M = \begin{bmatrix} K(x,x) & K(y,x) \\ K(x,y) & K(y,y) \end{bmatrix} .$$

Hence

$$\int_H [h(x).h(y)]^2.\mu(dh) = \frac{1}{\lambda^2}\{K(x,x)K(y,y)$$

$$+2[K(x,y)]^2\} \quad ,$$

yielding

$$B = \frac{1}{\lambda^2} \int_a^b \int_a^b \{K(x,x)K(y,y)+2[K(x,y)]^2\}q(x).$$

$$q(y).dxdy \quad . \qquad \text{———} \quad (13)$$

Furthermore,

$$C_{jk} = \int_H \left\{ \int_a^b [h(x)]^2 q(x).dx \right\} h(x_j)h(x_k).\mu(dh)$$

$$= \int_a^b \left\{ \int_H [h(x)]^2 h(x_j)h(x_k).\mu(dh) \right\} q(x).dx$$

so that

$$C_{jk} = \frac{1}{\lambda^2} \int_a^b \{K(x_j,x_k)K)x,x)+2K(x_j,x).K(x_k,x)\}$$

$$q(x).dx \quad , \qquad \text{———} \quad (14)$$

and

$$D_{jkrs} = \int_H h(x_j)h(x_k)h(x_r)h(x_s).\mu(dh) \qquad ,$$

which may be expressed in the form

$$D_{jkrs} = \frac{1}{\lambda^2} \{K(x_j,x_k)K(x_r,x_s)+K(x_j,x_r).K(x_k,x_s)$$
$$+K(x_j,x_s).K(x_k,x_r)\} \quad \text{—} \quad (15)$$

The integrals (13), (14) and (15) may be evaluated numerically, if not analytically.

We now specialize further to a "practical" example for which the previous theory reduces to a particularly simple form. Suppose a time-dependent voltage h(.), band-limited so that its component frequencies all lie in the real interval [-a,a], is applied to a unit resistance. The applied voltage is sampled at time intervals of $\frac{\pi}{a}$ and it is required to estimate the total power dissipated over an interval $(o,\frac{n\pi}{a})$ from a knowledge of the n voltage values measured during this interval.

To formulate this problem mathematically, we take H to be the Payley-Wiener-Hilbert space of band-limited functions, square-integrable on the real line, with inner product

$$(f,g) = \int_{-\infty}^{\infty} f(x)g(x).dx \quad , \qquad \forall f,g\in H \quad ,$$

(e.g. de Branges, 1968). This space possesses the reproducing kernel

$$K(x,y) = \frac{\text{Sin}[a(x-y)]}{\pi(x-y)} \quad , \qquad \forall x,y\in(-\infty,\infty) \, ,$$

which happens to be the Whittaker Cardinal Function (Whittaker, 1915).
We suppose that

$$h(j\tfrac{\pi}{a}) = \alpha_j \quad ; \qquad j = 1,2,\ldots,n,$$

and wish to estimate the value of $\int_0^{n\pi/a} [h(x)]^2.dx$, i.e. we take $q(x) \equiv 1$.

It is easily verified that

$$\hat{h}(x) = \sum_{j=1}^{n} \alpha_j \frac{\text{Sin}(ax-j\pi)}{a(x-j\pi/a)}$$

from which we obtain the Maximum Likelihood estimator

$$\int_0^{n\pi/a} [\hat{h}(x)]^2.dx = \frac{1}{a} \sum_{j=1}^{n} \sum_{k=1}^{n} \alpha_j \alpha_k \int_0^{n\pi} \frac{\text{Sin}(u-j\pi).\text{Sin}(u-k\pi)}{(u-j\pi)\ (u-k\pi)} .du \qquad (16)$$

Furthermore, since

$$K(x,x) = \frac{a}{\pi} \quad , \qquad \forall x \quad ,$$

we see from (11) that

$$\text{trace}(A_0) = n.$$

Also, since

$$K(j\frac{x}{a},k\frac{x}{a}) = \frac{a}{\pi}.\delta_{jk} \quad ,$$

δ_{jk} being the Kronecker-δ, the matrix A appearing in (12) is simply $\frac{\pi}{a}$ times the unit matrix. Hence

$$\text{trace}(PA_0P) = \frac{\pi}{a}\int_0^{n\pi/a} \sum_{j=1}^{n} \left[\frac{\text{Sin}(ax-j\pi)}{\pi(x-j\pi/a)}\right]^2 .dx \quad ,$$

so the correction factor β required in order to make (12) unbiased is given by

$$\beta = n\pi\left\{\sum_{j=1}^{n} \int_0^{n\pi} \frac{\text{Sin}^2(u-j\pi)}{(u-j\pi)^2}.du\right\}^{-1} . \qquad (17)$$

It may be verified that $\beta > 1$ and tends to that limit for large n.

The weight coefficients for the Minimum Variance estimator can be found if the quantities $\{C_{jk}\}$ and $\{D_{jkrs}\}$ are known. From (14) we find that

$$\lambda^2 C_{jk} = \frac{na}{\pi}.\delta_{jk} + \frac{2}{\pi^2}\int_0^{n\pi} \frac{\text{Sin}(u-j\pi).\text{Sin}(u-k\pi)}{(u-j\pi)\ (u-k\pi)}.du \quad , \tag{18}$$

and from (15),

$$\lambda^2 D_{jkrs} = \frac{a^2}{\pi^2}\left\{\delta_{jk}\delta_{rs}+\delta_{jr}\delta_{ks}+\delta_{js}\delta_{kr}\right\} \tag{19}$$

Hence, making use of index symmetry, we find that the weights $\{w_{jk}\}$ satisfy the equations

$$\delta_{jk}\sum_{r=1}^{n} w_{rr} + 2w_{jk} = \frac{n\pi}{a}.\delta_{jk} + \frac{2}{a}.\int_0^{n\pi} \frac{\text{Sin}(u-j\pi).\text{Sin}(u-k\pi)}{(u-j\pi)(u-k\pi)}.du \quad , \tag{20}$$

$$j,k = 1,2,\ldots,n.$$

Equations (20) give the off-diagonal weight coefficients explicitly, and it requires only a little algebraic manipulation to show that, for $j = 1,2,\ldots,n$,

$$w_{jj} = \frac{1}{a}.\int_0^{n\pi} \left[\frac{\mathrm{Sin}(u-j\pi)}{u-j\pi}\right]^2 .du + \frac{n\pi}{(n+2)a} .(1-\frac{1}{\beta}) \to$$

$$\frac{1}{a}\int_0^{n\pi} \left[\frac{\mathrm{Sin}(u-j\pi)}{u-j\pi}\right]^2 .du, \quad \text{as } n\to\infty. \quad \text{———} \quad (21).$$

Comparing (16) with (20) and (21) we see that, although the off-diagonal weight coefficients agree for the Minimum Variance and uncorrected Maximum Likelihood estimators, the diagonal coefficients exhibit agreement only in the limit of large n. The Minimum Variance estimate will always exceed the uncorrected Maximum Likelihood estimate for this problem.

6. References

1. N. Aronszajn, Theory of reproducing kernels, Trans. Amer. Math. Soc. 68 (1950), 337-404. MR 14,#479.

2. L. de Branges, Hilbert spaces of entire functions, Prentice-Hall, Englewood Cliffs, N.J., (1968). MR 37 #4590.

3. P.J. Davis, Interpolation and Approximation, Blaisdell, New York and London, (1963). MR 28 #393.

4. M. Golomb and H.F. Weinberger, Optimal approximation and error bounds, Proc. Sympos. on Numerical Approximation (Madison, Wis., 1958), Math. Res. Center, U.S. Army, Univ. of Wisconsin Press, Madison, Wis., (1959), pp. 117-190. MR 22 #12697.

5. D.C. Handscomb (Editor), Methods of numerical

approximation, Pergamon Press, New York, (1966).

6. J. Kuelbs, F.M. Larkin and J. Williamson, Weak probability distributions on reproducing kernel Hilbert spaces, Rocky Mt. J. Math.2 (1972), 369-378.

7. F.M. Larkin, Estimation of a non-negative function, Nordisk Tidskr. Informationbehandling 9 (1969), 30-52.

8. ———, Optimal approximation in Hilbert spaces with reproducing kernel functions, Math. Comp. 24 (1970), 911-921.

9. ———, Optimal estimation of bounded linear functionals from noisy data, Proc. I.F.I.P. Congress, Ljubljana, (1971).

10. ———, Gaussian Measure in Hilbert Space, and Applications in Numerical Analysis, Rocky Mt. J. Math. 2 (1972), 379-421.

11. ———, Probabilistic Error Estimates in Spline Interpolation and Quadrature, Proc. I.F.I.P. Congress, Stockholm, (1974).

12. A Sard, Integral representations of remainders, Duke Math. J. 15 (1948), 333-345. MR 10, 197.

13. ———, Best approximate integration formulas; Amer. J. Math. 71 (1949), 80-91, MR 10, 576.

14. ———, Smoothest approximation formulas, Ann. Math. Statist. 20 (1949), 612-615, MR 12, 84.

15. ———, Linear approximation, Math. Surveys, no. 9, Amer. Math. Soc., Providence, R.I., (1963), MR 28 #1429.

16. E.T. Whittaker, "On the functions which are represented by the expansions of the interpolation theory", Proc. Edinburgh Math. Soc. v. 35, (1915), pp. 181-194.

On Alternation in the Restricted Range Problem

WILLIAM H. LING

1. INTRODUCTION

Let $C[a,b]$ denote the space of all continuous real valued functions on $[a,b]$ and let f, l and u be elements in $C[a,b]$ with $l \leq u$ on $[a,b]$. Assume V is an n-dimensional extended Haar subspace of order n and define
$W = \{v \in V \mid l(x) \leq v(x) \leq u(x)$ for all x in $[a,b]\}$. (For the definition of extended Haar subspace, see [2,p.6].) The restricted range problem consists of finding a v_0 in W such that

$$\|f - v_0\| = \inf\{\|f - v\| \mid v \in W\} \doteq \rho,$$

where the supremum norm is employed. If $W \neq \phi$, the compactness of W ensures the existence of a v_0 which is then called a best restricted approximation to f.

In the case that $l < u$ holds on $[a,b]$, an alternation theory with "one exception" has been developed by Taylor for best restricted approximations. (See [7] and [8].) In the case that l and u agree at one or more points of $[a,b]$, Sippel [6], using a modified definition of alternation, has shown that best restricted approximations with "one exception" must alternate. In [9] Taylor gives similar results in the case that u and l fulfill special conditions.

By employing a modification of the original problem, it is shown here that an alternation property may be expected in the "exceptional case" as well. This approach can be easily applied to simultaneous approximation for the straddle point situation. These results should prove useful in developing a modified Remes algorithm for handling "all possibilities" in restricted range and simultaneous approximation problems.

2. RESTRICTED RANGE APPROXIMATION

We present first Sippel's main result on restricted range approximation. The following theory includes the situation $l < u$ on $[a,b]$ as a special case.

Let $l \leq u$ be given and suppose
$T = \{y_1, y_2, \cdots, y_s\} := \{x \in [a,b] \mid u(x) = l(x)\}$.
Further, assume there exist positive integers $\{w_j\}_{j=1}^{s}$ such that for each j
$u^{(k)}(y_j) = l^{(k)}(y_j)$ for $0 \leq k \leq w_j - 1$ and
$u^{(w_j)}(y_j) \neq l^{(w_j)}(y_j)$. (Derivatives are assumed to exist and be continuous in a neighborhood of each y_j). In addition we require that $\sum_{j=1}^{s} w_j < n$, where n is the dimension of V. Observe that under this assumption, T necessarily has less than n elements.

The following notation is used. For $v \in W$, define

$$X_{+1} = \{x \in [a,b] \mid v(x) = f(x) - \|f - v\|\}$$

$X_{-1} = \{x \in [a,b] \mid v(x) = f(x) + \|f - v\|\}$

$X_{+2} = \{x \in [a,b] \mid v(x) = l(x)\}$

$X_{-2} = \{x \in [a,b] \mid v(x) = u(x)\}$

$E_u = X_{-1} \cup X_{-2} \cup \{y_j \in T \mid v^{(w_j)}(y_j) = u^{(w_j)}(y_j)\}$ and

$E_l = X_{+1} \cup X_{+2} \cup \{y_j \in T \mid v^{(w_j)}(y_j) = l^{(w_j)}(y_j)\}$. We now introduce a modified definition of alternation.

DEFINITION 2.1: Let $f \in C[a,b]$ be given and set $k(n) = n - \sum_{j=1}^{s} w_j$. An approximation $v \in W$ will be said to alternate n times on [a,b] with respect to f, multiplicity counted, if there exist $k(n) + 1$ points $a \leq x_0 < x_1 < \cdots < x_{k(n)} \leq b$ which are elements of E_u and E_l alternately.

This definition incorporates counting of multiplicities as done by Rosman and Rosenbaum in the simultaneous problem [5]. The phrase "with respect to f" will be omitted where the context is clear.

Sippel's main result may be stated as follows.

THEOREM 2.1. Let $f \in C[a,b]$, $v_0 \in W$ and suppose $f(x) + \|f - v_0\| > l(x)$ and $f(x) - \|f - v_0\| < u(x)$ (1) hold for all x in [a,b]. Then v_0 is a best restricted approximation to f iff v_0 alternates n times on [a,b], multiplicity counted.

We now describe the "exceptional case" referred to in the introduction. If there exists an x_0 in [a,b] such that either $f(x_0) + \|f - v_0\| = l(x_0)$ (2) or $f(x_0) - \|f - v_0\| = u(x_0)$ (3), condition (1) in theorem 2.1 is violated. Then alternation of a best restricted approximation is no longer guaranteed. We call (2) and (3) the "exceptional case".

The following lemma translates (2) and (3) into the standard notation employed by Taylor, [8].

LEMMA 2.1: Let f, l, u and v_0 be given as above. Then

(a) there exists an x_0 with $f(x_0)+\|f-v_0\| = l(x_0)$ iff $X_{-1} \cap X_{+2} \neq \phi$ and

(b) there exists an x_0 with $f(x_0)-\|f-v_0\| = u(x_0)$ iff $X_{+1} \cap X_{-2} \neq \phi$.

Proof. The lemma follows from the fact that $l(x) \leq v_0(x) \leq f(x) + \|f - v_0\|$ and $f(x) - \|f - v_0\| \leq v_0(x) \leq u(x)$ hold on [a,b].

Due to lemma 2.1, the "exceptional case" may be referred to as a restricted range problem with either $X_{-1} \cap X_{+2}$ or $X_{+1} \cap X_{-2}$ nonempty.

For later use we introduce the following simple example to illustrate the "exceptional case".

EXAMPLE 1: Let $[a,b] = [-1,1]$, $V = P_2$, $n = 3$, $l(x) = -1$, $u(x) = |x|$ and $f(x) = 2$. It is easy to check that $\rho = 2$ and that f(x) has many best restricted range approximations. In particular observe that $v_0(x) = (1/2)x^2$ is a best approximation which does not alternate n = 3 times. However with a modification of the original problem, one may still find a best approximation to f which exhibits an alternation property. The development of this modification follows.

Let f, u, l and V be the given functions in the original restricted range problem. Define the quantity

$$\alpha = \max \{ \max_{x\in[a,b]} (f(x)-u(x)), \max_{x\in[a,b]} (l(x)-f(x)), 0\}$$

and let $l_1(x) = \max \{f(x) - \alpha, l(x)\}$,

$$u_1(x) = \min \{f(x) + \alpha, u(x)\},\ f_1(x) = (1/2)(l_1(x) + u_1(x)).$$

Observe that $l_1(x) \le f_1(x) \le u_1(x)$ holds on [a,b]. With these definitions, we now state the modified problem, namely: find a $v_0 \in V$ which minimizes $\|f_1 - v\|$ where $l_1(x) \le v(x) \le u_1(x)$ on [a,b]. We give next a result concerning this modified restricted range problem.

LEMMA 2.2: Let $X_{-1} \cap X_{+2}$, $X_{+1} \cap X_{-2}$, f, f_1, l_1, u_1, α and ρ be defined as above. The following three statements are equivalent.

(a) f_1 has a best restricted range approximation, with respect to l_1, u_1.

(b) f has a best restricted range approximation with respect to l, u and $\rho = \alpha$.

(c) At least one of the sets $X_{-1} \cap X_{+2}$, $X_{+1} \cap X_{-2}$, $X_{+1} \cap X_{-1}$ is nonempty.

Proof. The fact that (b) and (c) are equivalent follows from the definition of the sets X_{+1}, X_{+2}, X_{-1} and X_{-2}. We show that (a) and (b) are equivalent.

To see this, assume v_0 is a best approximation to f_1. In general $\rho \geq \alpha$ holds. However, since $l_1(x) \leq v_0(x) \leq u_1(x)$ on $[a,b]$, we have both

$$f(x) - \alpha \leq v_0(x) \leq f(x) + \alpha$$

and $l(x) \leq v_0(x) \leq u(x)$ satisfied on the interval. Thus $\rho = \alpha$ and v_0 is a best restricted approximation to $f(x)$.

To show (b) implies (a), let $v_0(x)$ be a best restricted range approximation to f with $\rho = \alpha$. Then the set

$$W_1 = \{v \in V | \; l_1(x) \leq v(x) \leq u_1(x)$$

for all $x \in [a,b]\}$ is nonempty and a best restricted approximation to f_1 exists.

We make two remarks which will be employed in the next theorem.

REMARK 1: Let f, l, u and V be given and assume $l(x) \leq f(x) \leq u(x)$ holds on $[a,b]$. It is easily shown that if either $X_{-1} \cap X_{+2} \neq \phi$ or $X_{+1} \cap X_{-2} \neq \phi$, then f is in V. Of course then the best approximation to f, v_0, is identical with f. In such a case we employ the convention that $v_0 = f$ alternates n times, multiplicity counted, since it trivially satisfies the requirements of definition 2.1.

REMARK 2: We comment on "overagreement". Recall in Sippel's main theorem, theorem 2.1, that the assumption $\sum_{j=1}^{s} w_j < n$ is made. Here n is both the dimension and order of the approximating family V. We shall say that "overagreement" occurs if $\sum_{j=1}^{s} w_j \geq n$. Observe that this includes the situation that $T = \{x \in [a,b] \mid u(x) = l(x)\}$ has n or more points. Using the fact that V is an extended Haar system of order n, theorem 4.3 in [2, p.24] implies that the set

$$W = \{v \in V \mid l(x) \leq v(x) \leq u(x)\}$$

is either empty or contains one element. Thus in the restricted range problem, f has either no best approximation or a unique best approximation. However if a best approximation does exist in the "overagreement" situation, note that it alternates n times, multiplicity counted, as stipulated in definition 2.1.

We may now state the main theorem.

THEOREM 2.2: Let f, l, u, f_1, l_1, u_1 and V be given as defined above. Assume a best restricted range approximation to f from V exists. Then there exists a best approximation to f from V which:

(a) alternates n times on [a,b] with respect to f, multiplicity counted or

(b) alternates n times on [a,b] with respect to f_1, multiplicity counted.

Proof. Let $\alpha = \max \{ \max_{x \in [a,b]} (f(x) - u(x)),$

$$\max_{x \in [a,b]} (l(x) - f(x)), 0\}.$$

Suppose v_0 is a best restricted range approximation to f. In case 1, we show that part (b) of the theorem holds.

Case 1. $\rho = \alpha$.

Since $\rho = \alpha$, lemma 2.2 implies that f_1 has a best restricted range approximation, v_1, with respect to l_1, u_1. From the proof of lemma 2.2, we know that v_1 is also a best restricted range approximation to f. We define sets Y_{+1}, Y_{-1}, Y_{+2}, Y_{-2} for the function f_1 in a similar manner to the sets X_{+1}, X_{-1}, X_{+2}, X_{-2} for f; for example

$$Y_{+1} = \{x \in [a,b] \mid v_1(x) = f_1(x) - \|f_1 - v_1\|\}.$$

We may now describe the two possibilities for v_1.

(a): $Y_{-1} \cap Y_{+2} \neq \phi$ or $Y_{+1} \cap Y_{-2} \neq \phi$.

In this situation, since $l_1 \le f_1 \le u_1$, holds on $[a,b]$, $v_1 \equiv f_1$ and thus v_1 alternates n times with respect to f_1 as noted in remark 1.

(b): $(Y_{-1} \cap Y_{+2}) \cup (Y_{+1} \cap Y_{-2}) = \phi$.

By lemma 2.1, condition (1) in theorem 2.1 holds with respect to f_1, and Sippel's theorem applies to v_1. In the case of "overagreement", remark 2 assures that v_1 alternates n times on [a,b] with respect to f_1, multiplicity counted.

In case 2, we show that part (a) of the theorem holds.

Case 2. $\rho > \alpha$.

Using parts (b) and (c) of lemma 2.2, $\rho > \alpha$

ensures that $X_{-1} \cap X_{+2}$ and $X_{+1} \cap X_{-2}$ are both empty. Thus condition (1) in theorem 2.1 holds with respect to f. Sippel's theorem implies that v_0 alternates n times on [a,b] with respect to f, multiplicity counted. Remark 2 on "overagreement" of course holds in this case as well.

REMARK 3: As an illustration of theorem 2.2, recall example 1. In this example $v_0(x) = (1/2)x^2$ is a best restricted approximation to $f(x) = 2$, $X_{+1} \cap X_{-2}$ is nonempty and $\rho = \alpha = 2$. Using previous notation, $l_1(x) = 0$, $u_1(x) = |x|$ and $f_1(x) = (1/2)|x|$. Now it is easy to check that $v_1(x) = [(1 + \sqrt{2})/4]x^2$ from $V = P_2$ is a best restricted range approximation to f_1 with respect to l_1, u_1. Since $v_1(x)$ is also a best restricted range approximation to f and it alternates n = 3 times on [-1,1] with respect to f_1, multiplicity counted, part (b) of theorem 2.2 holds.

REMARK 4: Observe that in example 1, the function $v_0(x) = x^2$ from P_2 is a best restricted range approximation to f. Further, if we consider the restraining curves to be $l_1(x)$ and $u_1(x)$, note that $v_0(x)$ alternates n = 3 times on [-1,1] <u>with respect to f</u>, multiplicity counted. The question arises: Consider a restricted range problem in which the exceptional case, $X_{-1} \cap X_{+2} \neq \phi$ or $X_{+1} \cap X_{-2} \neq \phi$, occurs. Can one always find a best approximation to f which alternates n times with respect to f, multiplicity counted, where the alternation is determined by l_1 and u_1, rather than by l and u? If such a best approximation could always be found, theorem 2.2 could be modified so that part (b) of the theorem might be omitted. Unfortunately, such a best approximation cannot always be found, as the following example demonstrates.

EXAMPLE 2: Let $[a,b] = [-1,0]$, $V = P_3$, $n = 4$, $f(x) = 2$, $u(x) = \exp(-x) - 1$ and $l(x) = 0$. Here $\rho = \alpha = 2$, $v_0(x) = 0$ is a best restricted approximation to f and $X_{+1} \cap X_{-2} \neq \phi$ occurs. It is easy to check that $u_1(x) = u(x)$, $l_1(x) = l(x)$ and $T = \{0\}$, $w_1 = 1$, $s = 1$, $k(4) = 4 - 1 = 3$. Using the fact that an element of P_3 has at most one inflection point, a short argument shows that there exists no v in P_3, $l_1(x) \leq v(x) \leq u_1(x)$, which alternates 4 times on $[-1,0]$ <u>with respect to f</u>, multiplicity counted.

In closing, we comment briefly on the possible numerical application of theorem 2.2. An approach for developing a modified Remes algorithm for restricted range approximation, including "all possibilities", might be as follows. (The computational details may of course require one to assume special forms for l, u, l_1, u_1 and f_1 as is done in [9, p. 72].) We remark that the application of an algorithm for computing best approximations in the "sum" norm ([4]) would be expanded if a general restricted range algorithm could be developed.

Let f, l, u and V be given.

Step 1. Compute α.

Step 2. If $\alpha = 0$ then $l(x) \le f(x) \le u(x)$ holds on [a,b]. Thus a "standard" algorithm based on Sippel's theorem 2.1 holds.

Step 3. If $\alpha > 0$, compute the best restricted range approximation to f_1. Since $l_1(x) \le f_1(x) \le u_1(x)$ holds, the "standard" algorithm of step 1 may be applied.

Step 4. If a best approximation to f_1 is found in step 3, we are done. If the algorithm finds no best approximation to f_1, then necessarily $\rho > \alpha$ holds and the sets $X_{-1} \cap X_{+2}$ and $X_{+1} \cap X_{-2}$ are empty. Hence the "standard" algorithm of step 1 applies to f.

REFERENCES

1. C. B. Dunham, Simultaneous Chebyshev Approximation of Functions on an Interval, Proc. Am. Math. Soc. 18 (1967), 472-477.

2. S. J. Karlin and W. J. Studden, "Tchebycheff Systems: with applications in analysis and statistics," Interscience Publications, New York, 1966.

3. W. H. Ling and J. E. Tornga, The Constant Error Curve Problem for Varisolvent Families, J. Approximation Theory 11 (1974), 54-72.

4. W. H. Ling, H. W. McLaughlin and M. L. Smith, Approximation of Random Functions, to appear.

5. B. H. Rosman and P. D. Rosenbaum, Chebyshev Approximation of Regulated Functions by Unisolvent Families, J. Approximation Theory 4 (1971), 113-119.

6. W. Sippel, Approximation by Functions with Restricted Ranges, to appear.

7. G. D. Taylor, On Approximation by Polynomials having Restricted Ranges, SIAM J. Numer. Anal. 5 (1968), 258-268.

8. G. D. Taylor, Approximation by Functions having Restricted Ranges, III., J. Math. Anal. Appl. 27 (1969), 241-248.

9. G. D. Taylor, Approximation by Functions having Restricted Ranges: Equality Case, Numer. Math. 14 (1969), 71-78.

10. G. D. Taylor and M. J. Winter, Calculation of Best Restricted Approximations, SIAM J. Numer. Anal. 7 (1970), 248-255.

Some Results On The Dual Of An Approximation Problem.

S.J. Poreda

<u>Introduction</u>. Let f be a complex valued function continuous on the unit circle $U = \{|z| = 1\}$, and for $n=0,1,2,\ldots$ let p_n denote the algebraic polynomial of degree n of best uniform approximation to f on U. A classical result tells us that the sequence $\{p_n\}_{n=0}^{\infty}$ converges uniformly to f on U if and only if $f \in A$ (the set of all functions analytic in $D = \{|z| < 1\}$ and continuous in $\overline{D}$). This sequence always converges uniformly on compact subsets of D to h_f, the unique best H^∞ approximation to f on U (with respect to the L^∞ norm). Whether or not these polynomials converge to h_f on U is still an open question.

If f is Dini-continuous, that is, if $\int_0^1 w_f(t)t^{-1}\,dt$ converges, where $w_f(t)$ is the modulus of continuity of f, then [1] $h_f \in A$ and thus the difference $d_f = (f-h_f)$ is continuous on U. Since all of the polynomials of best approximation to d_f on U are identically zero, it follows [4] that d_f has constant modulus on U

and its image of U, $d_f(U)$, has negative index or winding number with respect to the origin (assuming of course that $f \notin A$). Let the negative integer Δ_f denote this index.

The problem of approximating f by H^∞ functions on U is dual to the problem of finding the supremum of the linear functional

$$L_f(g) = \frac{1}{2\pi} \int_{-\pi}^{\pi} g(e^{i\theta})f(e^{i\theta})e^{i\theta}d\theta \quad \text{over all}$$

$g \in H^1$ with $\|g\|_1 = \frac{1}{2\pi} \int_{-\pi}^{\pi} |g(e^{i\theta})|d\theta \leq 1.$

Furthermore, since $\int_{-\pi}^{\pi} g(e^{i\theta})h(e^{i\theta})e^{i\theta}d\theta = 0$ for any $h \in H^\infty$, we have that for all $g \in H^1$,

$$L_f(g) = L_{(f-h_f)}(g),$$

and by Hölder's inequality that $|L_f(g)| \leq \|d_f\|_\infty = \operatorname{ess\,sup}_{z\in U} |d_f(z)|$. If we now apply the Hahn-Banach Theorem we get the classical result [1] that

$$\sup_{\substack{g\in H^1 \\ \|g\|_1\leq 1}} |L_f(g)| = \|d_f\|_\infty. \tag{1.}$$

Finally, it has been shown [2] that there exists at least one such function g for which equality is attained. Such a function will be referred to as an extremal of L_f.

Properties of the Extremals in the Dual Problem

DeLeeuw and Rudin [2] proved that there exists a positive integer n_f, that depends only on f, such that no extremal of L_f can have more that n_f zeros in D. Furthermore, if we assume that n_f is the least such integer, then they also showed that there exists a unique strong outer function $G(z)$ with $\|G\|_1 = 1$ and with the property that for any choice of points $a_1, a_2, \ldots, a_{n_f}$ in D there is a unique extremal g for L_f such that $g(a_1) = g(a_2) = \ldots = g(a_{n_f}) = 0$. This function g is of the form

$$g(z) = \lambda\, G(z) \prod_{j=1}^{n_f} (z-a_j)(1-\bar{a}_j z),$$

where λ is a positive constant.

Carleson and Jacobs [1] proved that if g_1 and g_2 are both extremals for the same L_f, then g_1/g_2 is a rational function and that a necessary and sufficient condition for L_f to have a unique extremal g is that $g^{-1} \in H^1$.

In what follows we present two theorems

which relate the properties of the extremals of L_f to the properties of the kernel f. We shall continue to use the notation and the definitions of the introduction and assume that f is Dini-continuous and not in A.

<u>Theorem 1</u>. If $g \in H^1$ is an extremal of L_f, then g has exactly $(-\Delta_f-1)$ zeros in $\overline{D}$ in the sense that g is of the form,

$$g(z) = h(z) \prod_{j=1}^{-\Delta_f-1} [(z-a_j)(1-\bar{a}_j z)],$$

where $|a_j| \leq 1$ for $j=1,2,\ldots,(-\Delta_f-1)$ and $h \in H^1$ is a strong outer function. Furthermore, g is unique if and only if $\Delta_f = -1$.

The proof of this theorem necessitates the following lemma.

<u>Lemma 1</u>. Suppose $\phi(\theta)$ is a real valued function continuous on $[-\pi,\pi]$, $\phi(\pi) = \phi(-\pi) + K2\pi$ for some positive integer K and $g \in H^p$ for some $p > 1$ with $\arg g(e^{i\theta}) = \phi(\theta)$ a.e. Then $g(z)$ is of the form

$$g(z) = h(z) \prod_{j=1}^{k} [(z-a_j)(1-\bar{a}_j z)]$$

where $|a_j| \leq 1$ and $h \in H^1$ is non vanishing in D.

Proof. (of Lemma 1.). Let us first consider the case $K = 0$. By means of a Poisson integral construct a function $v(z)$ harmonic in D, continuous in $\overline{D}$ and such that $v(e^{i\theta}) = -\phi(\theta)$ for $\theta \varepsilon [-\pi,\pi]$. Let $u(z)$ be the harmonic function conjugate to $v(z)$ and $T(z) = \exp(u(z)+iv(z))$. The function $T(z) \varepsilon H^p$ for $p < \infty$ [3]. Now suppose $g \varepsilon H^p$ for some $p > 1$ and $\arg g(e^{i\theta}) = \phi(\theta)$ a.e. on $[-\pi,\pi]$. It then follows that $T(e^{i\theta})g(e^{i\theta}) \geq 0$. a.e. on $[-\pi,\pi]$ and that $T(z)g(z) \varepsilon H^1$. This implies that $T(z)g(z)$ is identically constant and, in particular, that $g(z)$ has no zeros in D.

Now suppose that $K \neq 0$ and that as before, $g \varepsilon H^p$ for some $p > 1$ and $\arg g(e^{i\theta}) = \phi(\theta)$ a.e. on $[-\pi,\pi]$. We consider two cases.

Case 1. Suppose g has more than K zeros in D. We can then write

$$g(z) = B_k(z)g^*(z),$$

where $B_k(z)$ is a finite Blaschke product of degree K, $g^* \varepsilon H^p$ for some $p > 1$ and g^* has at least one zero in D. Let $\phi^*(\theta) = i \log[B_k(e^{i\theta})e^{i\phi(\theta)}]$. Then $\phi^*(\theta)$ is continuous on $[-\pi,\pi]$ with $\phi^*(\pi) = \phi^*(-\pi)$ and $\arg g^*(e^{i\theta}) = \phi^*(e^{i\theta})$ a.e., which implies that $g^*(z)$ has no zeros in D by the above and which is, of course, a contradiction.

Case 2. Suppose g has less than K zeros in D. As before we can write

$$g(z) = B_L(z)g^*(z)$$

where this time $B_L(z)$ is a Blaschke product of degree L, $0 \leq L < K$, $g^* \in H^p$ for some $p > 1$ and has no zeros in D. Now set $\phi^*(\theta) = i \log [e^{i(K-L)\theta}B_K(e^{i\theta})e^{-i\phi(\theta)}]$. As before, $\phi^*(\theta)$ is continuous on $[-\pi,\pi]$ with $\phi^*(\pi) = \phi^*(-\pi)$. Also as before, construct a function $T(z) \in H^p$ for all $p < \infty$ for which $\arg T(e^{i\theta}) = \phi^*(\theta)$ on $[-\pi,\pi]$. By multiplying $T(z)$ by an appropriate constant, we have $\|T\|_1 = 1$ and so we see that both $z^{K-L}T(z)$ and $g^*(z)$ are extremals for some linear functional L_{f^*} where $f^*(e^{i\theta}) = \exp-(i[\phi^*(\theta)+(K-L)\theta])$ and hence their quotient is a rational function (that is ≥ 0 on U), that is,

$$g^*(z)/T(z) = \lambda \prod_{j=1}^{K-L} [(z-a_j)(1-\bar{a}_j z)],$$

where $\lambda > 0$ and whereby our assumption on

g^*, $|a_j| = 1$ for $j=1,\ldots,K-L$. The proof of our lemma is now complete.

Proof. (of Theorem 1.). We may assume without loss of generality that $L_f(g) > 0$ and so may write

$$\|d_f\|_\infty = L_f(g)$$

$$= \frac{1}{2\pi} \int_{-\pi}^{\pi} g(e^{i\theta})d_f(e^{i\theta})e^{i\theta}d\theta$$

$$\leq \|d_f\|_\infty \|g\|_1$$

$$= \|d_f\|_\infty .$$

Thus we must have that

$$g(e^{i\theta})d_f(e^{i\theta})e^{i\theta} = \|d_f\|_\infty \qquad \text{a.e.}$$

If we let $\phi(\theta) = -\arg(d_f(e^{i\theta})e^{i\theta})$, then ϕ is continuous on $[-\pi,\pi]$ and $\phi(\pi) = \phi(-\pi) + (-\Delta_f-1)2\pi$ and so by lemma 1. we have that

$$g(z) = h(z) \prod_{j=1}^{-\Delta_f-1} [(z-a_j)(1-\bar{a}_j z)]$$

where $|a_j| \leq 1$ for $j=1,2,\ldots,(-\Delta_f-1)$ and $h \in H^1$ is nonvanishing in D. We need only show now that h is a strong outer function but this follows from the previously mentioned result [2] of DeLeeuw and Rudin and arguments similar to those used in the proof of lemma 1.

The second part of our theorem now

follows easily, for if $\Delta_f < -1$ we can easily construct infinitely many extremals for L_f starting with any given one (see [2]). On the other hand if $\Delta_f = -1$, then since the quotient of any two extremals must be a rational function and since any extremal is nonvanishing in D, there can be but only one extremal (and there must be at least one). The proof of Theorem 1. is now complete.

A natural question that arises here pertains to how the number $(-\Delta_f-1)$ is related to the corresponding linear functional. More precisely, given a particular linear functional L on H^1 is there any way of determining the number of zeros its extremal(s) will have. By appealing to known properties of d_f in the case where f is rational, we can derive such a relationship. Although a special case, the rational functions are nevertheless dense in the space of continuous functions. Let us first give the following theorem.

<u>Theorem 2</u>. Let $f(z) = \sum_{j=1}^{n} \frac{\alpha_j}{(z-a_j)^{\ell_j}}$ where $|a_j| < 1$, ℓ_j is a positive integer and α_j is a

complex constant for $j=1,2,\dots,n$. The best H^∞ deviation from f is then of the form

$$d_f(z) = \lambda_f \left[\prod_{j=1}^{n} \left(\frac{1-\bar{a}_j z}{z-a_j}\right)^{\ell_j}\right]\left[\prod_{k=1}^{k_f} \left(\frac{z-c_k}{1-\bar{c}_k z}\right)\right]$$

where λ_f is a prescribed root of an explicitly determined polynomial and where $k_f (< \sum_{j=1}^{n} \ell_j)$ and the c_k's can be directly calculated in terms of λ_f.

Proof. This theorem is a slight modification of a theorem given in [5] and its proof follows from it together with the remarks made in our introduction.

We can now apply this result to the question at hand.

Theorem 3. For $g \in H^1$ let the linear functional L be defined by

$$L(g) = \sum_{j=1}^{n} \alpha_j\, g^{(\ell_j)}(a_j) \quad \text{where}$$

$g^{(\ell_j)}(a_j)$ denotes the ℓ_jth derivative of g at a_j, $|a_j| < 1$ for $j=1,2,\dots,n$, and the α_j's are complex constants. Then $L \equiv L_f$ where

$$f(z) = \sum_{j=1}^{n} \frac{(\alpha_j/\ell_j!)}{(z-a_j)^{\ell_j}},$$ and it follows that any

function $g \in H^1$ which is an extremal for L has exactly $(n-k_f-1)$ zeros in $\overline{D}$ (in the sense of Theorem 1.) when k_f is as in Theorem 2.

Remarks. In lemma 1., we assumed that $g \in H^p$ for some $p > 1$. An open question is whether this lemma remains true for $p = 1$.

The author wishes to thank Profs. Lennart Carleson and David Tepper for their assistance.

REFERENCES

[1.] L. Carleson and S. Jacobs, Best uniform approximation by analytic functions, Arkiv fur Mathematik, Vol. 10, No. 2., pp. 219-229 (1972).

[2.] K. DeLeeuw and W. Rudin, Extreme points and extremum problems in H^1, Pacific J. of Math. 8, pp. 467-485 (1958).

[3.] G. M. Goluzin, Geometric Theory of functions of a complex variable, A.M.S. Translation, Vol. 26, p. 415 (1969).

[4.] S. J. Poreda, A characterization of badly approximable functions, Trans. A.M.S., Vol. 169, pp. 249-256, (1972).

[5.] ___________, On the convergence of best uniform deviations, Trans. A.M.S., Vol. 179, pp. 49-59 (1973).

BEST LEAST SQUARE APPROXIMATION BY MINIMUM PROPERTY OF FOURIER EXPANSIONS

Chung-Lie Wang

I. INTRODUCTION

The central problem under consideration here is the solution of the following least squares problem:

LSP. Let A be a given m×n real matrix with $m \geq n$ and of rank r, and β a given m-vector. Then determine an n-vector x_0 such that

$$||\beta - Ax_0|| = \min_x ||\beta - Ax|| \quad , \tag{1.1}$$

where $|| \ ||$ indicates the Euclidean norm.

It is known that this problem always has a solution. The solution is unique if r=n. If r<n there is an (n-r)-dimensional linear variety of solutions. In this solution variety there is a unique x_0 such that

$$||x_0|| = \min_x ||x||. \tag{1.2}$$

This is an old problem (e.g. see [4,5,6,15]). However, since high speed computers have been produced and new techniques of numerical analysis found, this useful and important problem has attracted, and continues to attract, considerable attention.

It is well known that x_0 satisfies normal equations:

$$A^TAx=A^T\beta ,$$

and the conditioning of A^TA is worse than that of A (e.g. see [7,19]). The problem of finding techniques in order to compute the solution precisely has concerned mathematicians for many years. Recently, Businger-Golub [3], Golub [7], Golub-Wilkinson [8] and Hanson-Lawson [9] have given algorithms for the solution of LSP based upon Householder orthogonal transformations (see [10]) with or without a modification. Their algorithms have favourable numerical properties (e.g. see [11]) due to the use of these orthogonal transformations and the avoidance of the formation of A^TA. However, Rutishauser [17] has been concerned with possible cancellation involved in the evaluation of Ax_0 by using already computed x_0, and he considered a too precise x_0 undesirable. Further, Bauer [1] has established an algorithm using modified Gaussian eliminations while Björck [2] has established an algorithm using modified Gram-Schmidt orthogonalization. Osborne [13] has solved the problem by using eigenvalues of A^TA. Their techniques have ensured some improvement.

This paper is to present exact solutions of (1.1) for the cases r=n and r<n (with (1.2)) by means of the minimum property of Fourier expansions. To this end, in §2, we introduce over a Hilbert space the Gram-Schmidt ortho- normalizing process, minimum property of Fourier expansions, and a recursive formula for least square approximation

established in [20] by the author. In §3, we generalize the indicated recursive formula for an Euclidean space. In §4, we give solutions of LSP using the results from the previous sections. In §5, we mention further applications over Hilbert spaces. In §6, we will conclude with several remarks.

II. PRELIMINARIES

Let H be a separable Hilbert space over F, where F denotes the field R of real numbers or the field C of complex numbers. The notation $\langle , \rangle$ and $\|\ \|$ will be adopted for the scalar product and usual norm in H respectively. We also refer to [5] for notation and terminology used throughout the paper, unless otherwise specified.

Let $\alpha_1, \alpha_2, \ldots$ be a sequence of linearly independent elements of H. For an arbitrary β in H, there is a best approximation $\sum_{j=1}^{n} x_j \alpha_j$ to β and the measure of best approximation is defined by means of

$$E_n(\beta) = \min_{x_j} \|\beta - \sum_{j=1}^{n} x_j \alpha_j\| . \tag{2.1}$$

For any finite or infinite linearly independent sequence of elements $\alpha_1, \alpha_2, \ldots$ of H, there always exists an orthonormal system $\alpha_1^* \alpha_2^*, \ldots$ such that (e.g. see [5,6])

$$\alpha_j^* = \beta_j / \|\beta_j\|, \; j=1,2,\ldots \tag{2.2}$$

where

$$\beta_1=\alpha_1, \beta_{j+1}=\alpha_{j+1}-\sum_{k=1}^{j} \langle\alpha_{j+1},\alpha_k^*\rangle\alpha_k^*, \ j=1,2,\ldots, \tag{2.3}$$

The recursive scheme (2.2)-(2.3) is generally called the Gram-Schmidt orthonormalizing process. With $\alpha_1^*,\ldots,\alpha_n^*$ given in (2.2), for any β in H

$$\min_{x_j}||\beta-\sum_{j=1}^{n} x_j\alpha_j^*||=||\beta-\sum_{j=1}^{n} a_j\alpha_j^*||, \tag{2.4}$$

where $a_j=\langle\beta,\alpha_j^*\rangle$, $j=1,\ldots,n$, are the Fourier coefficients of β. Note that (2.4) is known to be the minimum property of Fourier expansions (see [5]).

We now cite a recursive formula for least square approximation from [20] with proof (cf. [20] or see below) omitted as

<u>Theorem 2.1</u>. Let $\alpha_1,\ldots,\alpha_n$; $\beta_1,\ldots,\beta_n$ be given as in (2.2)-(2.3). Then for any β in H, the best approximation to β is given by $\sum_{j=1}^{n} b_j\alpha_j$, where

$$b_n=\langle\beta,\beta_n\rangle/||\beta_n||^2, \tag{2.5$_n$}$$

$$b_{n-j}=\langle\beta-\sum_{k=n-j+1}^{n} b_k\alpha_k,\beta_{n-j}\rangle/||\beta_{n-j}||^2,$$

$j=1,\ldots,n-1$.

III. AN EXTENSION

In this section we consider only the case $H=R^d$, $d\geq 1$. A natural extension of Theorem 2.1 is now in order by relaxing the linear independency of the elements $\alpha_1,\ldots,\alpha_n$. To this end, we introduce

Definition 3.1. Let S be a set which contains n elements $\alpha_1,\ldots,\alpha_n$ of R^m ($m \geq n$). If n-r elements $\alpha_{r+1},\ldots,\alpha_n$ are spanned by the r linearly independent elements $\alpha_1,\ldots,\alpha_r$ for some r ($0 \leq r \leq n$), then the integer r is called the rank of S.

Note 3.1. From Definiton 3.1 we have,

$$\begin{aligned} \alpha_{r+1} &= b_{r+1,1}\alpha_1+\ldots+b_{r+1,r}\alpha_r \\ &\vdots \\ \alpha_n &= b_{n1}\alpha_1+\ldots+b_{nr}\alpha_r \end{aligned} \tag{3.1}$$

where $b_{jk} \varepsilon R$, $j=r+1,\ldots,n$, $k=1,\ldots,r$, are uniquely determined. (See [21,23] for a simple method of doing so.)

With the above notation, we now state

Theorem 3.1. Let $S=\{\alpha_1,\ldots,\alpha_n\}$ be a subset of R^m whose rank is r. Then for any β in R^m, the best approximation to β is given by $\sum_{j=1}^{n} c_j\alpha_j$, where

$$c_r+\sum_{\ell=r+1}^{n} b_{\ell r}c_\ell = <\beta,\beta_r>/||\beta_r||^2,$$

$$c_{r-j}+\sum_{\ell=r+1}^{n} b_{\ell,r-j}c_\ell \tag{3.2}$$

$$= \frac{<\beta-\sum_{p=r-j+1}^{r}(c_p+\sum_{\ell=r+1}^{n} b_{\ell p}c_\ell)\alpha_p,\beta_{r-j}>}{||\beta_{r-j}||^2}$$

$j=1,\ldots,r-1$ and β_{r-j} is defined in (2.3), $j=0,1,\ldots,r-1$.

Proof. Using (3.1) we have,

$$||\beta-\sum_{j=1}^{n} c_j\alpha_j||=||\beta-\sum_{j=1}^{r} c_j\alpha_j-\sum_{j=r+1}^{n} c_j \times$$

$$\times\ (\sum_{\ell=1}^{r} b_{j\ell}\alpha_\ell)||$$

$$=||\beta-\sum_{j=1}^{r} (c_j+\sum_{\ell=r+1}^{n} b_{\ell j}c_\ell)\alpha_j|| \ . \tag{3.3}$$

By applying $(2.5)_r$ to (3.3), (3.2) follows immediately.

In the solution variety provided by (3.2), there is a unique vector

$$x_0=(c_1,\ldots,c_n)^T$$

under the proposed condition (1.2).

Now set

$$f=f(c_{r+1},\ldots,c_n)=||x_0||^2=\sum_{j=1}^{n} c_j^2 \tag{3.4}$$

where

$$c_j=c_j(c_{r+1},\ldots,c_n),\ j=1,\ldots,r.$$

Differentiating (3.4) with respect to c_ℓ and then setting $\partial f/\partial c_\ell=0$, we obtain (by using (3.2))

$$c_\ell=\sum_{j=1}^{r} b_{\ell j}c_j,\ \ell=r+1,\ldots,n. \tag{3.5}$$

Now in order to simplify the writing, we set

recursively

$$\langle \beta, \beta_r \rangle / ||\beta_r||^2 = t_r,$$
$$\langle \beta - \sum_{p=r-j+1}^{r} t_p \alpha_p, \beta_{r-j} \rangle / ||\beta_{r-j}||^2 = t_{r-j}, \tag{3.6}$$
$$j=1,\ldots,r-1.$$

Again setting $\mu_j=(b_{r+1,j},\ldots,b_{nj})^T$ and using $[\ ,\ ]$ to denote the inner product in R^{n-r} with (3.5) and (3.6), (3.2) becomes

$$U\,\tilde{c} = \tau, \tag{3.7}$$

where $U=I_r+G_r$, $G_r=G(\mu_1,\ldots,\mu_r)=([\mu_j,\mu_k])$ is the Gram matrix of $\mu_1,\ldots,\mu_r$, $\tilde{c}=(c_1,\ldots,c_r)^T$ and $\tau=(t_1,\ldots,t_r)^T$.

Since $\det U \geq 1$ (see below), $c_1,\ldots,c_r$ can be obtained by Cramer's rule:

$$c_j=\det U_j/\det U,\ j=1,\ldots,r, \tag{3.8}$$

where U_j is the matrix obtained from the matrix U by replacing the jth column with τ. (See [21,22, 23] also for a simple treatment.)

Now combining (3.8) with (3.5),

$$c_j = \begin{cases} \det U_j/\det U, & j=1,\ldots,r, \\ \sum_{k=1}^{r} b_{jk}\det U_k/\det U, & j=r+1,\ldots,n. \end{cases} \tag{3.9}$$

Note 3.2. A straightforward expansion yields

$$\det U = \det(I_r + G_r)$$

$$= 1 + \{\sum g(\mu_j) + \sum_{j \neq k} g(\mu_j, \mu_k) + \ldots + g(\mu_1, \ldots, \mu_r)\}, \tag{3.10}$$

where $g(\mu_j) = \det G(\mu_j)$, etc. There are $2^r - 1$ Gramians inside the braces on the left hand side of (3.10) which are all nonnegative. Thus $\det U \geq 1$.

IV. SOLUTIONS FOR LSP

Here again consider $H = R^d$, $d \geq 1$. Let $\alpha_1, \ldots, \alpha_n$ be the n column vectors of the m×n matrix A. (Recall that the rank of A is r.) Then (1.1) becomes (2.1):

$$\min_x ||\beta - Ax|| = \min_{x_j} ||\beta - \sum_{j=1}^{n} x_j \alpha_j|| = E_n(\beta). \tag{4.1}$$

4.1 If r=n, then orthonormalizing $\alpha_1, \ldots, \alpha_n$ by (2.2)-(2.3), we obtain recursively the corresponding orthogonal vectors $\beta_1, \ldots, \beta_n$ which are defined in (2.3). Consequently, by Theorem 2.1, we find the solution of (1.1)

$$x_0 = (b_1, \ldots, b_n)^T,$$

where b_n, $b_{n-1}, \ldots, b_1$ are recursively given by $(2.5)_n$.

4.2 If r<n, then $\alpha_1, \ldots, \alpha_n$ form a set of rank r(<n). Naturally, $\alpha_{r+1}, \ldots, \alpha_n$ can be expressed explicitly in terms of $\alpha_1, \ldots, \alpha_r$ by (3.1). In this case, following the same arguments given in §3, we find an (n-r)-dimensional variety of solutions

provided by (3.2). In this solution variety, by assuming (1.2), we find a unique vector

$$x_0=(c_1,\ldots,c_n)^T \tag{4.2}$$

whose components are given by (3.9).

Now let N(A) be the null space of A (see [18]). Since dimN(A)=n-r, we can let $y_1,\ldots,y_{n-r}$ be a basis of N(A). For any $\ell_1,\ldots,\ell_{n-r}\varepsilon R$,

$$\sum_{j=1}^{n-r} \ell_j y_j \tag{4.3}$$

is a general solution of Ax=0. Combining (4.3) with (4.2), we find that

$$x=x_o+\sum_{j=1}^{n-r} \ell_j y_j \tag{4.4}$$

is the complete solution of (1.1) with (1.2). On the other hand, by means of elementary matrices (see [18]), we can easily construct an n×n matrix H of rank n-r such that AH=0. Then for any $y\varepsilon R^n$

$$x=x_o+Hy \tag{4.5}$$

is an alternative form of the complete solution of (1.1) with (1.2).

Note 4.1. The extended Businger-Golub algorithm introduced in [9] by Hanson and Lawson can also be adopted to construct the indicated matrix H.

V. FURTHER APPLICATIONS

Our results established above have a wide range of applications. Here we consider H instead of R^d.

5.1 Approximation of linear functionals often arises in numerical analysis: Let $L_1, \ldots, L_n$ be n prescribed "elementary" linear functionals. It is desired to approximate a given functional L by linear combinations of L_j:

$$L \approx a_1 L_1 + \ldots + a_n L_n.$$

In a Hilbert space H, every bounded linear functional can be represented by a unique element of H (using Frechét-Riesz Representation Theorem - e.g. see page 218 of [5]). Naturally, our method can be adopted to treat the problem. (e.g. see pages 341-342 of [5] for a detailed analysis of the topic.)

5.2 In [12], S. Kaniel considered the method of least squares used for approximate computation of $Ax=\beta$ where A is a positive definite (symmetric) $n \times n$ matrix. It amounts to determining $b_1, \ldots, b_s$ so that

$$||\beta - A(\sum_{j=1}^{s} b_j A^{j-1}\beta)|| = \min_{a_j} ||\beta - A(\sum_{j=1}^{s} a_j A^{j-1}\beta)|| \tag{5.1}$$

where s is a predetermined integer (usually much smaller than n). Then $\gamma = \sum_{j=1}^{s} b_j A^{j-1}\beta$ is taken as an approximate solution.

By assuming $p(\lambda) = g(\lambda) - 1$ (where $g(\lambda) = \sum_{j=1}^{s} b_j \times \lambda^{j-1}$) a polynomial of degree at most s, Kaniel's technique for solving the indicated problem was a combination of the method of conjugate gradients with the minimax principle (see [4,15]). No explicit construction of $p(\lambda)$ was given in [12]. Particularly, the Kaniel method requires knowing a few largest (or smallest) eigenvalues of A. It is known that computation of eigenvalues (even a few of them) is a time-consuming task for a matrix A of large order.

However, setting $\alpha_j=A^j\beta$, $j=1,\dots,s$, (5.1) becomes

$$\min_{a_j}||\beta-\sum_{j=1}^{s}a_j\alpha_j||=E_s(\beta). \tag{5.2}$$

Now by our method (Theorem 2.1), $\sum_{j=1}^{s}b_j\alpha_j$ is a solution of (5.2), and, in turn, a solution of (5.1), where $b_s,\dots,b_1$ are recursively given by (2.5)s.

Note 5.1. If β, $A\beta,\dots,A^s\beta$ are linearly dependent, then evidently $p(A)\beta=0$. Thus $\gamma=g(A)\beta$ is the exact solution of $Ax=\beta$. (e.g., see [22] for an extensive discussion on the subject.)

Note 5.2. The essential assumption of positive definiteness of A in [12] is not needed for our method.

5.3 Using S. Kaniel's approach, we now have the following generalizations: Let T be any operator from H into H. Then consider

$$Tx=\beta \tag{5.3}$$

where β is a given element of H. Evidently replacing A in (5.1) by T, $\gamma=\sum_{j=1}^{s}b_jT^{j-1}\beta$ can be taken as an approximate solution of (5.3). Naturally setting $T^j\beta=\alpha_j$, $j=1,\dots,s$ and following the above process, $b_s,\ldots,b_1$ are determined by (2.5)s.

Note 5.3. For instance, T can be taken as linear Volterra and Fredholm integral operators (e.g. see [16,24]) or any integral operators.

VI CONCLUDING REMARKS

6.1 In view of the above, our results (or algorithms) (2.5)n, (3.2), (3.6) are suitable for use with desk calculators and computers of any kind. A subsequent report on subroutine for the indicated algorithms is in preparation (see also [2,14,20-23]). For a completion of our analysis of LSP, a detailed error analysis of our results is indeed necessary and important but it will not be carried out here.

6.2 It is natural to consider LSP with constraints for the case $r<n$ (e.g. see [7,9] for some details). Now with the relation (3.2) available, we can discuss LSP with constraints to any extent.

6.3 It is hoped that the results established here will stimulate further analysis of LSP, both theoretical and experimental.

ACKNOWLEDGEMENTS

The author, a 1975 Summer Research Institute Fellow at the University of Calgary, was supported (in part) by the N.R.C. of Canada (Grant No. A3116) and the President's N.R.C. Funds of the University of Regina.

REFERENCES

1. F.L. Bauer, Elimination with weighted row combinations for solving linear equations and least squares problems, Numer. Math 7(1965), pp.338-352.

2. A. Björck, Solving linear least squares problems by Gram-Schmidt orthogonalization, Nordisk. Tidskr., Informations-Behandling, 7(1967), pp.1-21.

3. P. Businger and G.H. Golub, Linear least squares solutions by Householder transformations, Numer. Math.7(1965),pp.269-276.

4. E.W. Cheney, Introduction to approximation Theory, McGraw-Hill, New York, 1966.

5. P.J. Davis, Interpolation and approximation, 2nd ed., Blaisdell, New York, 1965.

6. P.J. Davis and P. Rabinowitz, Advances in orthonormalizing computation. Advances in computers, F.L. Alt (Ed.), Vol. 2, Academic Press, New York, 1961,pp.55-133.

7. G. Golub, Numerical methods for solving linear least square problems, Numer. Math. 7(1965) pp.206-216.

8. G.H. Golub and J.H. Wilkinson, Note on the iterative refinement of least squares solution, Numer. Math. 9(1966),pp.139-148.

9. R.J. Hanson and C.L. Lawson, Extensions and applications of the Householder algorithm for solving linear least squares problems, Math. Comp. 23(1969),pp.787-812.

10. A.S. Householder, Unitary triangularization of a nonsymmetric matrix, J. Assoc. Comput. Mach. 5(1968),pp.339-342.

11. T.L. Jordan, Experiments on error growth associated with some linear least-squares procedures, Math. Comp.22(1968),pp.579-588.

12. S. Kaniel, Estimates for some computational techniques in linear algebra, Math. Comp. 20(1966), pp.369-378.

13. E.E. Osborne, On least squares solution of linear equations, J. Assoc. Comput. Mach. 8(1961), pp.628-636.

14. J.R. Rice, Experiments on Gram-Schmidt orthogonalization, Math. Comp. 20(1966), pp.325-328.

15. J.R. Rice, The approximation of functions, Vol. 1, Linear Theory, Addison-Wesley, Reading Ms., 1964.

16. F. Riesz and B. Sz-Nagy, Functional analysis (Translated by L.F. Baron), Ungar, New York, 1955.

17. H. Rutishauser, Once again: The least square problem, Linear Algebra and Its Applications, 1(1968),pp.479-488.

18. H. Schneider and G.P. Barker, Matrices and Linear Algebra, Holt, Rinehart and Winston, New York, 1968.

19. O. Taussky, Note on the conditon of matrices, Math. Tables Aids Comput.4(1950),pp. 111-112.

20. Chung-lie Wang, A recursive formula for least square approximation and its applications, submitted for publication.

21. Chung-lie Wang, A direct and simple method for solving a system of linear equations, submitted for publication.

22. Chung-lie Wang, A new method for solving a system of linear equations and its applications, submitted for publication.

23. Chung-lie Wang, Applications on the classical Gram-Schmidt orthonormalization, submitted for publication.

24. K. Yosida, Lectures on differential and integral equations, Interscience, New York, 1960.

Splines in Orlicz Spaces

J. Baumeister

§1 Introduction. In this paper we show with the help of Orlicz spaces that certain nonlinear classes of spline functions satisfy an extremal property similar to the well-known extremal property of polynomial splines.
Given a partition

$\Delta : t_o := a < t_1 < \ldots < t_{n+1} := b$ of $I := [a,b]$,

we define:

$S_3(\Delta) := \{f \in C^2(I) \mid f_{|[t_i,t_{i+1}]} \in P_3,\ i=o,\ldots,n\}$ 1)

$R_{2,1}(\Delta) := \{f \in C^2(I) \mid f_{|[t_i,t_{i+1}]} = p/q,\ p \in P_2, q \in P_1, i=o,\ldots,n\}$

$E(\Delta) := \{f \in C^2(I) \mid f(t) = a_i \exp(b_i t) + c_i t + d_i,\ t \in [t_i,t_{i+1}],\ i=o,\ldots,n\}.$

Let $y = (y_o', y_o, y_1, \ldots, y_{n+1}, y_{n+1}') \in \mathbb{R}^{n+4}$ and let

$V(y) := \{f \in C^2(I) \mid f(t_i) = y_i, i=o,\ldots,n+1, f'(a) = y_o', f'(b) = y_{n+1}'\}$

If f_o belongs to $V(y)$, then it is known that

$\int_I f_o''(t)^2 dt = \min\{\int_I f''(t)^2 dt \mid f \in V(y)\}$ if and only if f_o

is a cubic spline (i.e. $f_o \in S_3(\Delta)$).

It was shown in [1] that the classes $R_{2,1}(\Delta)$, $E(\Delta)$

---

1) $P_k := \{p \mid p$ is a polynomial of degree $\leq k\}$

also satisfy a best interpolation property:

(1.1) Suppose $f_o \in R_{2,1}(\Delta) \cap V(y)$ and $f_o''(t)>0$ for all $t \in I$.[2)]

Then the following equality holds:

$$\int_I f_o''(t)^{2/3}dt = \max\{\int_I f''(t)^{2/3}dt \mid f \in V(y), f''(t) \geq 0 \text{ for all } t \in I\}$$

(1.2) Suppose $f_o \in (E(\Delta)-g_o) \cap V(y)$ and $f_o''(t)>0$ for all $t \in I$, where $g_o(t) := t^2/2$, $t \in I$. Then the following equality holds:

$$\int_I A(f_o''(t))dt = \min\{\int_I A(f''(t))dt \mid f \in V(y), f''(t) \geq 0 \text{ for all } t \in I\},$$

where $A(u) := (1+|u|)\log(1+|u|)-|u|$.

The smoothness criterion in (1.1) can be interpreted as the Frechet-norm in $L_{2/3}$. The search for a similar interpretation for (1.2) leads one to Orlicz spaces.

§ 2 Orlicz spaces. A detailed theory of these spaces with applications to integral equations is given in the book of Krasnoselskii-Rutickii [5]. The facts that we need here can be found in the work of Luxemburg [6], where these spaces have been considered in a general framework.

To define Orlicz spaces, we need some notions from the theory of convex functions:

1. $Y := \{A \in C(\mathbb{R}) \mid A \text{ convex and symmetric}, A(o)=o, A \not\equiv o\}$
   The functions belonging to this set are called

2) This is no restriction since for $f \in R_{2,1}(\Delta)$, sign(f_o'') is constant

Young-functions.

2) Let $F : \mathbb{R} \to \mathbb{R} \cup \{\infty\}$. The map

$$F^* : \quad \mathbb{R} \ni v \mapsto \sup \{uv - F(u) \mid u \in \mathbb{R}\} \in [-\infty, \infty]$$

is called the convex conjugate of F.

3) $\bar{Y} := Y \cup Y^*$, where $Y^* := \{F \mid F = A^*, A \in Y\}$.

For $A \in \bar{Y}$ the following holds:

(2.1) $A^{**} := (A^*)^* = A$

Suppose now that I is a finite interval in $\mathbb{R}$.[3)]
We define

$$L_o := L_o(I) := \{f : I \to \mathbb{R} \cup \{-\infty, \infty\} \mid f \text{ Lebesgue measurable}\}.$$

Definition: Let $A \in \bar{Y}$. We set

$$P_A := \{f \in L_o \mid \rho_A(f) := \int_I A(f(t))dt < \infty\}$$ [4)]

Examples:

1) $A_p(u) := p^{-1}|u|^p, \ 1 \leq p < \infty \qquad (P_{A_p} = L_p)$

2) $A_\ell(u) := (1+|u|)\log(1+|u|) - |u|$

3) $A_\infty(u) := \begin{cases} 0 & , |u| \leq 1 \\ \infty & , |u| > 1 \end{cases} \qquad (P_{A_\infty} = \{f \in L_o \mid \|f\|_\infty \leq 1\})$

The example 3) shows that P_A is not necessarily a linear subspace of L_o. But for every $A \in \bar{Y}$, P_A is absolutely convex, and therefore $L_A := \bigcup_{n \in \mathbb{N}} nP_A$ is a linear subspace of L_o.

One can define in a natural way a seminorm over L_A:

$$\|f\|_A := \inf \{k > 0 \mid \rho_A(k^{-1}f) \leq 1\}.$$

---

3) Orlicz spaces can also be defined over general measure spaces.

4) The function $t \mapsto A(f(t))$ is measurable, since A is lower semicontinuous.

It follows that L_A is a Banach space. [5)]

In the following we give certain results concerning the dual space of L_A and the weak topologies in L_A. Let $A \in Y$ and $E_A := \bigcap_{k>0} kP_A$.

(2.2) E_A is the closure of the step functions in L_A, and hence E_A is separable.

(2.3) The following are equivalent:

a) $E_A = L_A$.

b) There exist $m>0$ and $u_o \geq 0$ with $A(2u) \leq mA(u)$ for all $u \geq u_o$.

(2.4) The dual space of E_A is topologically isomorphic to L_{A^*}, and hence we may identify L_{A^*} with $(E_A)^*$.

(2.5) ρ_{A^*} is $\sigma(L_{A^*}, E_A)$ - lower semicontinuous. [6)]

The result (2.5) follows from results of R.T. Rockafellar [8a,8b] concerning convex integrals.

Associated with an Orlicz space L_A, we may define a space of functions analogous to the Sobolev spaces $H_{p,m}$. Given a positive integer m, let

$$H_{A,m} := \{ f \in L_A \mid f^{(m-1)} \text{ exists and is absolutely continuous, } f^{(m)} \in L_A \}$$

This is a Banach space under a variety of equivalent norms, for example

5) We identify functions which are equal a.e. on I.
6) With $\sigma(L_{A^*}, E_A)$ we denote the topology in L_{A^*} which is induced by the finite subsets of E_A.

$$\|f\|_{A,m} := \|f^{(m)}\|_A + \sum_{i=o}^{m-1} |f^{(i)}(a)|.$$

It can be shown that functionals of the following kind (point evaluation functionals):

$$\mu_t : H_{A,m} \ni f \mapsto f(t) \in \mathbb{R},$$

are bounded linear functionals on $H_{A,m}$.

§3 Interpolating splines. Let us consider the following situation:
$A \in \overline{Y}$; $m \in \mathbb{N}$;
$K \subset H_{A,m}$, $K \neq \emptyset$, K convex;
$R : H_{A,m} \to L_A$ a bounded linear operator.

The usual definition of a spline through its extremal property can be given as follows: $k_o \in K$ is called a (K-interpolating) spline provided it is a solution of the constrained variational problem

(3.1) $\|Rk_o\|_A = \inf\{\|Rk\|_A \mid k \in K\}$.

We will denote the splines, defined in this sense, as N-splines (norm-splines).

Definition:

We call a function $k_o \in K$ a (K-interpolating) M-spline (modular-spline) provided it is a solution of the following constrained variational problem :

(3.2) $\rho_A(Rk_o) = \inf\{\rho_A(Rk) \mid k \in K \cap R^{-1}(P_A)\}$.

In the case of Young-functions $A_p, p \in [1,\infty)$, every M-spline is an N-spline and conversely; but in the

case of A_∞ the situation is quite different.
One can give sufficient conditions on K, R and A so that the uniqueness of N- or M-splines is ensured.
The general existence theorem can be formulated as follows:

THEOREM: Let the following conditions hold:

(V1) $A^* \in Y$;

(V2) $R(K)$ is $\sigma(L_A, E_{A^*})$-closed;

(V3) $R(K) \cap R^{-1}(P_A) \neq \emptyset$.

Then there exists at least one M-spline.
We give here only a sketch of the proof. Let $(Rk_n)_{n\in\mathbb{N}}$ be a minimal sequence in R(K) for (3.2). From the definition of the norm it follows that the sequence $(Rk_n)_{n\in\mathbb{N}}$ is bounded. Since L_A is the dual space of the separable Banach space E_{A^*} (see (2.2) and (2.4)), it follows that a subsequence of $(Rk_n)_{n\in\mathbb{N}}$ exists that converges in the $\sigma(L_A, E_{A^*})$-topology. From the condition (V2) and from (2.5) the result follows.

Remarks:

(1) Under the conditions (V1) and (V2) one can prove in an analogous way the existence of an N-spline.

(2) The above theorem is not applicable for $A = A_1$ since $A_1^* \notin Y$. On the other hand there exist simple examples which show that in this case the existence of splines is not ensured. Another way of tackling this situation is given by R. Holmes [3] and S.D. Fisher - J.W. Jerome [2].

In order to apply the above theorem to concrete cases one essentially needs to check the condition (V2). In the case of reflexive spaces ($L_A = E_A$, $L_{A^*} = E_{A^*}$) one needs only to show that R(K) is closed; sufficient conditions for this purpose have been given by J.W. Jerome [4]. But there exist interesting examples in which L_A is not reflexive (for example $A(u) := (1+|u|)\log(1+|u|)-|u|$). In this case we specify R and K as follows.

Let R be a nonsingular differential operator of the form

(3.3) $R=\sum_{j=o}^{m} a_j D^j, a_j \in C^j(I), j=o,\dots,m, a_m(t) \neq o$ for all $t \in I$,

and let K be described by

(3.4) $K=\{f \in H_{A,m} \mid l_\lambda \leq \langle\lambda, f\rangle \leq u_\lambda, \lambda \in \Lambda\}$, where $\Lambda \subset H^*_{A,m}$, $l_\lambda, u_\lambda \in \mathbb{R}$.

It is easy to show that R is surjective. Further there exists a map

$\kappa : I \times I \to \mathbb{R}$ with the properties

i) $\kappa(t,.)$ is measurable for all $t \in I$,

ii) for all $f \in H_{\Lambda,m}$ there exists $p \in R^{-1}(\theta)$ with

$$f(t) = p(t) + \int_I \kappa(t,s) Rf(s)\,ds \text{ for all } t \in I.$$

For $\mu \in \Lambda$ we define: $\mu_\kappa : I \ni s \mapsto \langle\mu, \kappa(.,s)\rangle$.

Analogous to the case of N-splines in L_∞ (see J.W. Jerome [4]) one can prove the

LEMMA: Let the following hold:

(1) There exist $\mu_1,\dots,\mu_m \in \Lambda$ which are linearly independent over $R^{-1}(\theta)$;

(2) $A^* \in Y$ and $\{\mu_\kappa | \mu \in \Lambda\} \subset E_{A^*}$;

(3) For all $\mu \in \Lambda$ and all $f \in H_{A,m}$ we have
$<\mu, \int_I \kappa(.,s) Rf(s) ds> = \int_I \mu_\kappa(s) Rf(s) ds.$

Then $R(K)$ is $\sigma(L_A, E_{A^*})$-closed.

The conditions (1), (2), (3) are fulfilled when, for example, Λ consists of a sufficiently large number of point evaluation functionals.

§4 Characterization. Suppose $A \in Y$. Then there exist for all $u \in \mathbb{R}$

$$A'_+(u) := \lim_{h \downarrow o} h^{-1}(A(u+h) - A(u)),$$

$$A'_-(u) := \lim_{h \downarrow o} h^{-1}(A(u) - A(u-h)),$$

and we denote: $\partial A(u) := [A'_-(u), A'_+(u)]$.

With the results of R.T. Rockafellar [8a,8b] concerning subdifferentials of convex integrals we can prove the following characterization theorem.

THEOREM: Let the following conditions hold:

(C1) $A \in Y$, $E_A = L_A$;

(C2) R is surjective.

Then for $k_o \in K$ the following are equivalent:

a) k_o is an M-spline;

b) There exists $g \in L_{A^*}$ with:

(1) $g(t) \in \partial A(Rk_o(t))$ a.e.,

(2) $<g, Rk_o> \leq <g, Rk>$ for all $k \in K$.

COROLLARY: In addition to (C1), (C2) assume:

(C3) $K = f_o + U$, where U is a linear subspace of $H_{A,m}$

and $f_o \in H_{A,m}$.

(C4) $A \in C^1(\mathbb{R})$.

Then for $k_o \in K$ the following are equivalent:

a) k_o is an M-spline.

b) $R^* g \in U^{\perp}$, where $g(t) := A'(Rk_o(t))$ for all $t \in I$.[7)]

From the above corollary it is clear that the extremal property of interpolating spline functions in $E(\Delta)$ is a special case of this theorem.

Remark:

Problems of the form (3.3), (3.4) have been intensively studied via optimal control in the case A_p, $p \in [1,\infty)$, by O.L. Mangasarian - L.L. Schumaker [7]. One can do this without the theory of optimal control for other classes of Young-functions in the same generality. In the characterization problem a theorem in nonlinear programming of R.T.Rockafellar [8a] is helpful.

7) R^* is the adjoint operator of R

References:

1. Baumeister, J.: Extremaleigenschaft nichtlinearer Splines. Dissertation, Universität München (1974).

2. Fisher, S.D. - Jerome, J.W.: Spline solutions to L^1 extremal problems in one and several variables. To appear.

3. Holmes, R.: R-splines in Banach spaces I. J. Math. Anal. Appl. 40 (1972), 574-593.

4. Jerome, J.W.: Minimization problems and linear and nonlinear spline functions I: Existence. SIAM J. Num. Anal. 10 (1973), 803-819.

5. Krasnoselskii, M.A. - Rutickii, Y.B.: Convex functions and Orlicz spaces. P. Nordhoff Ltd.-Groningen (1961).

6. Luxemburg, W.A.J.: Banach function spaces. Thesis, Technische Hogeschool te Delft (1955).

7. Mangasarian, O.L. -Schumaker, L.L.: Splines via optimal control. Approximations with Special Emphasis on Spline Functions, I. Schoenberg ed., Academic Press, New York (1969).

8. Rockafellar, R.T.:
   a) Convex functions, monotone operators and variational inequalities. Theory and Applications of Monotone Operators, Proc. NATO Advanced Study Institute, Venice (1969).
   b) Integrals which are convex functionals I. Pacific J. Math. 24 (1968), 525-539.
   c) Integrals which are convex functionals II. Pacific J. Math. 39 (1971), 439-469.

GEOMETRIC CONVERGENCE OF RATIONAL FUNCTIONS TO THE RECIPROCAL OF EXPONENTIAL SUMS ON $[0,\infty]$

Hans-Peter Blatt

For any nonnegative integer m, let Π_m denote the collection of all real polynomials of degree at most m, and for any nonnegative integers m and n, set

$$\Pi_{m,n} := \{ r_{m,n} = \frac{p_m}{q_n} \mid p_m \in \Pi_m, \; q_n \in \Pi_n \}.$$

For any closed interval $[a,b]$ of the extended real numbers $\overline{\mathbb{R}}$, let $||\cdot||_{[a,b]}$ denote the Chebyshev norm on $[a,b]$. Recently much attention has been paid to the problem of obtaining relations between the analyticity of a function f and the asymptotic behaviour of the minimal distances

$$\lambda_{m,n} := \inf_{r_{m,n} \in \Pi_{m,n}} \left|\left| \frac{1}{f} - r_{m,n} \right|\right|_{[0,\infty]} .$$

Cody, Meinardus and Varga [2], resp. Schönhage [6], showed for $f(x) = e^x$ that

$$\lim_{n\to\infty} \lambda_{0,n}^{1/n} = \frac{1}{3} .$$

Newman [4] proved that one cannot achieve more than geometric convergence for the minimal distances by the use of general rational functions, since

$$\lambda_{n,n} > 1280^{-n-1} \quad \text{holds for all } n = 1,2,\ldots .$$

In this note we prove the geometric convergence of $\{\lambda_{0,n}\}$ to 0 for certain exponential sums with polynomial coefficients. Furthermore, we generalize the result of Newman: For exponential sums with polynomial coefficients, the sequence $\{\lambda_{n,n}\}$ satisfies

$$\lambda_{n,n} > \alpha^{-n} \quad \text{with } \alpha > 0.$$

1. Geometric Convergence for Exponential Sums with Polynomial Coefficients

For given $r > 0$ and $s > 1$, let $\mathcal{E}(r,s)$ denote the closed ellipse in the complex plane with foci at $x = 0$ and $x = r$ whose sum of both axes is $r \cdot s$. If $f(z)$ is any entire fuction, set

$$M_f(r,s) := \max_{z \in \mathcal{E}(r,s)} |f(z)|.$$

We say, f has geometric convergence if

$$\overline{\lim_{n\to\infty}} \lambda_{0,n}^{1/n} < 1.$$

Then the following sufficient condition for geometric convergence generalizes Theorem 5 of Meinardus, Reddy, Taylor and Varga [3].

Theorem 1 [1]:

Let $f(z) = \sum_{\nu=0}^{\infty} a_\nu z^\nu$ be an entire function with real coefficients and $a_{\nu_0} > 0$, $a_\nu \geq 0$ for $\nu \geq \nu_0$. If there exist real numbers $K > 0$, $s > 1$, $\theta > 0$ and $r_0 > 0$ such that

$$M_f(r,s) \leq K(||f||_{[0,r]})^\theta \quad \text{for all } r \geq r_0,$$

then f has geometric convergence.

One can prove this theorem either directly as in [1] or by the use of Theorem 3 of Roulier and

Taylor [5]. For finding more functions with geometric convergence, we define some abbreviations: Let

$$0 \leq x_1 < x_2 < \ldots < x_L < \infty$$

with respective nonnegative integers $\beta_1, \beta_2, \ldots, \beta_L$. We set

$$N := \left\{ h(z) = \sum_{\nu=0}^{\infty} a_\nu z^\nu \;\middle|\; \begin{array}{l} a_\nu \in \mathbb{R},\ h \text{ is an entire} \\ \text{transcendental function} \\ \text{having zeros at preci-} \\ \text{sely } x_i \text{ with order } \beta_i \\ (1 \leq i \leq L),\ \lim_{x\to\infty} h(x) = \infty \end{array} \right\}$$

and

$$\tilde{N} := \left\{ h(z) = \sum_{\nu=0}^{\infty} a_\nu z^\nu \;\middle|\; \begin{array}{l} a_\nu \in \mathbb{R},\ h \text{ is an entire} \\ \text{function} \not\equiv 0 \text{ having} \\ \text{zeros at } x_i \text{ with} \\ \text{order} \geq \beta_i \ (1 \leq i \leq L) \end{array} \right\}$$

Then we can generalize Theorem 3 of Roulier and Taylor [5] in the following way.

Theorem 2 [1]:

Let $f \in N$ and assume that, for every $s > 1$, there exist constants $K(s) > 0$, $\theta(s) > 0$ and $r(s) > 0$ such that

$$M_f(r,s) \leq K(s)(\|f\|_{[0,r]})^{\theta(s)} \quad \text{for all } r \geq r(s).$$

Further, assume there exist entire functions f_1, $f_2 \in \tilde{N}$ such that

a) $f = f_1 + f_2$,

b) f_1 has geometric convergence and there exists a real number $r_0 > 0$ such that f_1 is nondecreasing for $r \geq r_0$,

c) there exists $B > 0$ such that $f_2(x) \geq -B$ for

all $x \geq 0$,

d) there exist $\psi > 0$ and $A > 0$ such that $f_2(x) \leq A(f_1(x))^{\psi}$ for all $x \geq r_o$,

e) there exists a sequence of positive integers $\{n_j\}$ for which $1 < n_{j+1}/n_j < \rho$ (ρ a fixed real number) and

$$f_2^{(n_j+1)}(x) \leq 0 \text{ for all } x \geq 0,\ j = 1,2,\ldots.$$

Then f has geometric convergence.

Now we are able to prove geometric convergence for exponential sums.

Theorem 3:

Let

$$f(x) = \sum_{\nu=1}^{k} p_\nu(x)\, e^{\lambda_\nu x} \quad \text{with } \lambda_\nu \in \mathbb{R},\ p_\nu(x) \text{ real}$$

polynomials in x, $p_\nu(x) \equiv p_\nu \in \mathbb{R}$ for $\lambda_\nu < 0$, $\lim_{x\to\infty} |f(x)| = \infty$. Then f has geometric convergence.

Proof: Let $\lambda_1 > \lambda_2 > \ldots > \lambda_k$ and let j be the number of the negative frequencies λ_ν. We may also assume $\lim_{x\to\infty} f(x) = \infty$. For $j = 0$ the power series of $f(z)$ has only a finite number of negative coefficients and the assertion follows from Theorem 1. We now proceed by induction on j:

Let the assertion be true for j-1. For $f \in N$ we fix a polynomial p interpolating the function

$$g(x) = \sum_{\nu=1}^{k-1} p_\nu(x)\, e^{\lambda_\nu x}$$

at x_i with order β_i $(1 \leq i \leq L)$ such that the coefficient of the highest power is positive. Then

$f = f_1 + f_2$ with $f_1(x) = g(x) - p(x)$. Since f_1 has geometric convergence the assertion follows from Theorem 2.

2. A Generalization of a Result of Newman

Let p be a polynomial and $f = g + p$ with

$$g(x) = \sum_{\nu=1}^{k} p_\nu(x)\, e^{\lambda_\nu x} .$$

For real x let f be a real-valued function. The functions p_ν ($\neq 0$) are polynomials of degree $m_\nu - 1$ and the λ_ν are k distinct complex numbers $\neq 0$. Further, let

$$\lim_{x\to\infty} |f(x)| = \infty \tag{2.1}$$

and

$$p(x) = o(f(x)) \text{ for } x \to \infty . \tag{2.2}$$

For abbreviation we define

$$m := \sum_{\nu=1}^{k} m_\nu \quad \text{and} \quad \lambda := \max_{1\leq\nu\leq k} |\lambda_\nu| .$$

Now we are able to generalize the result of Newman [4] to such functions f.

<u>Theorem 4</u>:

There exists a real number $\alpha = \alpha(f) > 0$ such that we have for any $P, Q \in \Pi_n$:

$$\left\| \frac{1}{f} - \frac{P}{Q} \right\|_{[0,\infty]} > \alpha^{-n} .$$

<u>Proof</u>: By (2.1) and (2.2), we can get by a suitable transformation that

$$|f(x)| > 0 \text{ and } |p(x)| \leq \frac{1}{4} |f(x)| \tag{2.3}$$

hold for all $x \geq 0$. Let α be a real number such that

$$\left| \frac{1}{f(x)} - \frac{P(x)}{Q(x)} \right| \leq \alpha^{-n} \tag{2.4}$$

for all $x \geq 0$. We want to reach a contradiction for α sufficiently large and may suppose that $\deg(Q) > \deg(p)$ and P/Q is irreducible. Then $Pp \neq Q$. Now we normalize P and Q so that

$$\| p P - Q \|_{[0,n]} = \frac{1}{2} . \tag{2.5}$$

Thus from (2.3), (2.4) follows for $x \in [0,n]$:

$$|Q(x)| \leq \frac{1}{2} + \frac{1}{4}|Q(x)|(1 + \alpha^{-n}|f(x)|) . \tag{2.6}$$

For fixed $\varepsilon > 0$ there exists $K > 0$ such that

$$|f(x)| \leq K^n e^{(\lambda+\varepsilon)x} \quad \text{for all } x \geq 0. \tag{2.7}$$

Assuming $\alpha \geq K e^{\lambda+\varepsilon}$ we get by (2.6):

$$\|Q\|_{[0,n]} \leq 1.$$

Throughout $[0,3n+nm]$ we can use as Newman the Chebyshev polynomials to estimate $\|Q\|_{[0,3n+nm]}$:

$$\|Q\|_{[0,3n+nm]} \leq T_n(5+2m) < (10+4m)^n.$$

Thus from (2.4), (2.7) we obtain for $x \in [0,3n+nm]$

$$|f(x) P(x) - Q(x)| \leq K_1^n \alpha^{-n} ,$$

where $K_1 := K e^{(\lambda+\varepsilon)(3+m)} \cdot (10+4m)$.

Now we consider the difference operator

$$D(E) = \prod_{\nu=1}^{k} (E - e^{\lambda_\nu})^{m_\nu} ,$$

where E is defined by $Ef(x) = f(x+1)$. Then $g(x)$ is a solution of the difference equation

$$D(E)y(x) = 0$$

and $g(x)P(x)$ is a solution of $D(E)^n y(x) = 0$.

Defining

$$R_o(x) := f(x)\ P(x) - Q(x)$$

and

$$R_i(x) := (E - e^{\lambda_i})^{nm_i}\ R_{i-1}(x) \text{ for } i = 1,2,\ldots,k,$$

we get

$$R_1(x) = \sum_{j=0}^{nm_1} (-1)^{nm_1-j} \binom{nm_1}{j} e^{\lambda_1(nm_1-j)} R_o(x+j)$$

and for $x \in [0, 3n+n(m-m_1)]$:

$$|R_1(x)| \leq K_1^n \alpha^{-n} (1 + e^{|\lambda_1|})^{nm_1}.$$

By induction it follows for $x \in [0,3n]$ that

$$|R_k(x)| \leq K_2^n\ \alpha^{-n}, \text{ where } K_2 := K_1 \prod_{\nu=1}^{k} (1+e^{|\lambda_\nu|})^{m_\nu}.$$

On the other hand,

$$R_k(x) = D(E)^n (p(x)\ P(x) - Q(x)).$$

Therefore R_k is a polynomial of degree $\leq$ 2n-1, since $\deg(p) \leq n-1$. Now defining

$$a_\nu := e^{\lambda_\nu} - 1 \quad \text{for } \nu = 1,\ldots,k,$$

$$\tilde{D}(\Delta) := D(\Delta+1)^n,$$

we consider the decomposition of $1/\tilde{D}(\Delta)$ into partial fractions:

$$\frac{1}{\tilde{D}(\Delta)} = \sum_{\nu=1}^{k} \sum_{j=1}^{nm_\nu} \frac{\alpha_{\nu,j}}{(\Delta - a_\nu)^j},$$

where $\alpha_{\nu,j} = \frac{1}{2\pi i} \int_{C_\nu} \frac{(\xi - a_\nu)^{j-1}}{\tilde{D}(\xi)} d\xi$ with

$$C_\nu = \{ z \mid |z - a_\nu| = d \} \text{ and}$$

$$d := \min (1, \tfrac{1}{2} \min_{\substack{\nu,\mu \\ \nu \neq \mu}} |a_\nu - a_\mu|).$$

A simple calculation gives us $|\alpha_{\nu,j}| \leq d^{-nm}$ for

$1 \leq \nu \leq k$ and $1 \leq j \leq nm_\nu$. Since

$$p(x)\ P(x) - Q(x) = D(E)^{-n}\ R_k(x)$$

$$= \sum_{\nu=1}^{k} \sum_{j=1}^{nm_\nu} \frac{\alpha_{\nu,j}}{(\Delta - a_\nu)^j} R_k(x)$$

$$= \sum_{\nu=1}^{k} \sum_{j=1}^{nm_\nu} \alpha_{\nu,j} (-1)^j \frac{1}{a_\nu^j} \sum_{i=0}^{\infty} \binom{j+i-1}{i} \left(\frac{\Delta}{a_\nu}\right)^i R_k(x)$$

and $\deg(R_k) \leq 2n-1$, it follows for $x \in [0,n]$ that

$$|p(x)\ P(x) - Q(x)|$$

$$< d^{-nm} K_2^n \alpha^{-n} \sum_{\nu=1}^{k} \sum_{j=1}^{nm_\nu} \sum_{i=0}^{2n} \binom{j+i-1}{i} \frac{2^i}{|a_\nu|^{i+j}} .$$

Now it is easy to see that there exists a constant $K_3 > 0$ (independent of n) such that

$$|p(x)\ P(x) - Q(x)| < K_3^n\ \alpha^{-n}$$

holds for $x \in [0,n]$.
If $\alpha \geq \max\ (K\ e^{\lambda+\varepsilon},\ 2K_3)$ we get

$$\|p\ P - Q\|_{[0,n]} < \frac{1}{2}\ ,$$ which contradicts (2.5).

References

[1] Blatt, H.-P., Rationale Tschebyscheff-Approximation über unbeschränkten Intervallen. Habilitationsschrift, Universität Erlangen-Nürnberg, 1974.

[2] Cody, W.J., G. Meinardus and R.S. Varga, Chebyshev Rational Approximations to e^{-x} in [0,+∞) and Applications to Heat-Conduction Problems. J. Approx. Theory 2 (1969), 50-65.

[3] Meinardus, G., A.R. Reddy, G.D. Taylor and R.S. Varga, Converse Theorems and Extensions in Chebyshev Rational Approximation to Certain Entire Functions in [0,∞). Trans. Amer. Math. Soc. 170 (1972), 171-185.

[4] Newman, D.J., Rational Approximation to e^{-x}. J. Approx. Theory 10 (1974), 301-303.

[5] Roulier, J.A. and G.D. Taylor, Rational Chebyshev Approximation on [0,∞). J. Approx. Theory 11 (1974), 208-215.

[6] Schönhage, A., Zur rationalen Approximierbarkeit über [0,∞). J. Approx. Theory 7 (1973), 395-398.

ON LOCAL LINEAR FUNCTIONALS WHICH VANISH AT ALL B-SPLINES BUT ONE

Carl de Boor*)

1. Introduction. Let $k \in \mathbb{N}$, $\underline{\underline{t}} := (t_i)$ nondecreasing (finite, infinite or biinfinite) with $t_i < t_{i+k}$, all i, and let (N_i) be the sequence of B-splines of order k for the knot sequence $\underline{\underline{t}}$. This means that $N_i = N_{i,k,\underline{\underline{t}}}$ is the B-spline of order k with knots $t_i, \ldots, t_{i+k}$, i.e., N_i is given by the rule

$$(1.1) \quad N_{i,k,\underline{\underline{t}}}(t) := ([t_{i+1}, \ldots, t_{i+k}] - [t_i, \ldots, t_{i+k-1}])(\ -t)_+^{k-1}$$

with $[t_j, \ldots, t_{j+r}]f$ the r-th divided difference of f at the points $t_j, \ldots, t_{j+r}$. In particular,

$$N_i(t) > 0 \quad \text{on} \quad [t_i, t_{i+k}] \quad \text{and} \quad = 0 \quad \text{off} \quad [t_i, t_{i+k}],$$

and, for $t \in [t_j, t_{j+1}]$,

$$(\sum_i N_i)(t) = \sum_{i=j-k+1}^{j} N_i(t) = 1,$$

i.e., such a B-spline sequence provides a partition of unity. For more information about B-splines, see Curry and Schoenberg's paper [9], and [5].

The present paper is concerned with linear functionals λ_i for which

---

*) Sponsored by the United States Army under Contract No. DAHC04-75-C-0024.

(1.2) $\quad \operatorname{supp} \lambda_i \subseteq [t_i, t_{i+k}], \; \lambda_i N_j = \delta_{ij}, \quad \text{all } j .$

The first such linear functional seems to have been constructed in [1], for the purpose of demonstrating the linear independence over an interval of all B-splines which do not vanish identically on that interval. Since then, such linear functionals have been constructed in various ways and for a variety of jobs [2] - [7], [11], [13], [16], some of which are listed in Section 2.

In particular, it was shown in [6] that there exists a smallest number D_k so that, for all $\underline{\underline{t}}$ and all i with $t_i < t_{i+k}$, an $h_i \in \mathbb{L}_\infty$ can be found with $\operatorname{supp} h_i \subseteq [t_i, t_{i+k}]$, $\|h_i\|_\infty \le D_k/(t_{i+k} - t_i)$, and $\int h_i N_j = \delta_{ij}$, all j. After a discussion in Section 3 as to how to construct linear functionals λ_i satisfying (1.2), it is shown in Section 4 that

$$(\pi/2)^k/2 \le D_k \le 2k\, 9^{k-1} .$$

Also, numerical evidence is presented to indicate that probably

$$D_k = O(2^k) .$$

In Section 5, the related constant

$$D_{k,\infty} := \sup_{\underline{\underline{t}}} \sup_i 1/\operatorname{dist}_{\infty,[t_{i+1}, t_{i+k-1}]}(N_i, \operatorname{span}(N_j)_{j \ne i})$$

is discussed. As pointed out in [2], this number is related to the condition number of the B-spline basis,

$$\operatorname{cond}_k := \sup_{\underline{\underline{t}}} \operatorname{cond}_{k,\underline{\underline{t}}}$$

since

$$\operatorname{cond}_{k,\underline{\underline{t}}} := \frac{\sup \|\Sigma\alpha_j N_j\|_\infty / \|\underline{\underline{\alpha}}\|_\infty}{\inf \|\Sigma\alpha_j N_j\|_\infty / \|\underline{\underline{\alpha}}\|_\infty} = \frac{1}{\inf_i \operatorname{dist}_\infty(N_i, \operatorname{span}(N_j)_{j\neq i})}$$

$$\leq D_{k,\infty} .$$

It is shown that

$$(\pi/2)^{k-1}/2 \leq D_{k,\infty} \leq D_k,$$

and numerical evidence is presented to suggest that

$$D_{k,\infty} \sim 2^{k-1}/\sqrt{2} .$$

2. Some results obtainable with the aid of such functionals. In this section, we list some results obtainable through the explicit construction and analysis of specific local linear functionals which vanish at all B-splines but one.

(1) [1], [4]. For any open $I \subseteq \mathbb{R}$, $\{N_i \mid \operatorname{supp} N_i \cap I \neq \emptyset\}$ is linearly independent on I.

(2) [2]. Let $\$ = \$_{k,\underline{\underline{t}}}$ denote the linear space of all splines of order k with knot sequence $\underline{\underline{t}}$, i.e.,

$$\$:= \$_{k,\underline{\underline{t}}} := \{\sum_j \alpha_j N_{j,k,\underline{\underline{t}}} \mid \alpha_j \in \mathbb{R}, \quad \text{all } j\}$$

with the sum taken pointwise in case $\underline{\underline{t}}$ is not finite. There exists a constant $D_{k,\infty}$ depending only on k so that

$$\operatorname{dist}_\infty(f, \$) \leq D_{k,\infty} \max_j \operatorname{dist}_{\infty,[t_{j+1-k}, t_{j+k}]}(f, \mathbb{P}_k) .$$

Here, $\mathbb{P}_k$ denotes the collection of all polynomials of order k or degree $< k$, and the number $D_{k,\infty}$ is found as $\max_j \|\lambda_j\|$, with (λ_j) a sequence of local linear functionals dual to the B-spline sequence, i.e., $\lambda_i N_j = \delta_{ij}$, all i, j. This example raises the question of just how small one can make the norm of such linear functionals, a question taken up again in Section 5.

(3) [2], [4], [6]. There exists a smallest number D_k (depending only on k) so that, for all $\underline{\underline{t}}$ and all i with $t_i < t_{i+k}$, an $h_i \in \mathbb{L}_\infty$ can be found satisfying

$$\text{supp}\, h_i \subseteq [t_i, t_{i+k}] \, ,$$

$$\|h_i\|_p \leq D_k/(t_{i+k}-t_i)^{1/q} \, , \qquad (1/p + 1/q = 1) \, ,$$

$$\int h_i N_j = \delta_{ij}, \quad \text{all} \quad j \, .$$

(Note that the constant D_k mentioned here is k times the number D_k mentioned in [6].)

This fact has many consequences, among them the following two.

(4) [4]. If $f = \Sigma \alpha_j N_j$, then $\alpha_i = \int h_i f \leq \|h_i\|_q \|f\|_{p,[t_i, t_{i+k}]}$, hence

$$(2.1) \qquad |\alpha_i|(t_{i+k}-t_i)^{1/p} \leq D_k \|f\|_{p,[t_i, t_{i+k}]} \, ,$$

therefore, with E the diagonal matrix given by

$$E := \ulcorner \ldots, (t_{i+k}-t_i)/k, \ldots \lrcorner \, ,$$

we have

$$\| E^{1/p} \underline{\underline{\alpha}} \|_p \leq D_k \| \Sigma \alpha_j N_j \|_p \, .$$

(5) [8]. In particular, with

$$\overset{2}{N}_j := N_j \Big/ \left(\frac{t_{j+k} - t_j}{k} \right)^{1/2} ,$$

we get

$$\| \underline{\underline{\beta}} \|_2 \leq D_k \| \Sigma \beta_j \overset{2}{N}_j \|_2 \, .$$

Let L be $\mathbb{L}_2$-approximation by elements of $\$$, i.e.,

$$Lf \in \$, \quad \text{and} \quad f - Lf \perp \$ \, .$$

Then $Lf = \Sigma \alpha_j \beta_j \overset{2}{N}_j$, with $G\underline{\underline{\beta}} = (\int \overset{2}{N}_i f)$ and $G := (\int \overset{2}{N}_i \overset{2}{N}_j)$. Let $G^{-1} =: (\alpha_{ij})$. Then G^{-1} decays exponentially away from the diagonal, i.e.,

$$|\alpha_{ij}| \leq \text{const} \; \lambda^{|i-j|}$$

with $\lambda := (1 - D_k^{-2})^{1/(2k-2)} \in \,]0,1[$ and $\text{const} := D_k^3 / \lambda^{k-1}$ both independent of $\underline{\underline{t}}$, as can be proved using a very nice idea of Douglas, Dupont and Wahlbin [10]. This implies that, as a map on $\mathbb{L}_\infty$,

$$\| L \| \leq \text{const}_k \, (M_{\underline{\underline{t}}}^{(k)})^{1/2} ,$$

a bound in terms of the global mesh ratio

$$M_{\underline{\underline{t}}}^{(k)} := \max_{i,j} \, (t_{i+k} - t_i)/(t_{j+k} - t_j) \, .$$

Finally, here are two applications which have, offhand, nothing to do with splines, but rather are concerned with the smooth interpolation of data.

(6) [6]. Suppose we are given $\underline{\underline{t}} = (t_i)$ nondecreasing with $t_i < t_{i+k}$, all i. For given f, let

$f|_{\underline{\underline{t}}} := (f_i)$, with $f_i = f^{(j)}(t_i)$, where $j := \max\{r \mid t_{i-r} = t_i\}$.

Then, given $\underline{\underline{\alpha}} = (\alpha_i)$, there is no difficulty in finding <u>some</u> smooth f so that $f|_{\underline{\underline{t}}} = \underline{\underline{\alpha}}$, let $f_{\underline{\underline{\alpha}}}$ be one such, but it is not at all clear a priori under what circumstances such an f can be found in $\mathbb{L}_p^{(k)}(\mathbb{R})$. But, using (3) above, one can show that there exists $f \in \mathbb{L}_p^{(k)}(\mathbb{R})$ so that $f|_{\underline{\underline{t}}} = \underline{\underline{\alpha}}$ if and only if $((t_{i+k}-t_i)^{1/p}[t_i, \ldots, t_{i+k}]f_{\underline{\underline{\alpha}}}) \in \ell_p$.

The argument is based on the observation that f, given by the conditions that it agree with $f_{\underline{\underline{\alpha}}}$ at k points and that

$$f^{(k)} := \sum_i c_i((t_{i+k}-t_i)/k)\, h_i$$

with

$$c_i := k![t_i, \ldots, t_{i+k}]f_{\underline{\underline{\alpha}}} = \frac{k}{t_{i+k}-t_i} \int N_i\, f_{\underline{\underline{\alpha}}}^{(k)},$$

necessarily agrees with $f_{\underline{\underline{\alpha}}}$ at $\underline{\underline{t}}$ since $\int N_j h_i = \delta_{ij}$, i.e., since f and $f_{\underline{\underline{\alpha}}}$ have the same k-th divided differences.

(7) [6], [7]. In particular, with f the interpolant just constructed and $t_j < t_{j+1}$, at most k of the h_i are nonzero on $[t_j, t_{j+1}]$, therefore, from (3),

$$\|f^{(k)}\|_{\infty,[t_j,t_{j+1}]} \le \Big\| \sum_{h_i|_{[t_j,t_{j+1}]}\neq 0} |c_i| \frac{t_{i+k}-t_i}{k} |h_i| \Big\|_\infty$$

$$\le \max_{h_i|_{[t_j,t_{j+1}]}\neq 0} |c_i| D_k .$$

This proves that, for given $\underline{\underline{t}}$ and given $\underline{\underline{\alpha}}$, there exists $f \in \mathbb{L}_\infty^{(k)}$ so that $f|_{\underline{\underline{t}}} = \underline{\underline{\alpha}}$ and, for all $t_j < t_{j+1}$,

$$\|f^{(k)}\|_{\infty,[t_j,t_{j+1}]} \le D_k \max_{[t_j,t_{j+1}]\subseteq[t_i,t_{i+k}]} k!\,|[t_i,\ldots,t_{i+k}]f_{\underline{\underline{\alpha}}}|,$$

a fact of interest when carrying out an a posteriori error analysis for a finite difference approximation to the solution of an ordinary differential equation.

3. Construction of λ_i. The following observations seem to have been made first in [1]. They are also used implicitly by Jerome and Schumaker in [11].

Let

$$\psi_i(t) := (t-t_{i+1})\cdot \;\ldots\; \cdot(t-t_{i+k-1})/(k-1)!\,,$$

$$\overset{+}{\psi}_i(t) := (t-t_{i+1})_+ \;\ldots\; \cdot(t-t_{i+k-1})_+/(k-1)!\,.$$

Then

$$[t_j,\ldots,t_{j+k}]\overset{+}{\psi}_i = \delta_{ij}/((t_{i+k}-t_i)(k-1)!)$$

since, for $j < i$, $\overset{+}{\psi}_i = 0$ on $t_j,\ldots,t_{j+k}$, while, for $j > i$, $\overset{+}{\psi}_i = \psi_i \in \mathbb{P}_k$ on $t_j,\ldots,t_{j+k}$, and, finally, for $j = i$, $\overset{+}{\psi}_i$

agrees on $t_j, \dots, t_{j+k}$ with $\psi_i(t)(t-t_i)/(t_{i+k}-t_i)$, a polynomial of exact degree k with leading coefficient $1/((t_{i+k}-t_i)(k-1)!)$. Consequently,

$$(k-1)!\,([t_{j+1}, \dots, t_{j+k}] - [t_j, \dots, t_{j+k-1}])\psi_i^+ = \delta_{ij}.$$

On the other hand, from Taylor's expansion with integral remainder,

$$([t_{j+1}, \dots, t_{j+k}]-[t_j, \dots, t_{j+k-1}])f = \int N_{j,k,\underline{\underline{t}}}(t)f^{(k)}(t)dt/(k-1)!$$

for $f \in \mathbb{L}_1^{(k)}$. This proves the following lemma.

Lemma 3.1 $\lambda \in \mathbb{L}_q \subseteq \mathbb{L}_p^*$ satisfies $\lambda N_j = \delta_{ij}$ iff $\lambda = f^{(k)}$ for some $f \in \mathbb{L}_q^{(k)}$ with $f = \psi_i^+$ on $\underline{\underline{t}}$.

If we require from λ, in addition, that $\operatorname{supp} \lambda \subseteq [t_i, t_{i+k}]$, then $f\big|_{t<t_i}$ and $f\big|_{t>t_{i+k}}$ are both polynomials of degree $< k$, hence then

$$f = \begin{cases} 0, & t<t_i \\ \psi_i, & t>t_{i+k} \end{cases},$$

at least for sufficiently long $\underline{\underline{t}}$.

Corollary. If $[a, b] \subseteq [t_i, t_{i+k}]$, and $f \in \mathbb{L}_q^{(k)}[a, b]$ with

$$f = \begin{cases} 0\ , & k\text{-fold at } a\ , \\ 0 = \psi_i & \text{at all } t_j \in\ ]a, b[\ , \\ \psi_i\ , & k\text{-fold at } b\ , \end{cases}$$

then $\lambda_i \in \mathbb{L}_p^*$ given by

$$\lambda_i g := \int_{t_i}^{t_{i+k}} g\, f^{(k)}$$

has support in $[a, b]$ and satisfies

$$\lambda_i N_j = \delta_{ij}, \quad \text{all} \quad i, j .$$

As a simple example, choose r so that $t_i \le t_r < t_{r+1} \le t_{i+k}$ and let $f \in \mathbb{L}_1^{(k)}[t_r, t_{r+1}]$ so that

$$f = \begin{cases} 0, & k\text{-fold at } t_r \\ \psi_i, & k\text{-fold at } t_{r+1} \end{cases} .$$

Then $\lambda_i := f^{(k)}$ satisfies $\lambda_i N_j = \delta_{ij}$, all j, by the Corollary. Now note that, by assumption,

$$\psi_i^{(m)}(t_{r+1}) = f^{(m)}(t_{r+1}) = \int_{t_r}^{t_{r+1}} (t_{r+1}-s)^{k-m-1} f^{(k)}(s)ds/(k-m-1)! ,$$

i.e.,

$$\lambda_i : (t_{r+1} - \cdot)^{k-m-1}/(k-m-1)! \mapsto \psi_i^{(m)}(t_{r+1}), \quad m = 0, \ldots, k-1 .$$

Since $p(s) = \sum_{m=0}^{k-1} (-)^{k-m-1} p^{(k-m-1)}(t_{r+1})(t_{r+1}-s)^{k-m-1}$, all $p \in \mathbb{P}_k$, this implies that

$$\lambda_i p = \sum_{m=0}^{k-1} (-)^{k-m-1} p^{(k-m-1)}(t_{r+1})\psi_i^{(m)}(t_{r+1}), \quad \text{all} \quad p \in \mathbb{P}_k .$$

But, for any $p, \psi \in \mathbb{P}$,

$$\begin{aligned}(d/d\tau) \sum_{m=0}^{k-1} & (-)^{k-m-1} p^{(k-m-1)}(\tau)\psi^{(m)}(\tau) \\ &= (-)^{k-1} p^{(k)}(\tau)\psi(\tau) + p(\tau)\psi^{(k)}(\tau) \\ &= 0 .\end{aligned}$$

Hence

$$\lambda_i p = \sum_{m=0}^{k-1} (-)^{k-m-1} p^{(k-m-1)}(\tau)\psi_i^{(m)}(\tau), \quad \text{all } \tau, \quad \text{all } p \in \mathbb{P}_k .$$

Further, for all j, $N_j\big|_{[t_r, t_{r+1}]} \in \mathbb{P}_k$. Therefore

$$\sum_{m=0}^{k-1} (-)^{k-m-1} N_j^{(k-m-1)}(\tau)\psi_i^{(m)}(\tau) = \delta_{ij}, \quad \text{all } \tau \in [t_i, t_{i+k}], \text{ all } j,$$

which is the identity on which the quasi-interpolant of [3] is based. The corresponding specific linear functional $\hat{\lambda}_i$ given by the rule

$$\hat{\lambda}_i g := \sum_{m=0}^{k-1} (-)^{k-m-1} g^{(k-m-1)}(\tau)\psi_i^{(m)}(\tau)$$

for some fixed $\tau \in [t_i, t_{i+k}]$ is the k-th derivative (in the weak sense) of the function

$$f := (\cdot - \tau)_+^0 \psi_i$$

which indeed agrees with $\overset{+}{\psi}_i$ at $\underline{\underline{t}}$.

4. <u>An estimate for</u> D_k. In this section, we get an estimate for the number D_k of (3) of Section 2 by constructing a specific linear functional λ_i with $\lambda_i N_j = \delta_{ij}$, all j, using Lemma 3.1 and its Corollary.

Let $a < b$ with

$$t_i \le a \le t_{i+1}, \quad t_{i+k-1} \le b \le t_{i+k}$$

and take $G \in \mathbb{L}_\infty^{(k)}$ to be such that

$$G = \begin{cases} 0 & k\text{-fold at } a \\ 1 & k\text{-fold at } b \end{cases} . \tag{4.1}$$

Then, as was observed by D. J. Newman, the function $f := G\psi_i$ on $[a,b]$ satisfies the assumptions of the Corollary to Lemma 3.1. Therefore, the function $h_i := f^{(k)}$ on $[a,b]$ is in $\mathbb{L}_\infty$ and satisfies

$$\operatorname{supp} h_i \subseteq [a,b] \subseteq [t_i, t_{i+k}]$$

$$\|h_i\|_p \leq \|h_i\|_\infty (b-a)^{1/p} \tag{4.2}$$

$$\int h_i N_j = \delta_{ij}, \quad \text{all } j .$$

Next, we estimate $\|h_i\|_\infty$. We have

$$\|h_i\|_\infty \leq \sum_{m=0}^{k-1} \binom{k}{m} \|\psi_i^{(m)}\|_\infty \|G^{(k-m)}\|_\infty$$

and

$$\|\psi_i^{(m)}\|_\infty \leq \frac{(k-1)\ldots(k-m)}{(k-1)!}(b-a)^{k-1-m}, \quad m = 0,\ldots,k . \tag{4.3}$$

Also,

$$G^{(k-m)}(t) = \int_a^b (t-s)_+^{m-1} G^{(k)}(s)ds/(m-1)! ,$$

hence

$$\delta_{mk} = G^{(k-m)}(b) = \int_a^b (b-s)^{m-1} G^{(k)}(s)ds/(m-1)! , \tag{4.4}$$

$$m = 1,\ldots,k ,$$

i.e., $G^{(k)}$ is orthogonal to $\mathbb{P}_{k-1}$ on $[a,b]$. This implies that

$$G^{(k-m)}(t) = \int_a^b [(t-s)_+^{m-1} - p(t,s)]G^{(k)}(s)ds/(m-1)! ,$$

$$\text{all } p(t,\cdot) \in \mathbb{P}_{k-1}$$

and, choosing $p(t,\cdot)$, e.g. by interpolation, so that

$$\int_a^b |(t-s)_+^{m-1} - p(t,s)|ds \le 4\left(\frac{b-a}{4}\right)^m ,$$

we conclude that

$$(4.5) \qquad \|G^{(k-m)}\|_\infty \le 4\left(\frac{b-a}{4}\right)^m/(m-1)!\ \|G^{(k)}\|_\infty .$$

Next, we choose $G \in \mathbb{L}_k^{(k)}[a,b]$ so as to minimize $\|G^{(k)}\|_\infty$ subject to the conditions (4.1), i.e., subject to the conditions (4.4). This problem has been solved by Louboutin ten years ago and a solution is described by Schoenberg in [13]. Here is a simple argument:

Conditions (4.4) describe $G^{(k)} \in \mathbb{L}_\infty[a,b]$ as an extension to all of $\mathbb{L}_1[a,b]$ of the linear functional μ on $\mathbb{P}_k$ given by the rule

$$\mu(b-\cdot)^{m-1}/(m-1)! = \delta_{mk},\ m = 1,\ldots,k ,$$

i.e.,

$$\mu p = (-)^{k-1}p^{(k-1)}, \quad \text{all } p \in \mathbb{P}_k .$$

Therefore, $\min\|G^{(k)}\|_\infty = \|\mu\|$, and $G^{(k)}$ is minimal iff $G^{(k)}$ takes its norm in $\mathbb{P}_k$. Let ч_k be the Chebyshev polynomial of degree k. Then, sign $\text{ч}_k^{(1)}$ is well known to be orthogonal to $\mathbb{P}_{k-1}$ on $[-1,1]$, while $\text{ч}_k^{(k)} = k!\,2^{k-1}$. Hence, with

$$\hat{\text{ч}}_k(t) := (-)^{k-1}\text{ч}_k(2\,\frac{t-a}{b-a} - 1) ,$$

sign $\hat{\text{ч}}_k^{(1)}$ is orthogonal to $\mathbb{P}_{k-1} = \ker\mu$ on $[a,b]$ while

$$\mu\hat{u}_k = (-)^{k-1}\hat{u}_k^{(k)} = \left(\frac{4}{b-a}\right)^k k!/2 .$$

It follows that

$$\hat{G}^{(k)} := \text{sign}\, \hat{u}_k^{(1)} \left(\frac{4}{b-a}\right)^k k!/(2\|\hat{u}_k^{(1)}\|_\infty) \in \mathbb{L}_\infty = \mathbb{L}_1^*$$

extends μ to $\mathbb{L}_1$ and takes on its norm in $\mathbb{P}_k$ (at the point $\hat{u}_k^{(1)} \in \mathbb{P}_k$), hence is minimal. Since

$$\|\hat{u}_k^{(1)}\|_\infty = \text{Var}_{[a,b]}\hat{u}_k = 2k ,$$

this shows the minimal $G^{(k)}$ to be

$$(4.6) \qquad \hat{G}^{(k)} := \text{sign}\, \hat{u}_k^{(1)} \left(\frac{4}{b-a}\right)^k \frac{(k-1)!}{4}$$

with

$$(4.7) \qquad \min\|G^{(k)}\|_\infty = \|\hat{G}^{(k)}\|_\infty = \left(\frac{4}{b-a}\right)^k \frac{(k-1)!}{4} .$$

Correspondingly, $\hat{G}^{(1)}$ is the perfect B-spline of order k with simple knots at the $k+1$ extrema of $\hat{u}_k$ in $[a,b]$ and normalized to have unit integral.

For this particular choice for G, (4.3), (4.5) and (4.7) give

$$\|\psi_i^{(m)}\|_\infty \|G^{(k-m)}\|_\infty \le \frac{(k-1)\ldots(k-m)}{(k-1)!}(b-a)^{k-1-m}\left(\frac{b-a}{4}\right)^{m-k}\frac{(k-1)!}{(m-1)!}$$

$$= \frac{(k-1)\ldots(k-m)}{(m-1)!}4^{k-m}/(b-a),\ m=1,\ldots,k-1,$$

and

$$\|\psi_i\|_\infty \|G^{(k)}\|_\infty \le 4^{k-1}/(b-a) ,$$

hence

$$(b-a)\|h_i\|_\infty \leq 4^{k-1} + \sum_{m=1}^{k-1} \binom{k}{m} \frac{(k-1)!}{(k-m-1)!(m-1)!} 4^{k-m}$$

$$< 4^{k-1} + 2(k-1)\left(\sum_{m=0}^{k} \binom{k}{m} 2^{k-m} - 2^k - 1\right) \sum_{m=1}^{k-1} \binom{k-2}{m-1} 2^{k-m-1}$$

$$= 4^{k-1} + 2(k-1)(3^k - 2^k - 1) 3^{k-2}$$

$$< 2k\, 9^{k-1} .$$

Theorem 4.1 Let D_k be the smallest number with the property that, for every $\underline{t}$, every i and every $a < b$ with

$$t_i \leq a \leq t_{i+1}, \quad t_{i+k-1} \leq b \leq t_{i+k} ,$$

there exists $h_i \in \mathbb{L}_\infty$ such that

(4.8) $\operatorname{supp} h_i \subseteq [a, b]$, $\|h_i\|_\infty \leq D_k/(b-a)$, $\int h_i N_j = \delta_{ij}$, all j.

Then

$$(\pi/2)^k/2 \leq D_k \leq 2k\, 9^{k-1} .$$

Proof. Only the first inequality still requires proof. For this, take Schoenberg's Euler spline [14], [16],

$$\mathcal{E}_k(t) := \gamma_k \sum_{j=-\infty}^{\infty} (-)^j N_{j,k+1,\mathbb{Z}}\left(t - \frac{k+1}{2}\right)$$

with

(4.9) $$\gamma_k = 1/\varphi_{k+1}(\pi) = \left(\frac{\pi}{2}\right)^{k+1} \Big/ \sum_j \left(\frac{(-1)^j}{2j+1}\right)^{k+1} \geq \left(\frac{\pi}{2}\right)^k /2$$

so chosen that $\mathcal{E}_k(\nu) = (-)^\nu$, all $\nu \in \mathbb{Z}$. Then

$$\mathcal{E}_k^{(1)}(t) = 2\gamma_k \sum_j (-)^j N_{j,k,\mathbb{Z}}(t - \frac{k+1}{2})$$

is a spline of order k, with knot sequence $\mathbb{Z} - s$ where $s := (k+1)/2$, hence, by (2.1), and since $\mathcal{E}_k$ is monotone between integers,

$$|2\gamma_k k| \leq D_k \|\mathcal{E}_k^{(1)}\|_{1,[s,s+k]} = D_k \operatorname{Var}_{[s,s+k]}\mathcal{E}_k = 2k\, D_k$$

and so

$$\left(\frac{\pi}{2}\right)^k /2 \leq \gamma_k \leq D_k \; ; \qquad \text{Q.E.D.}$$

It is possible to compute D_k for small k as follows. For $\underline{\underline{\sigma}} := (\sigma_i)_1^{3k-1}$ with

$$0 = \sigma_1 = \ldots = \sigma_k \leq \sigma_{k+1} \leq \ldots \leq \sigma_{2k} = \ldots = \sigma_{3k-1} = 1 ,$$

compute the norm of the linear functional $\mu_{\underline{\underline{\sigma}}}$ given on $\$_{k,\underline{\underline{\sigma}}} \subseteq \mathbb{L}_1[0,1]$ by the rule

$$\mu_{\underline{\underline{\sigma}}} N_{j,k,\underline{\underline{\sigma}}} = \delta_{jk}, \; j = 1, \ldots, 2k-1 .$$

Much as in the computations reported in [7], this amounts to constructing (by Newton's method, say) an absolutely constant step function g on $[0,1]$ with $\dim \$_{k,\underline{\underline{\sigma}}}$ steps so that

$$\int_0^1 g\, N_j = \delta_{jk}, \quad \text{all } j .$$

Then $\|\mu_{\underline{\underline{\sigma}}}\| = \|g\|_\infty$, and

$$D_k = \sup_{\underline{\underline{\sigma}}} \|\mu_{\underline{\underline{\sigma}}}\| .$$

Somewhat more explicitly, the construction of such a g proceeds as follows. With

$$s := \dim \$_{k,\underline{\underline{\sigma}}} ,$$

and $0 = \rho_0 < \ldots < \rho_s = 1$, one computes $(\beta_j)_1^s$ such that

$$(4.10) \qquad \sum_j \beta_j \int_{\rho_{j-1}}^{\rho_j} N_i = \delta_{ik}, \quad \text{all } i .$$

Now

$$\int^\rho N_{i,k} = \frac{\sigma_{i+k}-\sigma_i}{k} \int^\rho M_{i,k} = \frac{\sigma_{i+k}-\sigma_i}{k} \sum_{i \le n} N_{n,k+1}(\rho) ,$$

as one checks easily, therefore

$$\int_{\rho_{j-1}}^{\rho_j} N_i = \frac{\sigma_{i+k}-\sigma_i}{k} \sum_{i \le n} N_{n,k+1}(\rho_j) - N_{n,k+1}(\rho_{j-1}) .$$

Since $\sigma_{2k} - \sigma_k = 1$, this shows that (4.10) is equivalent to

$$\sum_j \beta_j \sum_{i \le n} (N_{n,k+1}(\rho_j) - N_{n,k+1}(\rho_{j-1})) = k\delta_{ik}, \quad \text{all } i .$$

But, subtracting in order each equation in this system from all its predecessors, starting with the last, we obtain the equivalent system

$$(4.11) \quad \sum_j \beta_j (N_{i,k+1}(\rho_j) - N_{i,k+1}(\rho_{j-1})) = \begin{cases} -k, & i = k-1 \\ k, & i = k \\ 0, & \text{otherwise} \end{cases} ,$$

which is very similar to the system dealt with in [7]. In particular, one proves that $\mu_{\underline{\underline{\sigma}}}$ <u>has exactly one extremal</u>, i.e., <u>there exists exactly one absolutely constant</u> g <u>with</u> s <u>steps on</u> $[0,1]$ <u>for which</u> $\int_0^1 gN_i = \delta_{ik}$, <u>all</u> i. This means that the nonlinear system for the β_j and ρ_j consisting of (4.11) and

$$\beta_{j-1} + \beta_j = 0, \; j = 2, \ldots, s \,, \tag{4.12}$$

has exactly one solution.

For all k considered, such computations show $\sup_{\underline{\underline{\sigma}}} \|\mu_{\underline{\underline{\sigma}}}\|$ to be taken on at the middle vertex of the simplex over which $\underline{\underline{\sigma}}$ varies, i.e., at the point $\underline{\underline{\sigma}} = (\sigma_j)$ with

$$\sigma_j = \begin{cases} 0, & j < k + k/2 \\ 1, & j \geq k + k/2 \end{cases} .$$

Computed values for D_k are

k	D_k	$\ln_2 D_k$
1	1	0
2	2.4142..	1.2715..
3	5.2044..	2.3797..
4	10.0290..	3.3261..
5	21.3201..	4.4141..
6	40.8972..	5.3539..
7	86.3688..	6.4324..
8	166.4052..	7.3785..
9	348.5582..	8.4452..
10	674.2949..	9.3972..
11	1402.9478..	10.4542..

These numbers strongly suggest that D_k grows like 2^k rather than like the upper bound 9^k established in

Theorem 4.1.

5. <u>An estimate for</u> $D_{k,\infty}$. If $a < b$ and $t_i \leq a \leq t_{i+1}$, $t_{i+k-1} \leq b \leq t_{i+k}$, then we can construct $h_i \in \mathbb{L}_\infty[a, b]$ so that $\int h_i N_j = \delta_{ij}$. In fact, such a function h_i with smallest possible ∞-norm can be constructed as a minimum norm extension to all of $\mathbb{L}_1[a, b]$ of the linear functional μ_i on $\$\big|_{[a,b]} \subseteq \mathbb{L}_1[a, b]$ given by the rule

$$\mu_i N_j = \delta_{ij}, \quad \text{all } j .$$

This fact was the basis for the computation of D_k reported in the preceding section.

In general, if we think of $\$\big|_{[a,b]}$ as a subspace of $\mathbb{L}_\infty[a, b]$, then a minimum norm extension of μ_i to all of $\mathbb{L}_\infty$ does not exist in the form $h_i \in \mathbb{L}_1$, i.e., in the form of a function on $[a, b]$. For this reason, it is more convenient to consider $\$\big|_{[a,b]}$ as a subspace of $C[a, b]$, - this requires the assumption

$$(5.1) \qquad t_j < t_{j+k-1}, \quad \text{all } j, -$$

and to consider a norm preserving extension of μ_i to all of $C[a, b]$ since the dual of $C[a, b]$, while still not representable by functions on $[a, b]$, is in some sense simpler than that of $\mathbb{L}_\infty$. In particular, it is always possible to find norm preserving extensions of μ_i of the form

$$(5.2) \qquad \sum_{m=1}^{s} \alpha_m [\rho_m]$$

with

$$s := \dim \$_{k,\underline{\underline{t}}}\Big|_{[a,b]}$$

and

$$[\rho]f := f(\rho) .$$

In this section, we estimate the number

$$(5.3)\quad D_{k,\infty} := \sup_{\underline{\underline{t}}} \sup_{i} 1/\text{dist}_{\infty,[t_{i+1}, t_{i+k-1}]}(N_i, \text{span}(N_j)_{j\neq i})$$

$$= \sup_{\underline{\underline{t}}} \sup_{i} \|\lambda_i^*\| ,$$

with

$$\lambda_i^* := \text{minimum of } \|\cdot\| \text{ over } \{\lambda_i \in C^*[t_{i+1}, t_{i+k-1}] \,|\, \lambda_i N_j = \delta_{ij}, \text{ all } j\} .$$

This number was shown to be finite in [2]. The argument relied on constructing explicitly a norm preserving extension of μ_i of the form $\Sigma\, \alpha_i[\rho_i]$ with $t_r \leq \rho_1 < \ldots < \rho_k \leq t_{r+1}$ and $[t_r, t_{r+1}]$ a largest interval of that form in $[t_{i+1}, t_{i+k-1}]$. But the resulting bound for $D_{k,\infty}$ seemed very pessimistic.

Theorem 5.1 The constant $D_{k,\infty}$ defined by (5.3) satisfies

$$(5.4)\qquad (\pi/2)^{k-1}/2 \leq D_{k,\infty} \leq D_k .$$

Proof. By Theorem 4.1, the linear functional μ_i on $\$_{k,\underline{\underline{t}}}$ given by $\mu_i N_j = \delta_{ij}$, all j, satisfies

$$|\mu_i f| \leq D_k \|f\|_{\infty,[a,b]}$$

for any $a < b$ with $t_i \leq a \leq t_{i+1} \leq t_{i+k-1} \leq b \leq t_{i+k}$.
Hence, for $t_{i+1} < t_{i+k-1}$,

$$\text{dist}_{\infty,[t_{i+1},t_{i+k-1}]}(N_i, \text{span}(N_j)_{j\neq i}) = 1/\|\mu_i\| \geq D_k^{-1}$$

with $\|\mu_i\|$ the norm of μ_i with respect to $\|\cdot\|_{\infty,[t_{i+1},t_{i+k-1}]}$. For $t_{i+1} = t_{i+k-1}$, $N_j(t_{i+1}) = \delta_{ij}$, hence then $\text{dist}_{\infty,[t_{i+1},t_{i+k-1}]}(N_i, \text{span}(N_j)_{j\neq i}) = 1/\|\mu_i\|$ $= 1$. This proves that $D_{k,\infty} \leq D_k$. The inequality $\gamma_{k-1} \leq D_{k,\infty}$ was already proved in [5], using Schoenberg's Euler spline. Q.E.D.

To be precise, it was shown in [5] that

$$\gamma_{k-1} = \text{cond}_{k,\mathbb{Z}}$$

with

$$\text{cond}_{k,\underline{\underline{t}}} := \frac{\sup \|\Sigma \alpha_j N_j\|_\infty / \|\underline{\underline{\alpha}}\|_\infty}{\inf \|\Sigma \alpha_j N_j\|_\infty / \|\underline{\underline{\alpha}}\|_\infty} = \frac{1}{\inf_i \text{dist}_\infty(N_i, \text{span}(N_j)_{j\neq i})}$$
$$\leq D_{k,\infty}$$

hence

$$\text{cond}_k := \sup_{\underline{\underline{t}}} \text{cond}_{k,\underline{\underline{t}}} \leq D_{k,\infty}.$$

It is, of course, possible to prove that $D_{k,\infty} = O(9^k)$ directly without reference to Theorem 4.1: Let $[a,b] = [t_{i+1}, t_{i+k-1}]$ with $a < b$ and consider λ_i of the form $(G\psi_i)^{(k)}$ with

$$G(t) := \begin{cases} 0, & t < a , \\ G^{(k)}\{(t-\cdot)_+^{k-1}/(k-1)!\}, & t \geq a , \end{cases}$$

and $G^{(k)} \in C^*[a,b]$ so that

$$G^{(k)}\{(b-\cdot)^{k-j}/(k-j)!\} = \delta_{1j}, \; j = 1,\ldots,k .$$

Then $G\psi_i$ agrees with $\overset{+}{\psi}_i$ at $\underline{\underline{t}}$, hence $\lambda_i N_j = \delta_{ij}$, all j, i.e., $\lambda_i \in C^*[a,b]$ and λ_i extends μ_i. Next, choose $G^{(k)}$ to have as small a norm as possible. This requires $G^{(k)}$ to be a norm preserving extension to all of $C[a,b]$ of the linear functional μ on $\mathbb{P}_k$ given by the rule

$$\mu(b-\cdot)^{k-j}/(k-j)! = \delta_{1j}, \; j = 1,\ldots,k ,$$

i.e.,

$$\mu p = (-)^{k-1} p^{(k-1)}, \quad \text{all} \quad p \in \mathbb{P}_k .$$

Hence, with $a \leq \rho_1 < \ldots < \rho_k \leq b$,

$$(-)^{k-1}[\rho_1,\ldots,\rho_k] = \Sigma \alpha_j [\rho_j]$$

is an extension of μ. This extension is norm preserving provided it takes its norm in $\mathbb{P}_k$. Since the coefficients $\alpha_1,\ldots,\alpha_k$ strictly alternate in sign, this will happen iff $\rho_1,\ldots,\rho_k$ are chosen as the extrema of the Chebyshev polynomial of degree $k-1$ adjusted to the interval $[a,b]$. The resulting minimal G is an old acquaintance, viz. the integral of the perfect B-spline of order $k-1$ with support equal to $[a,b]$ and unit integral. We record this curious fact in the following

Proposition. Let $G_k(t) := \int_a^t B_k(s)ds$ with $B_k(s) := k[\rho_0,\dots,\rho_k](\cdot - s)_+^{k-1}$ and

$$\rho_j = (a + b + (a-b)\cos \pi j/k)/2,\ j = 0,\dots,k$$

the extrema of the k-th degree Chebyshev polynomial for $[a,b]$. Then, not only is $G_k^{(k)}$ the unique norm preserving extension to all of $\mathbb{L}_1[a,b]$ of the linear functional μ_k on $\mathbb{P}_k$ given by

$$\mu_k p = (-)^{k-1} p^{(k-1)}, \quad \text{all } p \in \mathbb{P}_k,$$

and therefore $G_k^{(k)}$ is absolutely constant, hence B_k is perfect and

$$\|G_k^{(k)}\|_\infty = \|\mu_k\|_{\|\cdot\|_1} = \left(\frac{4}{b-a}\right)^k \frac{(k-1)!}{4}$$

- this much was shown already by Louboutin [15], - but also $G_k^{(k+1)}$ is the unique norm preserving extension of the form $\Sigma\alpha_j[\rho_j]$ to all of $C[a,b]$ of μ_{k+1}, therefore

$$\|G_k^{(k+1)}\|_{"1"} = \mathrm{Var}_{[a,b]} G_k^{(k)} = \|\mu_{k+1}\|_{\|\cdot\|_\infty} = \left(\frac{4}{b-a}\right)^k \frac{k!}{2}.$$

The rest of the argument for the estimate $D_{k,\infty} = O(9^k)$ now proceeds as in the proof of Theorem 4.1.

It is possible to compute $D_{k,\infty}$ for small k as

$$D_{k,\infty} = \sup_{\underline{\sigma}} \|\mu_{\underline{\sigma}}\|$$

with

(5.5) $0 = \sigma_1 = \dots = \sigma_k < \sigma_{k+1} \le \dots \le \sigma_n < \sigma_{n+1} = \dots \sigma_{n+k} = 1,$

$$n := 2k - 3,$$

and $\mu_{\underline{\underline{\sigma}}}$ the linear functional on $S := \$_{k,\underline{\underline{\sigma}}}\big|_{[0,1]} \subseteq C[0,1]$ given by

$$\mu_{\underline{\underline{\sigma}}} N_{j,k,\underline{\underline{\sigma}}} = \delta_{j,k-1} .$$

In order to compute $\|\mu_{\underline{\underline{\sigma}}}\|$, one constructs $\varphi \in S\backslash\{0\}$ and $0 = \rho_1 < \ldots < \rho_n = 1$ so that

$$(-)^j \varphi(\rho_j) = \|\varphi\|_\infty, \quad \text{all } j .$$

This is possible since (N_j) is a weak Chebyshev system (see, e.g., [12]). Next, one constructs the extension of $\mu_{\underline{\underline{\sigma}}}$ of the form $\Sigma \alpha_j [\rho_j]$ to all of $C[0,1]$. Then $\Sigma N_r(\rho_j)\alpha_j = \delta_{r,k-1}$, hence $\alpha_{j-1}\alpha_j \le 0$, all j, since $(N_r(\rho_j))$ is totally positive (see, e.g., [12]). Therefore

$$|\mu_{\underline{\underline{\sigma}}}\varphi| = |\Sigma \alpha_j \varphi(\rho_j)| = \Sigma |\alpha_j| \, \|\varphi\|_\infty ,$$

i.e.,

$$\|\mu_{\underline{\underline{\sigma}}}\| = \Sigma |\alpha_j| .$$

As with the earlier reported calculation of D_k, it appears from these computations that $\sup \|\mu_{\underline{\underline{\sigma}}}\|$ is taken on at the "middle" vertex of the simplex described by (5.5), i.e., at the point $\underline{\underline{\sigma}}$ with

$$\sigma_j = \begin{cases} 0, & j \le k + k/2 - 1 \\ 1, & j > k + k/2 - 1 \end{cases} .$$

This would mean that

$$D_{k,\infty} = \|(N_{j,k,\underline{\underline{\tau}}}(\rho_i))^{-1}\|_\infty \tag{5.6}$$

with $\underline{\underline{\tau}} := (\tau_i)_1^{2k}$ given by

$$0 = \tau_1 = \ldots = \tau_k, \ \tau_{k+1} = \ldots = \tau_{2k} = 1$$

and $0 = \rho_1 < \ldots < \rho_k = 1$ the extrema of the Chebyshev polynomial of degree $k-1$ for $[0,1]$. This gives the following values for $D_{k,\infty}$.

k	$D_{k,\infty}$	$\ln_2 D_{k,\infty}$
2	1	0
3	3	1.5849..
4	5	2.3219..
5	11 2/3	3.5443..
6	21	4.3923..
7	46 1/5	5.5298..
8	85 4/5	6.4229..
9	183 6/7	7.5224..
10	347 2/7	8.4399..
15	.1169 E 5	13.5128..
20	.3635 E 6	18.4715..
25	.1193 E 8	23.5075..
30	.3747 E 9	28.4813..
35	.1219 E11	33.5053..
40	.3850 E12	38.4861..

It is striking that the first few values of $D_{k,\infty}$ are such simple rational numbers and that these numbers conform so quickly to the pattern $D_{k,\infty} \sim 2^{k-1}/\sqrt{2}$, as can be seen by their logarithms to the base 2. This raises the hope that such a relation might be provable with a little effort.

References

1. C. de Boor, The method of projections etc., Ph.D. thesis, University of Michigan, Ann Arbor, Mich. 1966.
2. C. de Boor, On uniform approximation by splines, J. Approx. Theory 1 (1968) 219-235.
3. C. de Boor and G. Fix, Spline approximation by quasi-interpolants, J. Approx. Theory 8 (1973) 19-45.
4. C. de Boor, The quasi-interpolant as a tool in elementary polynomial spline theory, in "Approximation Theory", G. G. Lorentz ed., Academic Press, New York, 1973, 269-276.
5. C. de Boor, On calculating with B-splines, J. Approx. Theory 6 (1972) 50-62.
6. C. de Boor, How small can one make the derivatives of an interpolating function?, J. Approx. Theory 13 (1975) 105-116.
7. C. de Boor, A smooth and local interpolant with 'small' k-th derivative, in "Numerical solutions of boundary value problems for ordinary differential equations", A. K. Aziz ed., Academic Press, New York, 1975, 177-197.
8. C. de Boor, A bound on the $\mathbb{L}_\infty$-norm of $\mathbb{L}_2$-approximation by splines in terms of a global mesh ratio, MRC TSR #1597, 1975.
9. H. B. Curry and I. J. Schoenberg, On Polya frequency functions. IV., J. d'Anal. Math. 17 (1966) 71-107.

10. J. Douglas, Jr., J. Dupont and L. Wahlbin, Optimal $\mathbb{L}_\infty$-error estimates for Galerkin approximations to solutions of two point boundary value problems, Math. Comp. 29 (1975) 475-483.

11. J. W. Jerome and L. L. Schumaker, Characterizations of functions with higher order derivatives in $\mathbb{L}_p$, Trans. Amer. Math. Soc. 143 (1969) 363-371.

12. S. Karlin, "Total positivity, vol. I", Stanford University Press, Stanford, Calif., 1968.

13. Tom Lyche and L. L. Schumaker, Local spline approximation methods, MRC TSR #1417, 1974.

14. I. J. Schoenberg, Cardinal interpolation and spline functions, J. Approx. Theory 2 (1969) 167-206.

15. I. J. Schoenberg, The perfect B-splines and a time-optimal control problem, Israel J. Math. 10 (1971) 261-274.

16. I. J. Schoenberg, "Cardinal spline interpolation", CBMS vol. 12, SIAM, Philadelphia, PA., 1973.

SPLINE APPROXIMATION IN BANACH FUNCTION SPACES

Stephen Demko *

§1. Introduction

In this paper we extend the known L^p-error bounds for spline approximation to the more general class of rearrangement invariant norms. We had originally wanted to extend the L^p theory to the case of Orlicz norms by the techniques used in [6]. However, the proof in [6] uses an L^p inequality (due to de Boor) whose extension to Orlicz spaces is not obvious. Fortunately, we have been able to circumvent this difficulty and obtain more general results by using an interpolation theorem originally proved by Calderón, [3]. While other authors, cf. [7], have used interpolation theorems in spline theory, we believe that this is the first application of Calderón's theorem to splines. The remainder of this section is devoted to a brief exposition of the necessary results from the theory of rearrangement-invariant Banach function spaces (r.i. spaces, for short); our exposition follows that of C. Bennett [2].

If f is a Lebesgue measurable function on [0,1], then f* will denote its non-increasing rearrangement. That is, f*(x) is the non-increasing, right continuous function having the property that $m\{x:|f(x)| > a\} = m\{x:f^*(x) > a\}$ for all $a \in R$ (m(C) denotes the measure of the set C). A <u>rearrangement-invariant norm</u> is an extended-real valued functional

ρ defined on the space of non-negative (a.e.) measurable functions having the following properties:

$\rho(f) = 0 \iff f = 0$ a.e. ; $\rho(\lambda f) = \lambda\rho(f)\ \forall\lambda > 0$;

$\rho(f+g) \leq \rho(f) + \rho(g)$;

$f \leq g$ a.e. $\Rightarrow \rho(f) \leq \rho(g)$

$\rho(f) = \rho(f^*)$, $\rho(1) < \infty$

and

$f_n \uparrow f$ a.e. $\Rightarrow \rho(f_n) \uparrow \rho(f)$, (Fatou property).

The r.i. space L^ρ is defined to be the set of all equivalence classes of measurable functions for which $\rho(|f|) < \infty$. Because of the Fatou property, it is a Banach space with the norm $\|\cdot\| = \rho(|f|)$. We shall write $\rho(f)$ for $\rho(|f|)$. It is known that $L^\infty \subset L^\rho \subset L^1$ in the sense that the identity mappings are continuous. The result of Calderón alluded to above is

<u>Theorem 1</u>. Let L^ρ be any r.i. space. If T is a bounded linear operator from L^i into L^i , $i = 1,\infty$, then T is a bounded operator from L^ρ into itself. Furthermore, the norm of T as an operator on L^ρ is bounded by the maximum of its norms on L^1 and L^∞ .

Concrete examples of r.i. spaces are the separable Orlicz spaces, L^M , which we now define (see [8] for details). Let $M(x)$ be a non-decreasing, continuous convex function such that $M(0) = 0$, $M(\infty) = \infty$ and $\limsup_{x\to\infty} \frac{M(2x)}{M(x)} < \infty$. Then, $L^M \equiv \{f : \int_0^1 M(|f(x)|)dx < \infty\}$ is a separable r.i. space with the norm $\rho(f) \equiv \inf\{k>0 : \int_0^1 M(\frac{|f(x)|}{k})dx \leq 1\}$.

Finally, we note that for a general r.i. space L^ρ, the given norm $\rho(f)$ is equivalent to a norm of the form $\sigma(f) = \sup_{g \varepsilon C} \int_0^1 f(x)g(x)dx$ where C is a bounded convex subset of L^1 depending only upon ρ.

§2. Approximation Theoretic Background

In the remainder of this paper, L^ρ will always be an r.i. space.

Definition 2.1. The L^ρ-modulus of continuity is defined for $f \varepsilon L^\rho$ and $\delta > 0$ as $\omega_\rho(f;\delta) \equiv \sup_{|t| \leq \delta} \rho([f(x+t)-f(x)]\chi_t)$. Here, χ_t is the characteristic function of $[0,1-t]$ if $t > 0$ and of $[t,1]$ if $t < 0$.

Proposition 2.2. If $C[0,1]$ is dense in L^ρ, then $\omega_\rho(f;\delta) \to 0$ as $\delta \to 0$.

Proof. Let $\varepsilon > 0$ and $f \varepsilon L^\rho$. Choose a continuous function $g(x)$ such that $\rho(f-g) < \varepsilon/3$ and fix t. Now, $\rho(f(x+t)-f(x)) \leq \rho([f(x+t)-g(x+t)]\chi_t) + \rho([g(x)-f(x)]\chi_t) + \rho([g(x+t)-g(x)]\chi_t) < \frac{2\varepsilon}{3} + \|g(x+t)-g(x)\|_\infty$. The result now follows from the uniform continuity of g.

The natural function spaces for our work are the extensions of the usual Sobolev spaces defined as follows:

$W_\rho^k \equiv \{f : D^j f(x) \varepsilon C[0,1], 0 \leq j \leq k-1 ; D^{k-1}f(x)$ is absolutely continuous on $[0,1]$; and $D^k f \varepsilon L^\rho\}$.

We shall also make use of a piecewise version of these spaces which we now define. Let $\Delta : 0 = x_0 < x_1 < \cdots < x_N = 1$ be a partition of $[0,1]$. Then, $W_\rho^k(\Delta) \equiv \{f : D^j f \in C[x_{i-1}, x_i] , 0 \leq j \leq k-1 , 1 \leq i \leq N ; D^{k-1}f(x)$ is absolutely continuous on each subinterval $[x_{i-1}, x_i]$; and $D^k f \in L^\rho\}$. We shall allow the elements of $W_\rho^k(\Delta)$ to be double valued at the interior points of Δ. Note that if $f \in W_\rho^k(\Delta)$, then for $0 \leq j \leq k-1$, $D^j f(x)$ exists everywhere except at a finite number a points and is bounded; hence, $D^j f \in L^\rho$. Of course, if $k = 0$, then $W_\rho^k = W_\rho^k(\Delta) = L^\rho$.

Let us now recall the definition of the <u>Stekloff function</u> (cf. [1], p. 173), f_h, of a function $f \in L_1$: $f_h(x) = \frac{1}{h} \int_{-\frac{h}{2}}^{\frac{h}{2}} f(x+t)dt$ $(0 < h \leq 1)$.

The important properties of the Stekloff function are contained in the following result whose proof we leave to the reader.

<u>Proposition 2.3</u>. Let $f \in W_\rho^k$ for some $k \geq 0$. Then, there is a constant $K > 0$, independent of f, k, and h such that

(a) $f_h \in W_\rho^{k+1}$ and $(D^j f_h)(x) = (D^j f)_h(x)$

$$\forall x \in [0,1] , 0 \leq j \leq k .$$

(b) $\rho(D^j(f-f_h)) \leq K\omega_\rho(D^j f; {}^h/2)$

(c) $\rho(D^{k+1}f_h) \leq Kh^{-1}\omega_\rho(D^k f;h)$.

We conclude this section by defining the spaces of piecewise polynomials with which we shall work and discussing some of their properties. Let $\Delta : 0 = x_0 < \cdots < x_N = 1$ be any partition of $[0,1]$ and let k be a positive integer. Then, $S_k(\Delta) \equiv \{f : f \in C^{k-2}[0,1]$ and f is a polynomial of degree $\leq k-1$ on $[x_{i-1}, x_i]$, $1 \leq i \leq N\}$ and $P_k(\Delta) = \{f : f$ is a polynomial of degree k-1 on $[x_{i-1}, x_i]$, $1 \leq i \leq N\}$. Again, we allow the elements of $P_k(\Delta)$ to be double valued at the points x_i , $1 \leq i \leq N-1$. It is well known (cf. [4]) that $S_k(\Delta)$ has a basis $\{B_i\}_{i=1-k}^{N}$ with $B_i(x) \neq 0$ only if $x_i < x < x_{i+k}$ $(x_{1-k} = \cdots = x_{-1} = x_0)$. This basis, the so called B-spline basis, has an L^1-normalization, $\{B_{i,1}\}_i$, which has the property that for any collection of real scalars $\{a_i\}$

$$D_{k,1} \sum |a_i| \leq \|\sum a_i B_{i,1}\|_1 \leq \sum |a_i|$$

where $D_{k,1}$ is an absolute constant <u>independent of</u> $\underline{\Delta}$, (cf. [4]). From these inequalities and their proofs one can deduce the existence of bounded measurable functions, $\{\lambda_i\}$, such that (1) $\int_0^1 \lambda_i(x) B_{j,1}(x)dx = \delta_{i,j}$, (2) $\lambda_i(x) B_j(x) = 0$ if $i \neq j$ (i.e. the λ_i's have local supports), and (3) $\|\lambda_i\|_\infty \leq \frac{1}{D_{k,1}}$; for details, see [5]. It then follows that the operator on L^1 defined by $Pf = \sum_i \int_0^1 f(x)\lambda_i(x)dx \; B_{i,1}$ has norm no greater than $k D_{k,1}^{-1}$. We formalize this fact in

<u>Theorem 2.4</u>. Let L^ρ be an r.i. space. There exists a constant $C > 0$ such that for any Δ there

is a projection $P : L^\rho \to S_k(\Delta)$ with norm less than C.

§3. Main Results

In this section, we find upper bounds for the quantities $\inf\{\rho(D^j(f-g)) : g \in S_k(\Delta)\}$, $0 \le j \le r$, for functions $f \in W^r_\rho$, $0 \le r \le k-1$. We shall assume that $\Delta : 0 = x_0 < \cdots < x_N = 1$ is a fixed but arbitrary partition with $\bar{\Delta} \equiv \max\{x_i - x_{i-1} : 1 \le i \le N\}$.

__Lemma 3.1.__ Let $g \in W^1_\rho(\Delta)$ and suppose that there exist points $t_i \in (x_{i-1}, x_i)$, $1 \le i \le N$, at which g vanishes. Then, $\rho(g) \le \bar{\Delta}\, \rho(g')$.

__Proof.__ Define $T : L^i \to L^i$, $i = 1, \infty$, by the rule $(Tf)(x) = \int_{t_i}^{x} f(s)ds$ if $x_{i-1} \le x \le x_i$. Since T has norm no greater than $\bar{\Delta}$ in both L^1 and L^∞, it follows that $T : L^\rho \to L^\rho$ and $\rho(Tf) \le \bar{\Delta}\, \rho(f)$ $\forall f \in L^\rho$. Since $T(L^\rho) = \{f \in W^1_\rho(\Delta) : f(t_i) = 0, 1 \le i \le N\}$, we are done.

__Theorem 3.2.__ Let $f \in W^r_\rho$ for some $0 \le r \le k-1$. Then, there is a constant K independent of f and Δ such that $\inf\{\rho(D^j(f-g)) : g \in P_k(\Delta)\} \le K(\bar{\Delta})^{r-j}$ $\omega_\rho(D^r f; \bar{\Delta})$.

__Proof.__ Let $p \in P_n(\Delta)$ be defined by

$$p(x) = \sum_{m=0}^{r-1} \frac{D^m f(t_i)}{m!} (x-t_i)^m + \frac{D^r f_h(t_i)}{r!} (x-t_i)^r, \quad \text{if } x_{i-1} \le x \le x_i .$$

Here, we take $t_i = \frac{x_i + x_{i-1}}{2}$ and $h = \bar{\Delta}$. By the previous lemma, $\rho(D^j(f-p)) \leq (\bar{\Delta})^{r-j} \rho(D^r(f-p))$. Now, using the properties of the Stekloff function and the fact that $D^{r+1}p(x) = 0$ a.e., we have $\rho(D^r(f-p)) \leq \rho(D^r(f-f_h)) + \rho(D^r(f_h-p)) \leq K\omega_\rho(D^r f;\frac{h}{2}) + \bar{\Delta}\ \rho(D^{r+1}f_h) \leq K\omega_\rho(D^r f;\bar{\Delta})$.

Now, let P be a projection from L^ρ onto $S_k(\Delta)$ of the type described in the discussion preceding Theorem 2.4. That is, $Pf = \sum_i \lambda_i(f)B_{i,1}$, where $\{B_{i,1}\}$ is the L^ρ-normalized B-spline basis for $S_k(\Delta)$, $\lambda_i(f) = 0$ if f vanishes on the support of B_i , $\|\lambda_i\|_\infty \leq K < \infty\ \forall i$, and the norm of P is independent of Δ . Let $I_i \equiv [x_i,x_{i+1}]$ and $J_i \equiv [x_{i-k+1},x_{i+k}]$ so that if $x \in I_i$, (Pf)(x) depends only upon the behavior of f on J_i .

For each $0 \leq s \leq 2k-2$, let $A_s = \{i : 0 \leq i \leq N$ and $i \equiv s \mod(2k-2)\}$ and let Δ_s be the partition whose subintervals are $\{J_i\}_{i\epsilon A_s}$. Define a restriction operator, R_s , by $R_s f = \sum_{i\epsilon A_s} f\chi_{I_i}$, so that $\|R_s\| \leq 1$ and $f = \sum_{s=0}^{2k-2} R_s f$. Note that if $g \in P_k(\Delta_s)$, then $R_s Pg = g$.

Theorem 3.3. With P and R_s as above and $f \in L^\rho$, $\rho(R_s(f-Pf) \leq \{1 + \|P\|\} \inf_{g\epsilon P_k(\Delta_s)} \rho(f-g)$. Consequently, $\rho(f-Pf) \leq (2k-1)\{1 + \|P\|\} \max_{0\leq s\leq 2k-2} \{\inf_{g\epsilon P_k(\Delta_s)} \rho(g-f).\}$

Proof. Let $g \in P_k(\Delta_s)$. Then, $R_s(f-Pf) = R_s(f-g) + R_sP(g-f)$. The first inequality is now clear. The second is also obvious since $f-Pf = \sum_{s=0}^{2k-2} R_s(f-Pf)$.

Corollary 3.4. There is a constant $K > 0$, independent of Δ, such that for all $f \in W_\rho^r$, $0 \le r \le k-1$, $\inf\{\rho(D^j(g-f)) : g \in S_k(\Delta)\} \le K(\bar{\Delta})^{r-j}\, \omega_\rho(D^r f;\bar{\Delta})$, $0 \le j \le r$.

Proof. Apply the bounds of theorem 3.2 to the result of theorem 3.3 and note that $\inf\{\rho(D^j(f-g)) : g \in S_k(\Delta)\} = \inf\{\rho(D^j f-g) : g \in S_{k-j}(\Delta)\}$ and $\omega_\rho(f;M\delta) \le (1+M)\, \omega_\rho(f;\delta)$ $\forall M > 0$, (cf. [9]).

BIBLIOGRAPHY

[1] Achieser, N. I., Theory of Approximation, Ungar, 1956.

[2] Bennett, C., Banach function spaces and interpolation methods 1. the abstract theory, J. Func. Anal. 17 (1974), pp. 409-440.

[3] Calderón, A. P., Spaces between L^1 and L^∞ and the theorem of Marcinkiewicz, Studia Math. 26 (1966), pp. 273-299.

[4] de Boor, C., The quasi interpolant as a tool in elementary polynomial spline theory, in Approximation Theory (G. G. Lorentz, ed.) Academic Press, 1973, pp. 269-276.

[5] _________, How small can one make the derivatives of an interpolating function? J. Approx. Theory 13 (1975), pp. 105-116.

[6] Demko, S., Local mappings onto spline spaces, to appear.

[7] Hedstrom, G. W. and Varga, R. S., Applications of Besov spaces to spline approximation, J. Approx. Theory 4 (1971), pp. 295-327.

[8] Krasnosel'skii, M. A. and Rutickii, Ya. B., Convex functions and Orlicz spaces, Noordhoff, Groninger, 1961.

[9] Lorentz, G. G., Approximation of Functions, Holt, Rinehart and Winston, 1966.

---

* This work was supported in part by NSF GP-44017.

CHEBYSHEV NODES FOR INTERPOLATION ON
A CLASS OF ELLIPSES

K.O. Geddes*

Abstract Let ξ_ρ denote an ellipse with foci at ±1 and sum of semi-axes $\rho > 1$, let $A(\xi_\rho)$ denote the space of functions analytic in points interior to ξ_ρ and continuous onto the boundary, and let $T_n(z)$ denote the classical nth-degree Chebyshev polynomial of the first kind. A Lagrange interpolating projection I_n^ρ from the space $A(\xi_\rho)$ onto polynomials of degree n is defined by interpolation at the zeros of the polynomial

$$T_{n+1}(z)-i\ \sinh((n+1)\log \rho),$$

which are n+1 distinct points on ξ_ρ. By computing a bound on $\|I_n^\rho\|$ valid for all $\rho > 1$, for degrees in the range $1 \le n \le 10$, it is shown that $I_n^\rho f$ is a practical near-minimax polynomial approximation to f in $A(\xi_\rho)$. This extends a well-known result for interpolation in the limiting space $A(\xi_1)$, the space of continuous functions on the real interval [-1,1], in which case the interpolating points are the zeros of $T_{n+1}(z)$.

1. Introduction

It is well known that a good choice of nodes for degree-n interpolation in the space C[-1,1] is the set of zeros of the (n+1)th-degree Chebyshev polynomial of the first kind. If $I_n^1 f$ denotes this degree-n interpolating polynomial to $f \in C[-1,1]$ then it was shown in [8] that

* This work was supported by the National Research Council of Canada under Grant A8967.

$$(1.1) \qquad \|f-I_n^1 f\| \le (1+\phi_n)\|f-B_n f\|,$$

where $\|\cdot\|$ is the minimax norm on $[-1,1]$, $B_n f$ denotes the best degree-n minimax polynomial approximation to $f \in C[-1,1]$, and ϕ_n are constants which we shall call the Gronwall constants since they arise in a 1921 paper by T.H. Gronwall [6]. The Gronwall constants grow logarithmically with n and, in particular, it is found that $\phi_n < 5.5$ for all $n \le 1000$. Thus $I_n^1 f$ is a near-minimax polynomial approximation to $f \in C[-1,1]$.

The above result is of practical significance because of the computational complexity of computing $B_n f$. For approximation in a region of the complex plane, a similar result is of even greater practical significance since there is no known Remez-type algorithm for computing best minimax polynomial approximations (cf.[9]). It was shown in [5] that in the space A(C) of functions analytic in the unit circle C, if $F_n f$ denotes the polynomial interpolating f at the (n+1)th roots of unity then

$$(1.2) \qquad \|f-F_n f\| \le (1+\phi_n)\|f-B_n f\|,$$

where $\|\cdot\|$ is the minimax norm on the unit circle C, $B_n f$ denotes the best degree-n minimax polynomial approximation to $f \in A(C)$, and ϕ_n are the Gronwall constants appearing above. The purpose of this paper is to identify a set of interpolation nodes on a class of ellipses for which an inequality similar to (1.1) and (1.2) will hold.

2. The Function Space $A(\xi_\rho)$ as the Image of A(C)

A Chebyshev ellipse ξ_ρ, where $\rho > 1$, is an ellipse with foci at ±1 and sum of semi-axes ρ. The family of Chebyshev ellipses ξ_ρ, for all real values $\rho > 1$, covers the entire complex plane with the interval [-1,1] deleted. A Chebyshev ellipse ξ_ρ collapses to the real interval [-1,1] as $\rho \to 1$.

For fixed $\rho > 1$, the function space $A(\xi_\rho)$ is defined to be the linear space of functions f which are analytic in the interior $I(\xi_\rho)$ of the ellipse ξ_ρ and continuous in $\overline{I(\xi_\rho)}$, the closure of $I(\xi_\rho)$. The space $A(\xi_\rho)$ is normed in the minimax sense:

$$\|f\| = \sup\{|f(z)| : z \in \xi_\rho\}.$$

We shall be interested in the approximation subspace Π_n, the subspace of (complex) algebraic polynomials of degree not exceeding n, for $n \geq 0$.

The minimax mapping $B_n : A(\xi_\rho) \to \Pi_n$ maps $f \in A(\xi_\rho)$ onto its best minimax polynomial approximation. A mapping $M_n : A(\xi_\rho) \to \Pi_n$ is called a near-minimax mapping with relative factor μ_n if it satisfies the inequality:

$$\|f - M_n f\| \leq (1+\mu_n)\|f - B_n f\|, \text{ for all } f \in A(\xi_\rho).$$

A projection $P_n : A(\xi_\rho) \to \Pi_n$ is a mapping which is bounded, linear, and idempotent. A projection P_n is a near-minimax mapping with relative factor $\|P_n\|$ since it satisfies the following inequality [2]:

$$(2.1) \qquad \|f - P_n f\| \leq (1+\|P_n\|)\|f - B_n f\|, \quad \text{for all } f \in A(\xi_\rho).$$

In spaces normed in the minimax sense it is generally found that if $\{P_n\}$ is a sequence of projections onto polynomials (where n denotes the degree) then $\|P_n\| \to \infty$ as $n \to \infty$ (see Cheney [1], Chapter 6, for a proof of this fact in the space of continuous functions on the real interval [-1,1]). However, if we can identify a sequence $\{P_n\}$ such that

$\|P_n\| \sim c \log n$ for a reasonably small constant c then, by inequality (2.1), P_n will be a practical near-minimax approximation mapping. (For example, if $\|P_n\| < 9$ for all $n \le 1000$ then $P_n f$ is within one decimal place of the accuracy of $B_n f$ for practical values of n.)

A Lagrange interpolating projection for the space $A(\xi_\rho)$ is a particular projection $P_n^I: A(\xi_\rho) \to \Pi_n$ defined by interpolation at n+1 preassigned distinct points on the boundary ξ_ρ. If the interpolation points for P_n^I are $z_k \in \xi_\rho$, $0 \le k \le n$, then we have the following well-known formula (Lagrange form) for $P_n^I f$:

$$(2.2) \qquad (P_n^I f)(z) = \sum_{k=0}^{n} f(z_k)\ell_k(z),$$

where if $w_{n+1}(z) = \prod_{k=0}^{n} (z-z_k)$ then

$$\ell_k(z) = w_{n+1}(z)/[w'_{n+1}(z_k)(z-z_k)], \qquad 0 \le k \le n.$$

It follows immediately from formula (2.2) that

$$(2.3) \qquad \|P_n^I\| \le \|\Lambda_n\|$$

where $\Lambda_n(z) = \sum_{k=0}^{n} |\ell_k(z)|$ is the Lebesgue function.

The theory of polynomial approximation in $A(\xi_\rho)$ is intimately connected with a conformal mapping between Chebyshev ellipses and circles (cf. [7], Chapter 3). The mapping

$$(2.4) \qquad w = \phi(z) = z + \sqrt{z^2-1}$$

maps the entire complex plane with the interval [-1,1] deleted conformally onto the exterior of the unit circle, where the branch of $\sqrt{z^2-1}$ to be used is uniquely determined by the condition

$$(2.5)\qquad |z+\sqrt{z^2-1}|>1.$$

The inverse conformal mapping is $\psi=\phi^{-1}$, where

$$(2.6)\qquad z=\psi(w)=\tfrac{1}{2}(w+w^{-1}).$$

A Chebyshev ellipse ξ_ρ is mapped by ϕ onto the circle

$$C_\rho=\{w:|w|=\rho\}$$

and conversely ξ_ρ is the image of C_ρ under the mapping ψ.

It is shown in [4] that, for any $z\notin[-1,1]$, the branch of $\sqrt{z^2-1}$ satisfying condition (2.5) can be conveniently specified by:

$$(2.7)\qquad \sqrt{z^2-1}=+i\sin(\operatorname{Arccos} z),$$

where the single-valued branch of Arccos z to be used is specified by

$$(2.8)\qquad -\pi\le \operatorname{Re}(\operatorname{Arccos} z)<\pi;\ \operatorname{Im}(\operatorname{Arccos} z)<0.$$

Moreover, a Chebyshev ellipse ξ_ρ satisfies the parametric equation

$$(2.9)\qquad z=\tfrac{1}{2}(\rho+\rho^{-1})\cos\theta+i\,\tfrac{1}{2}(\rho-\rho^{-1})\sin\theta,\ -\pi\le\theta<\pi,$$

and if $z\in\xi_\rho$ then

$$(2.10)\qquad \operatorname{Arccos} z=\theta-i\log\rho,$$

where θ is the parameter of z in equation (2.9).

It is convenient to consider the conformal mapping between a Chebyshev ellipse ξ_ρ and the <u>unit</u> circle

$$C=\{t:|t|=1\}.$$

Under the mapping

$$(2.11)\qquad t=\eta(w)=\rho^{-1}w,$$

the circle C_ρ is mapped conformally onto the unit circle C and hence the mapping

$$(2.12) \qquad t = \Phi(z) = \rho^{-1}(z+\sqrt{z^2-1})$$

maps the ellipse ξ_ρ conformally onto the unit circle C. The inverse conformal mapping is

$$(2.13) \qquad z = \Psi(t) = \tfrac{1}{2}(\rho t+\rho^{-1}t^{-1}).$$

The above conformal mappings induce a relationship between the function space $A(\xi_\rho)$ and the corresponding function spaces $A(C_\rho)$ and $A(C)$. It is known (cf. [7], Chapter 3) that expansions in series of Chebyshev polynomials are the natural analogues for $A(\xi_\rho)$ of expansions in Taylor series for $A(C_\rho)$ and $A(C)$. We identify the function $h \in A(C)$ defined by the Taylor series expansion

$$h(t) = \sum_{k=0}^{\infty} a_k t^k$$

with the function $g \in A(C_\rho)$ defined by the Taylor series expansion

$$g(w) = \sum_{k=0}^{\infty} c_k w^k, \quad \text{where } c_k = \rho^{1-k} a_k,$$

and finally with the function $f \in A(\xi_\rho)$ defined by the Chebyshev series expansion

$$f(z) = \sum_{k=0}^{\infty} c_k T_k(z).$$

These relationships arise by applying the mapping (2.11) and its inverse to the function $t^k \in A(C)$, yielding the transformation:

$$t^k \to \eta^{-1}(t^k),$$

where $\quad t = \eta(w);$

i.e. $\quad t^k \to \rho(\rho^{-1}w)^k = \rho^{1-k}w^k;$

and similarly, applying (2.4) and (2.6) to the function $w^k \in A(C_\rho)$, yielding the transformation:

$$w^k \to \psi(w^k),$$

where $\quad w = \phi(z);$

i.e. $\quad w^k \to \frac{1}{2}[(z+\sqrt{z^2-1})^k + (z+\sqrt{z^2-1})^{-k}] = T_k(z)$

(the final equality here is readily proved using (2.7) and the the substitution $z = \cos s$). The validity of identifying h, g and f as above is based on the fact that if

$$\overline{\lim_{k\to\infty}} \sqrt[k]{|a_k|} = \frac{1}{R}$$

then h is analytic in $I(C_R)$, and then

$$\overline{\lim_{k\to\infty}} \sqrt[k]{|c_k|} = \rho^{-1} \overline{\lim_{k\to\infty}} \sqrt[k]{|a_k|} = \frac{1}{\rho R} \tag{2.14}$$

so that g is analytic in $I(C_{\rho R})$, and finally f is analytic in $I(\xi_{\rho R})$ by equation (2.14) (cf. [7], Chapter 3).

In view of these relationships, it is not surprising that truncated Chebyshev series yield near-minimax approximations in $A(\xi_\rho)$ [4] as the analogue of truncated Taylor series in A(C) [5]. For the case of interpolation in $A(\xi_\rho)$, it is natural to consider as nodes the images of the roots of unity under the mapping Ψ defined by (2.13), since interpolation at the roots of unity yields near-minimax approximations in the space A(C) [5].

3. Interpolation at Roots-of-Unity Images

It was proved in [5], for the space A(C) of functions analytic in the unit circle, that if $F_n : A(C) \to \Pi_n$ is the discrete Fourier projection defined by interpolation at the n+1 roots of $t^{n+1}-1$ (i.e. the (n+1)th roots of unity) then

$$\|F_n\| \le \phi_n,$$

where ϕ_n are the Gronwall constants (cf. section 1) which have asymptotic order $\frac{2}{\pi}\log n$. Moreover, it is conjectured that F_n is a minimal Lagrange interpolating projection (i.e. that $\|F_n\| \le \|P_n^I\|$ where $P_n^I : A(C) \to \Pi_n$ is any other Lagrange interpolating projection).

If ω denotes a primitive (n+1)th root of unity then the roots of $t^{n+1}-1$ are ω^k $(0 \le k \le n)$. As mentioned at the end of section 2, it is natural to consider the points $z_k \in \xi_\rho$ defined by

$$z_k = \Psi(\omega^k), \quad 0 \le k \le n,$$

where Ψ is the mapping (2.13), as interpolation points for the space $A(\xi_\rho)$. Explicitly, these points are:

$$z_k = \tfrac{1}{2}(\rho\omega^k + \rho^{-1}\omega^{-k}), \quad 0 \le k \le n. \tag{3.1}$$

<u>Theorem 3.1</u> The points z_k $(0 \le k \le n)$ defined by equation (3.1) are the n+1 roots of the polynomial

$$T_{n+1}(z) - \cosh((n+1)\log\rho).$$

<u>Proof</u> The mapping $z = \Psi(t)$ defined by (2.13) is the composition of the two mappings

$$w = \eta^{-1}(t) \text{ and } z = \psi(w)$$

defined by (2.11) and (2.6); i.e. $\Psi = \psi\eta^{-1}$. Now since

$$t^{n+1} = 1 \text{ if } t = \omega^k, \tag{3.2}$$

it follows that

(3.3) $\eta^{-1}(t^{n+1}) = \eta^{-1}(1)$ if $t = \omega^k$.

But it was seen in section 2 that under η^{-1}, the function t^{n+1} is transformed into the function $\rho^{-n}w^{n+1}$; i.e. substituting $t = \eta(w)$ into (3.3) yields

$$\rho^{-n}w^{n+1} = \rho \quad \text{if } w = \eta^{-1}(\omega^k) = \rho\omega^k.$$

Alternatively, we have shown that

(3.4) $w^{n+1} = \rho^{n+1}$ if $w = \rho\omega^k$.

(Parenthetically, it was shown in [3] that the interpolating projection for the space $A(C_\rho)$ based on the points $\rho\omega^k$ $(0 \le k \le n)$, which are the n+1 roots of the polynomial $w^{n+1}-\rho^{n+1}$, is the analogue of F_n in A(C); in particular, the norm is bounded by the Gronwall constants.)

Repeating the above argument by applying the mapping ψ to equation (3.4), we find

$$\psi(w^{n+1}) = \psi(\rho^{n+1}) \quad \text{if } w = \rho\omega^k,$$

from which it follows that

$$T_{n+1}(z) = \frac{1}{2}(\rho^{n+1}+\rho^{-(n+1)})$$

$$\text{if} \quad z = \psi(\rho\omega^k) = \frac{1}{2}(\rho\omega^k+\rho^{-1}\omega^{-k}).$$

Finally, since $\cosh((n+1)\log\rho) = \frac{1}{2}(\rho^{n+1}+\rho^{-(n+1)})$, the theorem is proved.

We define the Lagrange interpolating projection $L_n^\rho : A(\xi_\rho) \to \Pi_n$ by interpolation at the points z_k given by (3.1), i.e.

$$(L_n^\rho f)(z) = \sum_{k=0}^{n} f(z_k)\ell_k(z),$$

where $\ell_k(z) = \dfrac{T_{n+1}(z) - \cosh((n+1)\log\rho)}{T'_{n+1}(z_k)(z-z_k)}$. Hence, the bound on the norm is given by:

$$\|L_n^\rho\| \le \|\Lambda_n^\rho\|,$$

where

$$\Lambda_n^\rho(z) = \sum_{k=0}^{n} \frac{|T_{n+1}(z) - \cosh((n+1)\log\rho)|}{|T'_{n+1}(z_k)|\,|z-z_k|}.$$

Values of the constants

$$\lambda_n^\rho = \|\Lambda_n^\rho\|$$

for $1 \le n \le 10$ and for various values of ρ are presented in Table 1.

TABLE 1

Dependence of λ_n^ρ on n and ρ

n \ ρ	1.00	1.05	1.10	1.20	1.50	2.00	5.00	25.00	∞
1	1.000	1.001	1.005	1.016	1.071	1.166	1.361	1.412	1.414
2	∞	15.952	8.312	4.569	2.501	1.955	1.696	1.668	1.667
3	∞	8.306	4.537	2.765	1.957	1.855	1.848	1.848	1.848
4	∞	12.670	6.725	3.922	2.525	2.188	2.010	1.990	1.989
5	∞	8.164	4.469	2.849	2.284	2.184	2.115	2.105	2.104
6	∞	10.517	5.757	3.624	2.639	2.379	2.222	2.203	2.202
7	∞	7.310	4.144	2.929	2.521	2.400	2.301	2.288	2.287
8	∞	9.103	5.178	3.514	2.763	2.531	2.381	2.363	2.362
9	∞	6.648	3.931	3.035	2.701	2.558	2.444	2.429	2.429
10	∞	8.124	4.822	3.495	2.875	2.655	2.508	2.490	2.489

Recalling the parametric equation (2.9) for ξ_ρ and choosing $\omega = \exp(i\,2\pi/n+1)$, we find that the interpolation nodes (3.1) can be specified as

$$z_k = \tfrac{1}{2}(\rho+\rho^{-1})\cos\theta_k + i\,\tfrac{1}{2}(\rho-\rho^{-1})\sin\theta_k, \tag{3.5}$$

where $\theta_k = 2k\pi/(n+1)$. Thus, for arbitrarily large ρ, ξ_ρ approaches the circle $C_{\rho/2}$ and the points z_k $(0 \le k \le n)$

approach a set of equally-spaced points on the circumference. It is therefore not surprising to find numerical evidence that, for fixed n,

$$\lim_{\rho\to\infty} \lambda_n^\rho = \phi_n, \text{ the nth Gronwall constant.}$$

In the last column of Table 1 we have listed the Gronwall constants.

From equation (3.5), it follows that

$$\lim_{\rho\to 1} z_k = \cos(2k\pi/n+1), \quad 0 \le k \le n.$$

In other words, as ξ_ρ collapses to the interval [-1,1], the interpolation nodes collapse to fewer than n+1 distinct points if $n > 1$. Thus, L_n^ρ is not well-defined in the limiting case and we find numerical evidence that, for fixed $n > 1$,

$$\lim_{\rho\to 1} \lambda_n^\rho = \infty.$$

For large values of ρ, λ_n^ρ is not appreciably larger than ϕ_n and the interpolating projection L_n^ρ is acceptable. However, for fixed $n > 1$, λ_n^ρ appears to increase monotonically as ρ decreases and, in particular, is unbounded as $\rho \to 1$. The projection defined in the following section surmounts this problem by using interpolation nodes which collapse to distinct points on [-1,1] as $\rho \to 1$.

4. Interpolation at Roots-of-i Images

Any Lagrange interpolating projection for the space A(C) which uses n+1 equally-spaced nodes on C will have norm bounded by the Gronwall constants, by symmetry. In other words, the roots of unity are only one possible "good" choice of nodes. In particular, the projection defined by interpolation at the n+1 roots of $t^{n+1} - i$ is as good in the space A(C) as the projection F_n.

The (n+1)th roots of i can be specified by

(4.1) $$t_k = \exp(i\ \theta_k), \quad 0 \le k \le n,$$

where

(4.2) $$\theta_k = (4k+1)\pi/(2n+2).$$

The images on ξ_ρ of the points (4.1) under the mapping (2.13) are:

(4.3) $$z_k = \frac{1}{2}(\rho t_k + \rho^{-1} t_k^{-1}), \quad 0 \le k \le n.$$

In terms of the parametric equation (2.9), the nodes z_k $(0 \le k \le n)$ are given by:

(4.4) $$z_k = \frac{1}{2}(\rho+\rho^{-1})\cos\theta_k + i\ \frac{1}{2}(\rho-\rho^{-1})\sin\theta_k,$$

where θ_k is defined by (4.2).

Theorem 4.1 The points z_k $(0 \le k \le n)$ defined by equation (4.3) are the n+1 roots of the polynomial

$$T_{n+1}(z) - i\ \sinh((n+1)\log\rho).$$

Proof Using the same notation as in the proof of Theorem 3.1 and with t_k defined by (4.1), we find:

$$\begin{aligned}
t^{n+1} = i \Rightarrow \eta^{-1}(t^{n+1}) &= \eta^{-1}(i) && \text{if } t = t_k \\
\Rightarrow \rho^{-n} w^{n+1} &= i\rho && \text{if } w = \rho t_k \\
\Rightarrow w^{n+1} &= i\rho^{n+1} && \text{if } w = \rho t_k \\
\Rightarrow \psi(w^{n+1}) &= \psi(i\rho^{n+1}) && \text{if } w = \rho t_k \\
\Rightarrow T_{n+1}(z) &= \frac{1}{2}(i\rho^{n+1} + i^{-1}\rho^{-(n+1)}) \\
& \qquad \text{if } z = \frac{1}{2}(\rho t_k + \rho^{-1} t_k^{-1}).
\end{aligned}$$

Then since $i^{-1} = -i$ and

$$\sinh((n+1)\log\rho) = \frac{1}{2}(\rho^{n+1}-\rho^{-(n+1)}),$$

the theorem is proved.

We define the Lagrange interpolating projection $I_n^{\rho}:A(\xi_{\rho}) \to \Pi_n$ by interpolation at the points z_k given by (4.3); i.e.

$$(I_n^{\rho}f)(z) = \sum_{k=0}^{n} f(z_k)\ell_k(z),$$

where

$$\ell_k(z) = \frac{T_{n+1}(z) - i\sinh((n+1)\log\rho)}{T'_{n+1}(z_k)(z-z_k)}.$$

Hence, the bound on the norm is given by:

$$\|I_n^{\rho}\| \le \|\Lambda_n^{\rho}\|,$$

where

$$\Lambda_n^{\rho}(z) = \sum_{k=0}^{n} \frac{|T_{n+1}(z) - i\sinh((n+1)\log\rho)|}{|T'_{n+1}(z_k)|\,|z-z_k|}.$$

Values of the constants

$$\iota_n^{\rho} = \|\Lambda_n^{\rho}\|$$

for $1 \le n \le 10$ and for various values of ρ are presented in Table 2.

As was the case in section 3 for λ_n^{ρ}, we find that

$$\lim_{\rho\to\infty} \iota_n^{\rho} = \phi_n, \quad \text{the nth Gronwall constant.}$$

Also as in section 3, we find that for fixed $n > 1$ the value of ι_n^{ρ} increases as ρ decreases from ∞. However, in this case ι_n^{ρ} does not increase indefinitely but rather achieves a

TABLE 2

Dependence of ι_n^ρ on n and ρ

n \ ρ	1.00	1.05	1.10	1.20	1.50	2.00	5.00	25.00	∞
1	1.414	1.414	1.414	1.414	1.414	1.414	1.414	1.414	1.414
2	1.677	1.668	1.673	1.687	1.737	1.742	1.680	1.667	1.667
3	1.848	1.853	1.868	1.917	2.020	1.968	1.864	1.848	1.848
4	1.989	2.001	2.034	2.136	2.237	2.129	2.006	1.989	1.989
5	2.104	2.126	2.185	2.351	2.399	2.253	2.122	2.105	2.104
6	2.202	2.237	2.330	2.553	2.523	2.355	2.220	2.203	2.202
7	2.287	2.338	2.474	2.732	2.623	2.442	2.305	2.288	2.287
8	2.362	2.433	2.619	2.884	2.706	2.518	2.380	2.363	2.362
9	2.429	2.524	2.765	3.013	2.779	2.585	2.447	2.430	2.429
10	2.489	2.613	2.911	3.120	2.842	2.647	2.508	2.490	2.489

maximum value and then decreases to ϕ_n as $\rho \to 1$. The limiting value ϕ_n as $\rho \to 1$ is to be expected since, from equation (4.4), it is clear that

$$\lim_{\rho\to 1} z_k = \cos((4k+1)\pi/(2n+2)), \quad 0 \le k \le n,$$

which are in fact the n+1 distinct roots of $T_{n+1}(z)$. Thus, as $\rho \to 1$, the projection I_n^ρ collapses "gracefully" to the standard projection on [-1,1] based on the Chebyshev nodes (cf. section 1), which we may denote by I_n^1. In this case, we are able to compute a bound on $\|I_n^\rho\|$ over all $\rho \ge 1$. For $1 \le n \le 10$, the constants

$$\iota_n = \max_{1<\rho<\infty} \iota_n^\rho \tag{4.5}$$

were computed numerically and are presented in Table 3 along with the values of ρ for which the maximum in (4.5) was achieved (the values of ρ are given to 2 decimal places). Thus we have the bound

$$\|I_n^\rho\| \le \iota_n, \text{ for any } \rho \ge 1,$$

and from Table 3 we may conclude that I_n^ρ is a practical near-minimax approximation mapping. Specifically, we have shown that for $1 \le n \le 10$ and for any $\rho \ge 1$,

$$\|f - I_n^\rho f\| < 5\|f - B_n f\|,$$

where $B_n : A(\xi_\rho) \to \Pi_n$ is the minimax mapping.

TABLE 3

Dependence of ι_n on n plus values of ρ such that $\iota_n = \iota_n^\rho$

n	ι_n	ρ
1	1.414	all
2	1.749	1.74
3	2.020	1.52
4	2.247	1.41
5	2.440	1.34
6	2.609	1.29
7	2.758	1.25
8	2.892	1.23
9	3.013	1.20
10	3.123	1.19

References

[1] E.W. Cheney, *Introduction to Approximation Theory*, McGraw-Hill, New York, 1966.

[2] E.W. Cheney and K.H. Price, *Minimal projections*, Approximation Theory, A. Talbot, ed., Academic Press, London, 1970, pp.261-289.

[3] K.O. Geddes, *Algorithms for analytic approximation*, Tech. Rep. 56, Dept. of Computer Science, Univ. of Toronto, 1973.

[4] K.O. Geddes, *Near-minimax polynomial approximation in an elliptical region*, submitted to SIAM J. Numer. Anal., 1975.

[5] K.O. Geddes and J.C. Mason, *Polynomial approximation by projections on the unit circle*, SIAM J. Numer. Anal., 12 (1975), pp.111-120.

[6] T.H. Gronwall, *A sequence of polynomials connected with the nth roots of unity*, Bull. Amer. Math. Soc., 27 (1921), pp.275-279.

[7] A.I. Markushevich, *Theory of functions of a complex variable*, vol.3, Translated by R.A. Silverman, Prentice-Hall, Englewood Cliffs, N.J., 1967.

[8] M.J.D. Powell, *On the maximum errors of polynomial approximations defined by interpolation and by least squares criteria*, Comput. J., 9 (1967), pp.404-407.

[9] J. Williams, *Numerical Chebyshev approximation in the complex plane*, SIAM J. Numer. Anal., 9 (1972), pp.638-694.

RRAS APPROXIMATION OF DIFFERENTIABLE FUNCTIONS

Darell J. Johnson

The restricted range approximation scheme introduced by G.D. Taylor [6] has aroused the interest both of mathematicians and engineers. While questions of existence, uniqueness, characterization and non-triviality of restricted range polynomial approximation has been dealt with by various authors, many open questions still remain the subject of research activity. One such question can loosely be phrased: what about the imposition of side conditions on the basic restricted range approximation scheme?

Some contributions to this question have been made, especially for the characterization of polynomials of best restricted range approximation satisfying certain special linear side conditions [5]. Another contribution to this question has been towards the determination of linear functional side conditions one may impose on the restricted range approximation scheme and still retain a Weierstrass-type theorem [2-4]. The present note will continue the latter approach to answering the above question by investigating what n-tuples of linear functionals one may impose in the OSAS [2] and RRAS [4] schemes when the continuous function being approximated is actually differentiable.

§1. APPROXIMATION OF DIFFERENTIABLE FUNCTIONS.

Proposition 1.1[2]. Suppose M is a dense subspace of $C[a,b]$ which contains the constants. Suppose $x_1^*,\ldots,x_n^*$ are span indefinite on $C[a,b]$. Then there exists an $m \in M$ such that $x_i^*m = 0$ $(i=1,\ldots,n)$ and $m \geq 1$.

Corollary 1.1. Suppose M is a dense subspace of $C[a,b]$ which contains the constants. Suppose $x_1^*,\ldots,x_n^*$ are span indefinite on $C[a,b]$. Suppose $\sigma = (\sigma_1,\ldots,\sigma_n) \in \{-1,1\}^n$ is arbitrary. Then there exists a $\nu > 0$, independent of σ, such that given $\eta = (\eta_1,\ldots,\eta_n) \in \mathbb{R}^n$ satisfying $\|\eta\|_{\ell_1} = \sum_{i=1}^n |\eta_i| < \nu$ there exists an $m_\sigma \in M$ for which $\sigma_i x_i^*(m -f) > \nu$ $(i=1,\ldots,n)$ while $0 < m_\sigma(x) \leq 1$ $(x \in [a,b])$.

Proof. Since $x_1^*,\ldots,x_n^*$ are linearly independent, let $g_i \in C[a,b]$ be such that $x_j^*g_i = \delta_{ij}$. Let $\mu_i = \inf\{g_i(x); x \in [a,b]\}$. Set $h_{i,1} = g_i + (\mu_i+1)m$, the $m \in M$ being chosen by Proposition 1.1. Let $\nu_i = \inf\{-g_i(x) : x \in [a,b]\}$ and set $h_{i,-1} = -g_i + (\nu_i+1)m$. For a given $\sigma \in \{-1,1\}^n$, let $h_\sigma = \sum_{i=1}^n h_{i,\sigma_i}$, and set $m_\sigma = h_\sigma/\|h_\sigma\|$.

Corollary 1.2. Suppose $x_1^*,\ldots,x_n^*$ are span indefinite linear functionals on $C[a,b]$. Suppose $\sigma = (\sigma_1,\ldots,\sigma_n) \in \{-1,1\}^n$ and $\varepsilon > 0$ are arbitrary. Then there exists a $\nu = \nu(\varepsilon) > 0$ such that given $\eta \in \mathbb{R}^n$ with $\|\eta\|_{\ell_1} < \nu$ there exists a $g_\sigma \varepsilon \Pi$ for which $\sigma_i x_i^*(m - f) > \nu$ $(i=1,\ldots,n)$, $\|f-g_\sigma\| < \varepsilon$, and $g_\sigma \geq f$ on $[a,b]$.

Proof. Let $p_\varepsilon \varepsilon \Pi$ be such that $x_i^* p_\varepsilon = x_i^* f$ $(i=1,\ldots,n)$, $\|f-p_\varepsilon\| < \varepsilon/2$, and $p_\varepsilon \geq f$ on $[a,b]$. Choose $\nu' > 0$ according to Corollary 1.1, and set $\nu = \varepsilon\nu'/2$. For $\sigma \varepsilon \{-1,1\}^n$ arbitrary, choose $m_\sigma \varepsilon \Pi$ by Corollary 1.1. Set $g_\sigma = p_\varepsilon + \varepsilon m_\sigma/2$.

Proposition 1.2[1]. Suppose $f \varepsilon C[a,b]$ and ℓ, u are permissible bounding functions such that $\ell \leq f \leq u$ on $[a,b]$. Then given $\varepsilon > 0$ arbitrary there exists a polynomial $p_\varepsilon \varepsilon \Pi$ for which $\|f-p_\varepsilon\| < \varepsilon$ while $\ell \leq p_\varepsilon \leq u$ on $[a,b]$ also.

Definition 1.1. Suppose $f \varepsilon C[a,b]$, $t \varepsilon [a,b]$. We say f satisfies the OSA necessary condition (from above) at t in case the pair of functions f, u are permissible bounding functions, u being $f(t)$ at t and $+\infty$ elsewhere.

Theorem 1.1. Suppose $x_1^*,\ldots,x_n^*$ are span indefinite linear functionals on $C[a,b]$. Suppose T is a finite subset of $[a,b]$, and set $B = [a,b]\setminus T$. Suppose $f \varepsilon C[a,b]$ satisfies the OSA necessary condition at each $t \varepsilon T$. If the linear functionals $x_1^* \circ \chi_B,\ldots,x_n^* \circ \chi_B$ are also span indefinite on $C[a,b]$, then given $\varepsilon > 0$ arbitrary there exists a polynomial $p_\varepsilon \varepsilon \Pi$ such that
i) $x_i^* p_\varepsilon = x_i^* f$ $(i=1,\ldots,n)$, ii) $p_\varepsilon(t) = f(t)$ $(t \varepsilon T)$, iii) $\|f-p_\varepsilon\| < \varepsilon$, and iv) $p_\varepsilon \geq f$ on $[a,b]$.

Proof. For $\delta > 0$ arbitrary, set $C_\delta = [a,b]\setminus N_\delta(T)$, $v_i^* = x_i^* \circ \chi_B$, and $u_{i,\delta}^* =$

$x_i^* \circ \chi_{C_\delta}$. Given $\varepsilon > 0$ arbitrary choose $\nu = \nu(\varepsilon/2)$ by Corollary 1.2 (for the span indefinite functionals $v_1^*,\ldots,v_n^*$). Since

$$\|u_{i,\delta}^* - v_i^*\| \to 0$$

as $\delta \to 0^+$, choose $\delta_0 > 0$ so that $0 < \delta < \delta_0$ implies $\|u_{i,\delta}^* - v_i^*\| < 10^{-6}n^{-1}\nu$ $(i=1,\ldots,n)$. For $\sigma \in \{-1,1\}^n$ arbitrary, let $g_\sigma \in \Pi$ be such that $\sigma_i x_i^*(g_\sigma - f) > \nu$ $(i=1,\ldots,n)$, $\|f-g_\sigma\| < \varepsilon/2$, and $g_\sigma \geq f$ on $[a,b]$. Fix a δ so that $0 < \delta < \delta_0$ and choose an $h_\sigma \in C[a,b]$ so that i) $h_\sigma(x) = g_\sigma(x)$ if $x \in C_\delta$, ii) $h_\sigma(x) = f(x)$ if $x \in N_{\delta/2}(T)$, and iii) elsewhere so h_σ is both continuous and $\|f-h_\sigma\| \leq \|f-g_\sigma\| < \varepsilon/2$. Define bounding functions

$$\ell(x) = f(x), \quad u(x) = \begin{cases} f(x), & \text{if } x \in T \\ +\infty, & \text{otherwise.} \end{cases}$$

By hypothesis ℓ, u are permissible, so by Proposition 1.2. there exists a sequence of polynomials $\{p_{\sigma,\zeta}\}_{\zeta>0}$ such that $\|h_\sigma - p_{\sigma,\zeta}\| < \zeta$, $p_{\sigma,\zeta}(x) \geq f(x)$, and $p_{\sigma,\zeta}(t) = f(t)$ $(t \in T)$. Fix ζ so that $0 < \zeta < \min\{4^{-1}\nu, \varepsilon/2\}$. Then $\sigma_i x_i^*(h_\sigma - f) > 2^{-1}\nu$ implies $\sigma_i x_i^*(p_{\sigma,\zeta} - f) > 4^{-1}\nu$, so we may as usual construct a $p \in \Pi$ for which i) $x_i^* p = x_i^* f$ $(i=1,\ldots,n)$, ii) $p(t) = f(t)$ $(t \in T)$, iii) $p(x) \geq f(x)$ on $[a,b]$, and iv) $\|f-p\| \leq \|f-p_{\sigma,\zeta}\| < \varepsilon$.

<u>Theorem 1.2</u>. Suppose $x_1^*,\ldots,x_n^*$ is a RRAS sequence on $C[a,b]$. Suppose T is a finite subset of $[a,b]$. Suppose $f \in C[a,b]$ satisfies the OSA necessary condition (from above and below) at every

$t \in T$. Suppose ℓ, u are permissible bounding functions such that $\ell \leq f \leq u$. Then given $\varepsilon > 0$ arbitrary there exists a polynomial p_ε such that i) $x_i^* p_\varepsilon = x_i^* f$ $(i=1,\ldots,n)$, ii) $p_\varepsilon(t) = f(t)$ $(t \in T)$, iii) $\|f-p\| < \varepsilon$, and iv) $\ell \leq p_\varepsilon \leq u$ on $[a,b]$.

<u>Remark 1.2</u>. Suppose $f \in C[a,b]$ is fixed. If x^* is neither purely finitely atomic nor a RRAS functional on $C[a,b]$, we showed in [4] how we may construct permissible bounding functions ℓ, u for which $\ell \leq f \leq u$ but f cannot be approximated arbitrarily closely in the RRAS scheme (the point being, for non-finitely purely atomic, non-RRAS functionals it is possible to find permissible bounding functions ℓ_f, u_f for any specified $f \in C[a,b]$ so that f cannot be approximated arbitrarily closely in the RRAS scheme when the bounding ℓ_f, u_f are used). We thus have characterized completely those linear functionals $x_1^*,\ldots,x_n^*$ for which some given $f \in C[a,b]$ may be approximated arbitrarily closely in the RRAS scheme where the permissible bounding functions ℓ, u are restrained only by the condition that $\ell \leq f \leq u$ on $[a,b]$ (f not varying):

<u>Corollary 1.3</u>. Suppose $f \in C[a,b]$ is arbitrary. Let $T = \{t \in [a,b] : f$ satisfies the OSA necessary condition at $t\}$. Suppose $x_1^*,\ldots,x_n^*$ is a finite set of bounded linear functionals on $C[a,b]$. Then given $\varepsilon > 0$ arbitrarily small and ℓ, u permissible for which $\ell \leq f \leq u$ there is a $p_\varepsilon \in \Pi$ for which i) $x_i^* p_\varepsilon = x_i^* f$ $(i=1,\ldots,n)$,

ii) $\|f-p_\varepsilon\| < \varepsilon$, and iii) $\ell \leq f \leq u$ on $[a,b]$ if and only if every $x^* \in \langle x_1^*,\ldots,x_n^*\rangle$ is either i) a RRAS functional or ii) purely atomic with support in S.

In particular, since any differentiable function satisfies the OSA necessary condition at every point of its range, we have now shown

Theorem 1.3. Suppose $x_1^*,\ldots,x_n^*$ is a finite set of bounded linear functionals on $C[a,b]$. In order that any continuously differentiable function $f \in C^1[a,b]$ may be approximated arbitrarily closely in

i) the OSAS scheme, and

ii) the RRAS scheme for arbitrary permissible bounding functions ℓ,u restrained only so that $\ell \leq f \leq u$,

it is necessary and sufficient that (respectively)

i') any $x^* \in \langle x_1^*,\ldots,x_n^*\rangle$ is either (span) indefinite or purely finitely atomic, and

ii') any $x^* \in \langle x_1^*,\ldots,x_n^*\rangle$ is either a RRAS functional or else purely finitely atomic.

§2. RRAS FOR FIXED ℓ,u.

For $f \in C[a,b], \|f\| = 1$, SAIN approximation may be viewed as RRAS when the ℓ,u are fixed at $-\mathbb{1},\mathbb{1}$ while the continuous f are arbitrary. Those n-tuples of linear functionals $x_1^*,\ldots,x_n^*$ for which SAIN approximation is always possible have been determined [3]; we now consider determining those n-tuples of linear functionals so that one has a Weierstrass theorem holding in the RRAS scheme for fixed (but arbitrary) permissible bounding functions ℓ,u and arbitrary $f \in C[a,b]$ subjected only to the restriction that $\ell \leq f \leq u$.

Definition 2.1. Suppose ℓ,u are fixed permissible bounding functions. A linear functional x^* on $C[a,b]$ is said to be a RRAS (ℓ,u) functional in case given $f \in C[a,b]$ for which $\ell \leq f \leq u$ and $\varepsilon > 0$ arbitrary there always exists a polynomial p_ε for which $\ell \leq p_\varepsilon \leq u$, $|f-p_\varepsilon| < \varepsilon$, and $x^*p_\varepsilon = x^*f$. $x_1^*,\ldots,x_n^*$ is a RRAS (ℓ,u) sequence in case any $x^* \in \langle x_1^*,\ldots,x_n^*\rangle$ is a RRAS(ℓ,u) functional.

We call those point t where $\ell(t) = u(t)$ the nodes of ℓ,u.

Theorem 2.1. Suppose ℓ,u are permissible bounding functions with nodes T. Then x^* is a RRAS(ℓ,u) functional if and only if either

i) x^* is finitely purely atomic, and ℓ satisfies the OSA necessary condition from above, u the OSA necessary condition from below at every atom of x^*, or

ii) $(\text{supp } x^{+*} \cap \text{supp } x^{-*})\backslash T \neq \phi$,

with the exception that if ℓ and u are both polynomials, then x^* is also a RRAS (ℓ,u) functional if it is a positive (or negative) linear functional with support $[a,b]$.

Example. Consider $\ell(x) = -u(x) = -|x|$, $[a,b] = [-1,1]$, $x^* = \int_0^1 \cdot\, dx - e_{3/4}$, $y^* = \int_{-1}^1 \cdot\, dx - e_0$. Then x^* is but y^* is not a RRAS(ℓ,u) functional.

Theorem 2.2 Suppose $x_1^*,\ldots,x_n^*$ is a set of bounded linear functionals on $C[a,b]$. Suppose

ℓ, u are fixed permissible bounding functions. Then given $f \in C[a,b]$ for which $\ell \leq f \leq u$ and $\varepsilon > 0$ arbitrary, in order that there always exists a polynomial p_ε for which $\ell \leq p_\varepsilon \leq u$, $\|f-p_\varepsilon\| < \varepsilon$, and $x_i^* p_\varepsilon = x_i^* f$ $(i=1,\ldots,n)$ it is necessary and sufficient that the $x_1^*,\ldots,x_n^*$ be a RRAS (ℓ,u) sequence on $C[a,b]$.

Proof. Necessity is clear. For sufficiency let T denote the nodes of ℓ,u, and for $\delta > 0$ set $B_\delta = [a,b] \backslash N_\delta(T)$. If ℓ,u are not both polynomials, then an examination of [4] will show it is possible to choose a $\delta_0 > 0$ for which $v_{1,\delta}^*,\ldots,v_{n,\delta}^*$ is a span indefinite sequence on $C[a,b]$ for any $0 < \delta < \delta_0$, where $v_{i,\delta}^* = x_i^* \circ \chi_{B_\delta}$. If ℓ,u are both polynomials and f is not ℓ or u, then careful consideration will show f is non-extremal for positive linear functionals having support $[a,b]$, and that if any such functionals are inherent in the $x_1^*,\ldots,x_n^*$ it is still possible to choose the B_δ so the $v_{1,\delta}^*,\ldots,v_{n,\delta}^*$ less the above inherent functionals become span indefinite on $C[a,b]$, and together with the inherent non-extremal (for f) such functionals will yield a lemma analogous to Corollary 1.1 or 1.2 (cf.[3]).

Corollary 3.1. Suppose ℓ,u are fixed permissible bounding functions. Suppose $x_1^*,\ldots,x_n^*$ is a finite set of bounded linear functionals on $C[a,b]$. In order that any continuously differentiable function $f \in C^1[a,b]$ may be approximated arbitrarily closely in the RRAS (ℓ,u) scheme it is necessary and sufficient that any $x^* \varepsilon \langle x_1^*,\ldots,x_n^*\rangle$ is either finitely purely atomic or else a RRAS (ℓ,u) functional.

REFERENCES

1. D.J. Johnson, On the nontriviality of restricted range polynomial approximation, SIAM J. Numer. Anal. 12 (1975).
2. ____________, One-sided approximation with side conditions, J. Approximation Theory, to appear.
3. ____________, SAIN approximation in C[a,b], J. Approximation Theory, submitted.
4. ____________, Restricted range approximation with side conditions, submitted.
5. E. Kimchi and N. Richter-Dyn, Properties of best approximation with interpolatory and restricted range side conditions, preprint.
6. G.D. Taylor, On approximation by polynomials having restricted ranges, SIAM J. Numer. Anal. 5 (1968), 258-268.

On the Expansion of Exponential Type Integrals in Series of Chebyshev Polynomials

Yudell L. Luke*

I. INTRODUCTION

Salient reasons for the efficiency of (at least formal) representations for functions $p(x)$ in the form

$$h(x) = \sum_{k=0}^{\infty} a_k x^k$$ are that integral powers,

derivatives and integrals of $h(x)$ are again of the form of $h(x)$. Further, for computational purposes, one deals with $h(x)$ suitably truncated, and evaluation of this polynomial is readily accomplished with the aid of a recursion formula used in the backward direction.

Now representations for functions $p(x)$ in the form

$$f(x) = \sum_{k=0}^{\infty} b_k T_k^*(x/\lambda),$$ where λ is fixed,

$0 \le x \le \lambda$ and $T_k^*(x)$ is the shifted Chebyshev polynomial of the first kind of order k, can be as easily managed as the form for $h(x)$ without the necessity for converting $T_k^*(x)$ to ordinary

*This work was supported by the United States Air Force Office of Scientific Research under Grant 73-2520.

powers in order to first achieve a form like $h(x)$.

Representation of functions by the form $f(x)$ has many advantages over the form $h(x)$. If $p(x)$ is analytic with singularities (if any) sufficiently far removed from the segment $0 \leq x \leq \lambda$, and if $f(x)$ is truncated after $(n+1)$ terms, then the error is very nearly $c_{n+1} T^*_{n+1}(x/\lambda)$ and so is very nearly uniform over the range $0 \leq x \leq \lambda$. In such an event, the difference between this approximation and the 'best' approximation in the sense of Chebyshev is insignificant. A striking advantage of the form $f(x)$ is that for a large class of functions, it converges and rapidly so, while its counterpart $h(x)$ diverges but is asymptotic. Further examination of these points will be found in volumes by Luke [1,2] and the reader is referred to these sources for more details.

Suppose we are given the coefficients b_k in the expansion

$$f(x) = \sum_{k=0}^{\infty} b_k T^*_k(x/\lambda), \quad 0 \leq x \leq \lambda. \tag{1.1}$$

We suppose that (1.1) converges even though its power series counterpart might be divergent but asymptotic. In this paper we show how to get the coefficients g_k for the expansion of

$$g(x) = e^{-ax} x^{-(u+1)} \int_0^x e^{at} t^u f(t)\,dt \tag{1.2}$$

in the form

$$g(x) = \sum_{k=0}^{\infty} g_k T^*_k(x/\lambda), \quad 0 \leq x \leq \lambda, \tag{1.3}$$

which converges so long as $t^u f(t)$ is integrable. These developments are given in Section II.

Similar developments for

$$F(x) = \sum_{k=0}^{\infty} c_k T^*_k(\lambda/x), \quad 0 < \lambda \leq x, \tag{1.4}$$

$$G(x) = e^{bx} x^{-u} \int_x^{\infty} e^{-bt} t^u F(t)\,dt, \quad \mathcal{R}(b) > 0, \tag{1.5}$$

$$G(x) = \sum_{k=0}^{\infty} h_k T_k^* (\lambda/x), \quad 0 < \lambda \le x \tag{1.6}$$

are presented in Section III. Here we assume (1.4) is convergent even though its power series counterpart might be divergent but asymptotic. We further suppose that the parameters in (1.5) (and also in (1.2)) are such that the integrals have meaning. In this connection, we must point out that the restrictions on (1.6) might be less severe than those on (1.5) in view of analytic continuation. Also in Section III, we consider (1.4) - (1.6) except that the limits of integration in (1.5) are 0 and x. In Section IV, we consider the situations akin to those of Sections II and III where the integrals are of the form $\int e^{at}t^u(\ln t)f(t)dt$ and $\int e^{-at}t^u(\ln t)F(t)dt$ respectively. Some applications are treated in Section V.

II. THE REPRESENTATION FOR $g(x)$

It is convenient to reproduce equations (1.1) - (1.3). Thus

$$f(x) = \sum_{k=0}^{\infty} b_k T_k^* (x/\lambda), \quad 0 \le x \le \lambda, \tag{2.1}$$

$$g(x) = e^{-ax}x^{-u-1}\int_0^x e^{at}t^u f(t)dt, \tag{2.2}$$

$$g(x) = \sum_{k=0}^{\infty} g_k T_k^* (x/\lambda), \quad 0 \le x \le \lambda. \tag{2.3}$$

We assume that (formally at least)

$$f(x) = \sum_{k=0}^{\infty} d_k x^k. \tag{2.4}$$

We also suppose that the Chebyshev series (2.1) is convergent. Now $T_k^*(x) = \cos k\theta$ if $x = \cos^2(\theta/2)$. So if $t^u f(t)$ is integrable, then in virtue of the theorem that the integral of a Fourier series is convergent and converges to the integral of the function to which the Fourier series corresponds, see Titchmarsh [3,p.419], it follows that (2.3) is convergent.

We now turn to developments for obtaining the g_k's. The ideas are as follows. From (2.2), we readily derive a differential equation for $g(x)$ which is then converted to an integral equation. Put (2.3) in the latter and make use of properties of integrals of Chebyshev series and the like to get a recursion formula for the g_k's. We then show how to use the recursion formula in the backward direction to produce numerical values for the g_k's. This technique for the solution of differential equations and certain types of integral equations is fairly standard. See Luke [1,2] for details and further references. See also Wimp [4] and Wimp and Luke [5].

Differentiation of (2.2) produces the differential equation

$$xg'(x) + (ax + u+1)g(x) = f(x), \tag{2.5}$$

and this is equivalent to the integral equation

$$xg(x) + a\int_0^x t\,g(t)dt + u\int_0^x g(t)dt = \int_0^x f(t)dt. \tag{2.6}$$

Given $g(x)$ in the form (2.3), the coefficients in like forms for $g'(x)$, $x^v g(x)$, $v = 0,1,2$ and $\int_0^x t^w g(t)dt$, $w = 0,1$, expressed in terms of the g_k's follow from the data given in [1,v.1,pp.314-316]. Making the appropriate substitutions and equating like coefficients of $T_k^*(x/\lambda)$ in (2.6), we get

$$2g_0 + g_1 + 2uG + a\lambda H = 2B ,$$

$$G = g_0 - \frac{g_1}{4} - \sum_{n=2}^{\infty} \frac{(-)^n g_n}{n^2-1} ,$$

$$H = \frac{3g_0}{4} + \frac{g_1}{12} - \frac{19g_2}{48} + 3\sum_{n=3}^{\infty} \frac{(-)^n g_n}{(n^2-1)(n^2-4)} ,$$

$$B = b_0 - \frac{b_1}{4} - \sum_{n=2}^{\infty} \frac{(-)^n b_n}{n^2-1} \quad , \tag{2.7}$$

$$2(1+u)g_0 + 2g_1 + (1-u)g_2 + (a\lambda/4)(4g_0+g_1-2g_2-g_3)$$

$$= 2b_0 - b_2 \quad , \tag{2.8}$$

$$(k+u)g_{k-1} + 2k\ g_k + (k-u)g_{k+1}$$

$$+ (a\lambda/4)((2g_{k-2}/\epsilon_{k-2}) + 2g_{k-1} - 2g_{k+1} - g_{k+2})$$

$$= b_{k-1} - b_{k+1} \ , \ k \geq 2 \ , \tag{2.9}$$

where

$$\epsilon_k = 1 \quad \text{for} \quad k = 0 \ , \ \epsilon_k = 2 \quad \text{for} \quad k > 0. \tag{2.10}$$

We can get a difference equation of one order less than (2.9) by antidifferencing. Thus if

$$v_k = (-)^k((k+u)g_{k-1} + (k-1-u)g_k - (b_{k-1} - b_k)) \ , \tag{2.11}$$

then

$$\Delta v_k = v_{k+1} - v_k =$$

$$(-)^{k+1}((k+u)g_{k-1}+2k\ g_k+(k-u)g_{k+1} - (b_{k-1} - b_{k+1})). \tag{2.12}$$

Similarly, for $k \geq 2$, we have the pair,

$$v_k = (-)^k((2g_{k-2}/\epsilon_{k-2}) + g_{k-1} - g_k - g_{k+1}), \tag{2.13}$$

$$\Delta v_k = (-)^{k+1}((2g_{k-2}/\epsilon_{k-2}) + 2g_{k-1} - 2g_{k+1} - g_{k+2}). \tag{2.14}$$

It follows that

$$(k+u)g_{k-1} + (k-1-u)g_k$$

$$+ (a\lambda/4)((2g_{k-2}/\epsilon_{k-2}) + g_{k+1} - g_k - g_{k+1})$$

$$= b_{k-1} - b_k \ k \geq 2. \tag{2.15}$$

Before discussing the determination of the g_k's, we first set down some relations useful for check and other purposes. If in (2.1) we replace $T_k^*(x/\lambda)$ by its polynomial representation, then by collecting the coefficients of $x^r, r = 0,1,2,$ and equating to (2.4), we have

$$\sum_{k=0}^{\infty} (-)^k b_k = d_0 \tag{2.16}$$

$$\sum_{k=1}^{\infty} (-)^k k^2 b_k = -(\lambda d_1/2) , \tag{2.17}$$

$$\sum_{k=2}^{\infty} (-)^k k^2(k^2-1) b_k = (3\lambda^2 d_2/2). \tag{2.18}$$

Now by solution of (2.5) or otherwise, we have (formally at least)

$$g(x) = \{d_0/(u+1)\} + \{d_1 - ad_0/(u+1)\}\{x/(u+2)\}$$

$$+ \{d_2 - [ad_1/(u+2)] + a^2 d_0/(u+2)(u+1)\}\{x^2/(u+3)\} + \dots , \tag{2.19}$$

whence

$$\sum_{k=0}^{\infty} (-)^k g_k = \{d_0/(u+1)\} , \tag{2.20}$$

$$\sum_{k=1}^{\infty} (-)^k k^2 g_k = -\{\lambda/2(u+2)\}\{d_1 - ad_0/(u+1)\}, \tag{2.21}$$

$$\sum_{k=2}^{\infty} (-)^k k^2(k^2-1) g_k = \{3\lambda^2/2(u+3)\}\{d_2-[ad_1/(u+2)] + a^2 d_0/(u+2)(u+1)\}. \tag{2.22}$$

We now turn to the evaluation of the g_k's. We first suppose that $a \neq 0$. For convenience, we rewrite (2.15) in the form

$$L\{g_k\} \equiv (2g_k/\epsilon_k) + \{(4/a\lambda)(k+u+2)+1\}g_{k+1} + \{(4/a\lambda)(k+1-u)-1\}g_{k+2} - g_{k+3} ,$$

$$L\{g_k\} = (4/a\lambda)(b_{k+1} - b_{k+2}),\ k \ge 0. \quad (2.23)$$

Let n be a large positive integer. Put $A_{k,n} = 0$ for $k \ge n+1$ and evaluate $A_{k,n}$ from

$$L\{A_{k,n}\} = (4/a\lambda)(b_{k+1} - b_{k+2}),\ 0 \le k \le n. \quad (2.24)$$

Again put $B_{k,n} = 0,\ k \ge n+2,\ B_{n+1,n} = 1$ and evaluate $B_{k,n}$ from

$$L\{B_{k,n}\} = 0,\ 0 \le k \le n. \quad (2.25)$$

Now define

$$g_{k,n} = A_{k,n} + B_{k,n}\left(\frac{d_0/(u+1) - \sum_{m=0}^{n} (-)^m A_{m,n}}{\sum_{m=0}^{n+1} (-)^m B_{m,n}}\right). \quad (2.26)$$

Then by using some results of Wimp [4] and following a study by Wimp and Luke [5], we have

$$g_k = \lim_{n\to\infty} g_{k,n},\ k = 0,1,\ldots. \quad (2.27)$$

However, we hasten to remark that though the algorithm possesses excellent theoretical convergence properties, its application can give rise to rather serious numerical irritabilities since one may be in the position of determining a linear combination of two large quantities which is small. The details are quite lengthy and will not be taken up here. The reader will find the discussion in Wimp and Luke [5] very illuminating. In this connection, the noted instabilities can be avoided if one determines the g_k's by solving systems of linear equations. Further details are given in the latter reference.

For checks on the solution, we can use (2.7), (2.8), (2.21) and (2.22) with g_k replaced by $g_{k,n}$.

If $f(x) = 1$, the analysis simplifies. For then $b_0 = 1$ and $b_k = 0$ if $k > 0$, whence $A_{k,n} = 0$ for all $k \ge 0$.

Next we consider the case $a = 0$. Then from (2.8) and (2.15), we have

$$(2g_k/\epsilon_k) = \{(u-k)g_{k+1} + (2b_k/\epsilon_k) - b_{k+1}\}/(k+u+1). \tag{2.28}$$

Let n be as before. Put $g_{n+1,n} = 0$ and evaluate $g_{k,n}$ from (2.28) with g_k replaced by $g_{k,n}$ for $0 \le k \le n$. Then

$$g_k = \lim_{n\to\infty} g_{k,n}, \quad k = 0,1,\ldots. \tag{2.29}$$

The backward recurrence technique for this situation has been previously analyzed. See [1, v.1,p.316].

III. THE REPRESENTATION FOR $G(x)$.

For convenience, we reproduce (1.4) - (1.6). Thus

$$F(x) = \sum_{k=0}^{\infty} c_k T_k^*(\lambda/x), \quad 0 < \lambda \le x, \tag{3.1}$$

$$G(x) = e^{bx} x^{-u} \int_x^{\infty} e^{-bt} t^u F(t)\,dt, \tag{3.2}$$

$$G(x) = \sum_{k=0}^{\infty} h_k T_k^*(\lambda/x), \quad 0 < \lambda \le x. \tag{3.3}$$

It is convenient to suppose that $b \neq 0$ and that u is not a positive integer or zero. If $b = 0$, then upon replacing x, λ, and t by their reciprocals, (3.2) with $b = 0$ becomes (2.2) if in the latter u and $f(t)$ are replaced by $-(u+2)$ and $F(1/t)$, respectively. The situation when u is a positive integer or zero is treated at the end of this section. We also assume that (formally at least)

$$F(x) = \sum_{k=0}^{\infty} t_k x^{-k} \tag{3.4}$$

The analysis is much akin to that of the previous section. From (3.2), we get

$$x\,G'(x) + (u-b)G(x) = -x\,F(x). \tag{3.5}$$

For convenience, let

$$x = 1/y, \quad G(x) = G^*(y), \quad F(x) = F^*(y). \tag{3.6}$$

Then

$$y^2 G^{*\prime}(y) + (a-uy)G^*(y) = F^*(y) \tag{3.7}$$

and by integration

$$y^2 G^*(y) + b\int_o^y G^*(t)dt - (u+2)\int_o^y tG^*(t)dt$$
$$= \int_o^y F^*(t)dt\ . \tag{3.8}$$

Proceeding as in (2.6) - (2.10), we find

$$b\lambda G - (u+2)H/2 + (6h_0+4h_1+h_2)/8 = \lambda B\ , \tag{3.9}$$

where G, H and B are defined in (2.7) with b_k and g_k replaced by c_k and h_k respectively,

$$(2b\lambda-u)h_0 + \tfrac{1}{4}(5-u)h_1 + (2+\tfrac{1}{2}u-b\lambda)h_2 + \tfrac{1}{4}(u+3)h_3$$
$$= \lambda(2c_0 - c_2)\ , \tag{3.10}$$

$$(2kh_{k-2}/\varepsilon_{k-2}) + 4kh_{k-1} + 6kh_k + 4kh_{k+1} + kh_{k+2}$$
$$+ 4a\lambda(h_{k-1} - h_{k+1}) - (u+2)\{(2h_{k-2}/\varepsilon_{k-2}) + 2h_{k-1}$$
$$- 2h_{k+1} - h_{k+2}\} = 4\lambda(c_{k-1} - c_{k+1})\ ,\ k \ge 2\ . \tag{3.11}$$

We simplify the latter by anti-differencing. Thus if $k \ge 2$,

$$v_k = (-)^k\{(2kh_{k-2}/\varepsilon_{k-2})$$
$$+ (3k-1)h_{k-1} + (3k-2)h_k + (k-1)h_{k+1}\}, \tag{3.12}$$

then

$$\Delta v_k = (-)^{k+1}\{(2kh_{k-2}/\varepsilon_{k-2})$$
$$+ 4kh_{k-1} + 6kh_k + 4kh_{k+1} + kh_{k+2}\}. \tag{3.13}$$

We also have the pairs

$$v_k = (-)^k(h_{k-1}-h_k),\ \Delta v_k = (-)^{k+1}(h_{k-1}-h_{k+1}), \tag{3.14}$$

$$v_k = (-)^k \{(2kh_{k-2}/\varepsilon_{k-2}) + h_{k-1}-h_k-h_{k+1}\},$$

$$\Delta v_k = (-)^{k+1} \{(2kh_{k-2}/\varepsilon_{k-2}) + 2h_{k-1}-2h_{k+1}-h_{k+2}\},$$

$$k \geq 2. \tag{3.15}$$

It follows that

$$M\{h_k\} \equiv (k-u)(2h_k/\varepsilon_k) + (3k+3+4b\lambda-u)h_{k+1}$$

$$+ (3k+6-4b\lambda+u)h_{k+2} + (k+3+u)h_{k+3},$$

$$M\{h_k\} = 4\lambda(c_{k+1}-c_{k+2}), \ k \geq 0. \tag{3.16}$$

The relations analogous to (2.16) - (2.18) are

$$\sum_{k=0}^{\infty} (-)^k b_k = t_0 , \tag{3.17}$$

$$\sum_{k=1}^{\infty} (-)^k k^2 b_k = -(t_1/2\lambda) , \tag{3.18}$$

$$\sum_{k=2}^{\infty} (-)^k k^2(k^2-1) b_k = (3t_2/2\lambda^2). \tag{3.19}$$

Further, (at least formally),

$$G(x) = (t_0/b) + \{t_1 + (ut_0/b)\}/(bx)\}$$

$$+ \{t_2 + (u-1)(t_1/b) + u(u-1)(t_0/b^2)\}/(bx^2) + \ldots, \tag{3.20}$$

and so

$$\sum_{k=0}^{\infty} (-)^k h_k = (t_0/b) ,$$

$$\sum_{k=1}^{\infty} (-)^k k^2 h_k = -\{t_1 + (ut_0/b)\}/(2\lambda b) , \tag{3.22}$$

$$\sum_{k=2}^{\infty} (-)^k k^2(k^2-1) h_k = \{t_2 + (u-1)(t_1/b)$$

$$+ u(u-1)(t_0/b^2)\}. \tag{3.23}$$

For the determination of the h_k's, we use (3.16) and (3.21) and proceed as in the developments (2.24) - (2.26). Thus let n be a large positive integer. Put $C_{k,n} = 0$ for $k \geq n+1$ and compute $C_{k,n}$ from

$$M\{C_{k,n}\} = 4\lambda(c_{k+1}-c_{k+2}), \quad 0 \leq k \leq n. \tag{3.24}$$

Again, let

$$D_{k,n} = 0 \quad \text{for} \quad k \geq n+2, \; D_{n+1,n} = 1 \quad \text{and}$$

$$\text{get} \quad D_{k,n} \quad \text{from} \quad M\{D_{k,n}\} = 0, \; 0 \leq k \leq n. \tag{3.25}$$

Define

$$h_{k,n} = C_{k,n} + D_{k,n}\left(\frac{(t_0/b) - \sum_{k=0}^{n} (-)^k C_{k,n}}{\sum_{k=0}^{n+1} (-)^k D_{k,n}}\right) \tag{3.26}$$

Then following techniques discussed in references [4,5], we find

$$h_k = \lim_{n\to\infty} h_{k,n}, \quad |\arg b| < \pi . \tag{3.27}$$

As a function of $\arg b$, the convergence of the backward recurrence scheme weakens as $|\arg b| \to \pi$ and fails when $|\arg b| = \pi$. In these situations, the h_k's can be efficiently obtained provided the backward recurrence scheme is modified as outlined in references [1,2,4]. We briefly describe this procedure. Let n_1, n_2 be two large distinct positive integers. Evaluate h_{k,n_i}, $i = 1,2$ from (3.26) with $n = n_i$. Let $N = N\{h_k\}$ be an operator acting on the h_k's which produces a known numerical value. We determine μ from

$$\mu N_1 + (1-\mu)N_2 = N, N_j = N\{h_k, n_j\}, \; j = 1,2, \tag{3.28}$$

and evaluate

$$h_{k,n_1,n_2} = \mu h_{k,n_1} + (1-\mu) h_{k,n_2} \tag{3.29}$$

Then under certain conditions noted in reference [4], we have

$$\lim_{n_1,n_2\to\infty} h_{k,n_1,n_2} = h_k, k = 0,1,\cdots, \quad n_1 \neq n_2. \tag{3.30}$$

These conditions are very difficult to apply in terms of the general formulation given here as they depend on the growth properties of the c_k's and the nature of the operator N. A common choice for N is often based on knowledge of $G(x)$ for a particular value of x, say $x = \lambda$. Then in this event

$$N = G(\lambda) = \sum_{k=0}^{\infty} h_k . \tag{3.31}$$

This was the procedure to get numerous sets of coefficients in Luke [1].

Other possible choices of N could be based on (3.10), (3.11), (3.22) or (3.23). As remarked, in practice, the hypotheses noted might be difficult to verify, and some experimentation might be necessary. Indeed, in some recent exploratory work, we found that the scheme worked if N is given by (3.31), but failed when N is based on any of the four possible choices named above. In particular, when (3.22) was employed, we found that N_1 and N_2 were much nearer to each other than h_{k,n_1} and h_{k,n_2}. Notice that if $N_1 = N_2$, any value of μ satisfies (3.28). The advantage of having several possible forms for N is that those not used in (3.28) are available as checks. In practice, it is usually best to circumvent the evaluation of μ per se since considerable loss of significant figures occurs if N_1 is near N_2. It is readily verified that alternative and preferred forms are

$$h_{k,n_1,n_2} = h_{k,n_1} + (N-N_1)(h_{k,n_2} - h_{k,n_1})/(N_2 - N_1) \tag{3.32}$$

or the right hand side of this formula with the

subscripts 1 and 2 interchanged.

In some applications, we need

$$G(x) = e^{ax}x^{-u} \int_0^x e^{-at}t^u F(t)dt \qquad (3.33)$$

where $F(x)$ and $G(x)$ are represented by (3.1) and (3.2) respectively. In this event, (3.4) is an asymptotic expansion for $F(x)$ in some domain. Since

$$xG'(x) + (\mu - ax)G(x) = xF(x) \; , \qquad (3.34)$$

the corresponding asymptotic expansion for $G(x)$ is given by (3.20) with t_k replaced by $-t_k$ for all k. Comparing (3.5) and (3.34), we see that for the present situation, (3.9) - (3.16), (3.24) - (3.32) are valid provided we replace c_k by $-c_k$ for all k.

We now consider the situation in (3.2) when u is a positive integer or zero. First suppose $u = 0$. Then use of (3.16) to compute the elements outlined in (3.24), (3.25) will fail for $k = 0$. In this case, designate the left hand side of (3.10) by $L^*(h)$. Then in (3.24), (3.25) replace the $k = 0$ computations by $L^*(C) = \lambda(2c_0 - c_2)$ and $L^*(D) = 0$ and proceed in the usual fashion. Next suppose that u is a positive integer. Then use of (3.16) to compute the coefficients outlined in (3.24), (3.25) will fail for $k = u$. In this event, notice from (3.4) that

$$x^u F(x) = F_1(x) + F_2(x) \; ,$$

$$F_1(x) = \sum_{k=0}^{u-1} t_{u-1-k} x^{k+1} \; , \; F_2(x) = \sum_{k=0}^{\infty} t_{k+u} x^{-k} . \qquad (3.35)$$

Now $e^{bx} \int_x^\infty e^{-bt} F_1(t)dt$ is a polynomial in x of degree $u-1$ and no Chebyshev expansions are needed. We can easily convert $F_2(x)$ into a form like (3.1), see Luke [1,Vol.1,pp.290-293)]. The problem is then reduced to the form (3.2) with $u = 0$ and $F(t)$ replaced by $F_2(t)$.

IV. EXPONENTIAL TYPE INTEGRALS INVOLVING LOGARITHMS

From (2.2),

$$\frac{\partial g(x)}{\partial u} =$$

$$-(\ln x)g(x) + e^{-ax}x^{-u-1}\int_0^x e^{at}t^u(\ln t)f(t)dt, \quad (4.1)$$

$$\frac{\partial g(x)}{\partial u} = \sum_{k=0}^{\infty} p_k T_k^*(x/\lambda) \;, \; p_k = \frac{\partial g_k}{\partial u} . \quad (4.2)$$

Then for the present situation, (2.7), (2.8), (2.9) and (2.15) remain valid provided that in the left hand sides of these equations g_k is replaced by p_k and that the right hand sides of these equations are replaced by

$$-2G,\; -2g_0 + g_2 \;,\; -(g_{k-1} - g_{k+1}) \;\text{ and }\; -(g_{k-1} - g_k) \quad (4.3)$$

respectively, G as in (2.7). The analogs of equations (2.19) - (2.23) are

$$\frac{\partial g(x)}{\partial u} = w_0+w_1x+w_2x^2+\ldots,\; w_0 = -d_0/(u+1)^2,$$

$$w_1 = -\{d_1 - ad_0(2u+3)/(u+2)^2\}\{x/(u+1)^2\},$$

$$w_2 = -\{d_2 - [ad_1(2u+5)/(u+2)^2]$$

$$- a^2d_0(3u^2+12u+11)/(u+1)^2(u+2)^2\}\{x^2/(u+3)^2\} \;, \quad (4.4)$$

$$\sum_{k=0}^{\infty} (-)^k p_k = w_0,\; \sum_{k=1}^{\infty} (-)^k k^2 p_k = -(\lambda w_1/2),$$

$$\sum_{k=2}^{\infty} (-)^k k^2(k^2-1)p_k = (3\lambda^2 w_2/2). \quad (4.5)$$

In place of (2.23), we have

$$L\{p_k\} \equiv (2p_k/\varepsilon_k) + \{(4/a\lambda)(k+u+2) + 1\}p_{k+1}$$

$$+ \{(4/a\lambda)(k+1-u) - 1\}p_{k+2} - p_{k+3} ,$$

$$L\{p_k\} = -(4/a\lambda)(g_{k-1} - g_k). \tag{4.6}$$

Equations (2.24) - (2.26) can stand except that for (2.24), put

$$L\{A_{k,n}\} = -(4/a\lambda)(g_{k-1} - g_k) \tag{4.7}$$

and in (2.26) replace $g_{k,n}$ by $p_{k,n}$. Then

$$p_k = \lim_{n\to\infty} p_{k,n} . \tag{4.8}$$

The situation when $a = 0$ follows readily enough from (2.28) and we omit the details.

Analysis of $\partial G(x)/\partial u$ where $G(x)$ is given by (3.2) or (3.33) is much akin to the above developments and we dispense with specific results, save that the backward recursion process converges so long as $|\arg b| < \pi$. It will converge more slowly as $|\arg b| \to \pi$ and fails when $|\arg b| = \pi$. Here, the pertinent coefficients can be easily determined if the backward recurrence scheme is modified after the manner of discussion surrounding (3.28) - (3.32).

V. NUMERICAL EXAMPLES

In this section, we describe developments to get Chebyshev coefficients for

$$x^{-1}F(x) = x^{-1}\int_0^x t^{-1}\{Ei(t)-\gamma-\ln t\}dt \tag{5.1}$$

and

$$x^{-1}e^xP(x) = x^{-1}e^x\int_0^x t^{-1}e^{-t}\{Ei(t)-\gamma-\ln t\}dt \tag{5.2}$$

where γ is Euler's constant and

$$Ei(x) = -P.V.\int_{-x}^{\infty} t^{-1}e^{-t}dt. \tag{5.3}$$

Here P.V. stands for Cauchy principal value. It is known that, see Luke [1,Vol.1,p.222],

$$Ei(x) - (\gamma + \ln x) = \int_0^x t^{-1}(e^t - 1)dt. \quad (5.4)$$

In all of the numerics, we take $\lambda = 8$.

To get the coefficients for $x^{-1}F(x)$, we obtain in all three sets of coefficients based on the theory in Section II. Pass 1 is the nomenclature employed to describe the input and output data for the computer to achieve the first set of coefficients by use of the backward recursion process described by (2.23) - (2.29). Similarly, we speak of Pass 2 and Pass 3. We also need three passes for $x^{-1}e^xP(x)$, but the first two passes to get this function and $x^{-1}F(x)$ are identical.

The data for the passes are based on (2.2) and are given in the following table.

Pass No.	a	u	f(x)	g(x)
1	-1	0	1	$U(x) = x^{-1}(e^x - 1)$
2	0	0	U(x)	$V(x) = x^{-1}\int_0^x U(t)dt$
3A	0	0	V(x)	$x^{-1}F(x) = x^{-1}\int_0^x V(t)dt$
3B	-1	0	V(x)	$x^{-1}e^xP(x) = x^{-1}e^x\int_0^x e^{-t}V(t)dt$

(5.5)

To describe the Chebyshev expansions, some notation changes are necessary to avoid confusion with the notation of Section II, and to correspond to the notation of the computer. We use the nomenclature of (5.5) and write

$$U(x) = \sum_{K=0}^{\infty} B(K)T_K^*(x/8), \quad (5.6)$$

$$V(x) = \sum_{K=0}^{\infty} G(K)T_K^*(x/8), \quad (5.7)$$

$$x^{-1}F(x) = \sum_{K=0}^{\infty} H(K)T_K^*(x/8) \ , \qquad (5.8)$$

$$x^{-1}e^{x}P(x) = \sum_{K=0}^{\infty} E(K)T_K^*(x/8) \ , \qquad (5.9)$$

where the range of validity is $0 \le x \le 8$. The coefficients B(K) and G(K) are presented to 20D in Table 1, and the coefficients H(K) and E(K) are given to 20D in Table 2.

In the numerics quadruple precision was employed. Several values of n (see (2.24)) were used to insure that the final coefficients would be accurate to about 28D. For each n, the coefficients were checked using (2.7), (2.21) and (2.28). These are called Test 1, Test 2 and Test 3, respectively. Test 1, for example, is the difference between the left and right hand sides of (2.7) with g_k replaced by the computed g_k. Numerous other checks were made. The data in Tables 1 and 2 are rounded 20D values from the $n = 40$ computations. The test values for the $n = 40$ unrounded data are presented below each corresponding set of rounded coefficients.

Using the data in Table 2, we calculated F(x) and P(x) for $x = 0(0.1)2(0.2)5(0.5)8$. Chipman [6] tabulated these functions to 12 significant figures for the same values of x as well as for many values of $x > 8$. For the x values stated, the data given by Chipman are in perfect agreement with our findings.

The FORTRAN program for computations based on the theory of Section 2 is available from the author on request.

VI. ACKNOWLEDGMENT

I am indebted to Mr. Rick Whitaker of the Computation Laboratory, University of Missouri, Kansas City, Missouri for the development of the FORTRAN program.

Table 1

Chebyshev Coefficients For U(x) And V(x)

(See (5.5)-(5.7) for definitions)

K	B(K)	G(K)
0	84.21528770042713795950	14.41456779944782689066
1	139.60143980195862213768	21.39840922840560632498
2	85.21528770042713795950	11.58933364472027152821
3	40.60764385021356897975	4.91982145802637719756
4	15.77850825131791519843	1.71661658893004833037
5	5.16137322191878098368	0.50851302118722314073
6	1.45576180748035323106	0.13090665746301778165
7	0.36059548129911692234	0.02980328732335197286
8	0.07958677802143021666	0.00608320067012441754
9	0.01583486248656648906	0.00112538868796799622
10	0.00286744579127871872	0.00019039220173420091
11	0.00047637622313130745	0.00002967553490712013
12	0.00007309908702441011	0.00000428824702013234
13	0.00001042060996351576	0.00000057760548326449
14	0.00000138697475244150	0.00000007285834195164
15	0.00000017311660000426	0.00000000864164451161
16	0.00000002034127261190	0.00000000096726768044
17	0.00000000225773521050	0.00000000010249917713
18	0.00000000023743852051	0.00000000001031244128
19	0.00000000002372498310	0.00000000000098761962
20	0.00000000000225795513	0.00000000000009024398
21	0.00000000000020514367	0.00000000000000788439
22	0.00000000000001782899	0.00000000000000065990
23	0.00000000000000148504	0.00000000000000005301
24	0.00000000000000011875	0.00000000000000000409
25	0.00000000000000000913	0.00000000000000000030
26	0.00000000000000000068	0.00000000000000000002
27	0.00000000000000000005	
	TEST1 = 0.31554Q-29	TEST1 = 0.13805Q-29
	TEST2 = 0.12572Q-29	TEST2 = 0.17218Q-30
	TEST3 = 0.29582Q-30	TEST3 = 0.49304Q-31

Table 2

Chebyshev Coefficients For $x^{-1}F(x)$ And $x^{-1}e^{x}P(x)$

(See (5.5)-(5.9) for definitions)

K	H(K)	E(K)
0	3.71536318524502372817	125.71108767544436730357
1	4.01002445709397614210	210.08614599238009058502
2	1.78902666949738251257	129.42645086068939103174
3	0.65121608910087339647	62.18703685169731545791
4	0.19944683756427842710	24.33360295198771131866
5	0.05271734498035826353	8.00700906357096171078
6	0.01226045876841115558	2.26965701687427594241
7	0.00254669312679795329	0.56458844764792477821
8	0.00047807737697770414	0.12506441586340564631
9	0.00008188944866963551	0.02496184587178997702
10	0.00001290022217082669	0.00453266875813334169
11	0.00000188142229479871	0.00075485244962925663
12	0.00000025547457721847	0.00011608009582794006
13	0.00000003245600275231	0.00001657953104894205
14	0.00000000387408482927	0.00000221052600142585
15	0.00000000043610178578	0.00000027633809425030
16	0.00000000004644988391	0.00000003251563439163
17	0.00000000000469502980	0.00000000361363776047
18	0.00000000000045154114	0.00000000038048072047
19	0.00000000000004141888	0.00000000003805883127
20	0.00000000000000363147	0.00000000000362573457
21	0.00000000000000030494	0.00000000000032971390
22	0.00000000000000002457	0.00000000000002867974
23	0.00000000000000000190	0.00000000000000239073
24	0.00000000000000000014	0.00000000000000019132
25	0.00000000000000000001	0.00000000000000001472
26		0.00000000000000000109
27		0.00000000000000000008
28		0.00000000000000000001

H(K)	E(K)
TEST1 = 0.89363Q-31	TEST1 = 0.17392Q-28
TEST2 =-0.26193Q-31	TEST2 = 0.17873Q-30
TEST3 = 0.0	TEST3 = 0.24652Q-30

REFERENCES

1. Y.L. Luke, "The Special Functions and Their Approximations," Vols. 1 and 2, Academic Press, New York, 1969.
2. Y.L. Luke, "Mathematical Functions and Their Approximations," Academic Press, New York, 1975.
3. E.C. Titchmarsh, "Theory of Functions," Oxford University Press, London, 1939.
4. J. Wimp, "On Recursive Computation," Office of Aerospace Research, United States Air Force, Wright-Patterson Air Force Base, Dayton, Ohio ARL Report 60-0186, 1969, Clearinghouse, U.S. Department of Commerce, Springfield, Virginia 22151.
5. J. Wimp and Y.L. Luke, "An Algorithm for Generating Sequences Defined by Nonhomogeneous Difference Equations," Rend. Circ. Mat. Palermo 18 (1969), 251-275.
6. D.M. Chipman, "The Numerical Computation of Two Transcendental Functions Related to the Exponential Integral," Math. Comp. 26 (1972), 241-249.

NEVILLE - AITKEN Algorithms for Interpolation by Functions of ČEBYŠEV-Systems in the Sense of NEWTON and in a Generalized Sense of HERMITE.

G. Mühlbach

0. Introduction

The algorithms of NEWTON and NEVILLE-AITKEN solving the classical interpolation problem with algebraic polynomials are well-known. While the first allows a simple calculation of the coefficients of the interpolating polynomial by means of a well-known recurrence relation for the ordinary divided differences, the algorithms of NEVILLE and AITKEN are used to compute the interpolant at a prescribed point by recurrence. Although evaluating the interpolant with one of these algorithms is not more economical than computing first the coefficients of the interpolant by NEWTON's algorithm and then its value using nested multiplication, the NEVILLE-AITKEN algorithms are often used in numerical analysis to accelerate convergence of numerical methods.

All these algorithms can be established likewise for complete ČEBYŠEV-systems replacing the algebraic polynomials, but the classical proofs do not carry over in an obvious manner. In this note,recurrence relations of NEVILLE-AITKEN-type for interpolation by linear combinations of functions of complete or extended complete ČEBYŠEV-systems respectively are proved (including "RICHARDSON's extrapolation method" as a special case). Here interpolation is meant in the sense of NEWTON as well as in the usual or a generalized sense of HERMITE. These formulas lead to an evaluation method of the interpolant by recurrence nearly as simple as in the case of algebraic polynomials.

1. Definitions

Let I be a nonempty real interval. An (m+1)-tuple of functions

(1) $(f_0, \ldots, f_m)$, $f_i \in C(I)$

will be called a ČEBYŠEV-system (of order m) over I when

$$V\binom{f_0, \dots, f_m}{x_0, \dots, x_m} := \det f_i(x_j) \neq 0$$

for all choices of pairwise distinct "knots" $x_0, \dots, x_m \in I$. The system (1) will be referred to as a complete ČEBYŠEV-system (of order m) over I if $(f_0, \dots, f_k)$ is a ČEBYŠEV-system for each $k=0,\dots,m$.

Now let I be a nonempty real open interval. Corresponding to two given systems of functions of C(I),

$$(2) \quad \begin{array}{r} w_0, w_1, \dots, w_m \in C(I) \text{ each strictly positive on } I, \\ v_1, \dots, v_m \in C(I) \text{ each strictly monotonic on } I, \end{array}$$

we define generalized differential operators as follows:

$$(3) \quad \left.\begin{array}{ll} D_j f(x) & := \lim\limits_{h \to 0} \dfrac{f(x+h)-f(x)}{v_j(x+h)-v_j(x)} \\ d_j f & := D_j \left(\dfrac{f}{w_{j-1}}\right) \\ d^j & := d_j d_{j-1} \dots d_1 \\ \delta^j f & := \dfrac{d^j f}{w_j}. \end{array}\right\} (j=1,\dots,m)$$ [1]

A system (1) will be called an extended ČEBYŠEV-system (of order m) over I, extended with respect to the system of differential operators (3), provided for all systems of possibly multiple knots $(x_0, \dots, x_m)^*$ [2] which can be reordered in the form

$$(4) \quad (\underbrace{y_0, \dots, y_0}_{r_0}, \underbrace{y_1, \dots, y_1}_{r_1}, y_2, \dots, y_{\mu-1}, \underbrace{y_\mu, \dots, y_\mu}_{r_\mu})$$

where the points y_i $(i=0,\dots,\mu)$ are pairwise dis-

[1] d^0 will always denote the identity; $\delta^0 f := f/w_0$.

[2] here and in the following, a star * will indicate that knots may be repeated.

tinct and $r_0 + r_1 + \ldots + r_\mu = m+1$, $0 \leq \mu \leq m$, $1 \leq r_i \leq m+1$, the following determinant is nonzero: [3)]

$$V\binom{f_0, \ldots, f_m}{x_0, \ldots, x_m}^* =$$

$$\begin{vmatrix} f_0(y_0) & d^1 f_0(y_0) & \cdots & d^{r_0-1} f_0(y_0) & f_0(y_1) & d^1 f_0(y_1) & \cdots & d^{r_1-1} f_0(y_1) & \cdots & d^{r_\mu-1} f_0(y_\mu) \\ \vdots & \vdots & & \vdots & \vdots & \vdots & & \vdots & & \vdots \\ f_m(y_0) & d^1 f_m(y_0) & \cdots & d^{r_0-1} f_m(y_0) & f_m(y_1) & d^1 f_m(y_1) & \cdots & d^{r_1-1} f_m(y_1) & \cdots & d^{r_\mu-1} f_m(y_\mu) \end{vmatrix}.$$

A complete ČEBYŠEV-system $(f_0, \ldots, f_m)$ over I will be referred to as an extended complete ČEBYŠEV-system over I, extended with respect to the differential operators (3), if for $k=0,\ldots,m$ $(f_0, \ldots, f_k)$ is an extended ČEBYŠEV-system (of order k) over I, extended with respect to the differential operators (3) with m replaced by k.

Examples:

(i) Let I be a nonempty open interval of the real axis. To given systems (2) we define for $k=0,\ldots,m$ $v_{k,k+1} := w_k$, and by repeated integration in the sense of STIELTJES for $k=1,\ldots,m$ and $i=k,k-1,\ldots,1$,

$$(5) \quad v_{k,i}(x) = w_{i-1}(x) \int_c^x v_{k,i+1}(t)\,dv_i(t),$$

where c denotes an arbitrary but fixed point of I. Then the system

3) in the case $\mu = 0$, $r_0 = m+1$ this determinant can be regarded as a generalization of the WRONSKIAN determinant of the functions $(f_0, \ldots, f_m)$.

(6) $(v_{0,1}, v_{1,1}, \dots, v_{m,1})$

is an extended complete ČEBYŠEV-system of order m with respect to the differential operators (3). A proof can be found in [2] p.542.

(ii) It is important that in the case when $v_j = \pm id_I$ $(j=1,\dots,m)$ [4] and when the functions w_k are endowed with differential properties in the usual sense, $w_k \in C^{m-k}(I)$ $(k=0,\dots,m)$, the system (6) is an extended complete ČEBYŠEV-system, extended with respect to another system of differential operators of type (3), the operators of ordinary repeated differentiation

(7) $$\tilde{d}^j f := \frac{d^j}{dx^j} f = f^{(j)} \qquad (j=1,\dots,m).$$

Conversely, to each extended complete ČEBYŠEV-system $(u_0,\dots,u_m)$, extended with respect to the differential operators (7), satisfying the initial conditions $u_k^{(p)}(c) = 0$ $(p=0,\dots,k-1;\ k=1,\dots,m)$ for some point $c \in I$, there exist systems $(w_0,\dots,w_m)$ and $(v_1,\dots,v_m)$ of strictly positive functions $w_k \in C^{m-k}(I)$ and strictly monotonic functions $v_j = \pm\, id_I$ $(j=1,\dots,m)$, such that $(u_0,\dots,u_m)$ is identical with the system (6) generated with the functions w_k and v_j according to (5) ([1], theorem 1.2, p.379).

2. NEWTON- and HERMITE-Interpolation Formulas; Divided Differences

Let I be a nonvoid real open interval and $(f_0,\dots,f_m)$ be an extended complete ČEBYŠEV-system over I, extended with respect to the differential operators

[4] id_I denotes the identical mapping of I

(3). To each system $(x_0,\ldots,x_m)^*$ of possibly multiple knots of I, which is assumed to be identical with (4) after suitable reordering, and to each real function f defined on I such that $d^m f \in C(I)$, there exists a unique linear combination $H_m^* f$ of $(f_0,\ldots,f_m)$ that interpolates f in the following generalized sense of HERMITE:

$$(8)\quad d^{s_i} H_m^* f(y_i) = d^{s_i} f(y_i) \quad (i=0,\ldots,\mu;\ s_i=0,\ldots,r_i-1).^{5)}$$

In view of CRAMER's rule this is an immediate consequence of the definition of an extended complete ČEBYŠEV-system. It is easily proved that $H_m^* f$ can be represented by means of the so-called divided differences of f,

$$[x_0,\ldots,x_k|f]^* := \begin{bmatrix} f_0,\ldots,f_k \\ x_0,\ldots,x_k \end{bmatrix} f \Big]^* := \frac{V\begin{pmatrix} f_0,\ldots,f_{k-1},f \\ x_0,\ldots,x_{k-1},x_k \end{pmatrix}^*}{V\begin{pmatrix} f_0,\ldots,f_{k-1},f_k \\ x_0,\ldots,x_{k-1},x_k \end{pmatrix}^*},$$

in the following way, which generalizes the classical NEWTON-HERMITE-interpolation formula:

$$(9)\quad H_m^* f(x) = \sum_{k=0}^{m} \left[\begin{matrix} f_0,\ldots,f_k \\ x_0,\ldots,x_k \end{matrix} \middle| f \right]^* \cdot g_k^*(x)$$

For $x \notin \{x_0,\ldots,x_m\}$ and $k=0,\ldots,m+1$ the abbreviations $g_0^*(x) := f_0(x)$ and

$$g_k^*(x) := \frac{V\begin{pmatrix} f_0,\ldots,f_{k-1},f_k \\ x_0,\ldots,x_{k-1},x \end{pmatrix}^*}{V\begin{pmatrix} f_0,\ldots,f_{k-1} \\ x_0,\ldots,x_{k-1} \end{pmatrix}^*} = f_k(x) - H_{k-1}^* f_k(x)$$

are used. When $(f_0,\ldots,f_{m+1})$ is an extended complete

5) d^0 will always denote the identity, $\delta^0 f := f/w_0$.

ČEBYŠEV-system, extended with respect to the differential operators (3) with m replaced by m+1, the remainder term $r_m^* f(x) := f(x) - H_m^* f(x)$ for $x \notin \{x_0, \ldots, x_m\}$ can be represented in the form

$$r_m^* f(x) = \frac{V\binom{f_0, \ldots, f_m, f}{x_0, \ldots, x_m, x}^*}{V\binom{f_0, \ldots, f_m}{x_0, \ldots, x_m}^*} = \left[\begin{matrix} f_0, f_1, \ldots, f_{m+1} \\ x, x_0, \ldots, x_m \end{matrix} \middle| f \right]^* \cdot g_{m+1}^*(x).$$

It should be mentioned that in the case of the system (6), where m is to be replaced by m+1 and when $d^{m+1} f \in C(I)$, there exists a point z in the convex hull of the set $\{x, x_0, \ldots, x_m\}$ such that $r_m^* f(x) = \delta^{m+1} f(z) \cdot g_{m+1}^*(x)$. Proofs of these representations and of the following recurrence formula can be found in [2].

For practical purposes it is important that the divided differences can be computed by recurrence:

$$(10) \quad \left[\begin{matrix} f_0, \ldots, f_k \\ x_0, \ldots, x_k \end{matrix} \middle| f \right]^* = \begin{cases} \dfrac{V\binom{f_0, \ldots, f_{k-1}, f}{x_0, \ldots, x_0, x_0}^*}{V\binom{f_0, \ldots, f_{k-1}, f_k}{x_0, \ldots, x_0, x_0}^*} \quad {}^{6)} \\[2ex] \dfrac{\left[\begin{matrix} f_0, \ldots, f_{k-1} \\ x_1, \ldots, x_k \end{matrix} \middle| f \right]^* - \left[\begin{matrix} f_0, \ldots, f_{k-1} \\ x_0, \ldots, x_{k-1} \end{matrix} \middle| f \right]^*}{\left[\begin{matrix} f_0, \ldots, f_{k-1} \\ x_1, \ldots, x_k \end{matrix} \middle| f_k \right]^* - \left[\begin{matrix} f_0, \ldots, f_{k-1} \\ x_0, \ldots, x_{k-1} \end{matrix} \middle| f_k \right]^*} \end{cases}$$

provided all knots are equal in the first case and $x_0 \neq x_k$ in the second. A proof of this formula when all knots are different can also be found in [3]. Clearly, in this case it is sufficient to assume the system $(f_0, \ldots, f_m)$ to be a complete ČEBYŠEV-system only, and to omit the stars in the above formulas.

6) $= \delta^k f(x_0)$ in the case of the system (6)

3. NEVILLE-AITKEN Algorithms for the Interpolation by ČEBYŠEV-Systems in the Sense of NEWTON-HERMITE

Let $(f_0,\dots,f_{m+1})$ be a complete ČEBYŠEV-system of order m+1 over a nonvoid real interval I, let f denote a real function defined on I and let $x_0,\dots,x_m$ be a system of pairwise distinct knots of I. For an arbitrary but fixed point $x \in I$ such that $x \notin \{x_0,\dots,x_m\}$, and for an arbitrary but fixed subsystem $\{x_{i_j}, x_{i_{j+1}},\dots,x_{i_k}\} \subseteq \{x_0,\dots,x_m\}$, we use the following notations [7]:

$$g_{j,k+1} := \frac{V\begin{pmatrix} f_0, f_1, \dots, f_{k-j}, f_{k+1-j} \\ x_{i_j}, x_{i_{j+1}}, \dots, x_{i_k}, x \end{pmatrix}}{V\begin{pmatrix} f_0, f_1, \dots, f_{k-j} \\ x_{i_j}, x_{i_{j+1}}, \dots, x_{i_k} \end{pmatrix}} \qquad (0 \le j < k \le m)$$

$$g_{j,j} := f_0(x)$$

$$H_{j,k} f := \sum_{\ell=j}^{k} \begin{bmatrix} f_0, \dots, f_{\ell-j} \\ x_{i_j}, x_{i_{j+1}}, \dots, x_{i_\ell} \end{bmatrix} f \, g_{j,\ell}(x) \qquad (0 \le j \le k \le m)$$

$$r_{j,k} f := f(x) - H_{j,k} f$$

$$= \begin{bmatrix} f_0, f_1, \dots, f_{k+1-j} \\ x, x_{i_j}, \dots, x_{i_k} \end{bmatrix} f \, g_{j,k+1}(x) \qquad (0 \le j < k \le m)$$

$$r_{j,j} f := f(x) - H_{j,j} f(x) = f(x) - \frac{f(x_{i_j})}{f_0(x_{i_j})} \cdot f_0(x).$$

Theorem 1: For $0 \le j < k \le m$ there holds

$$H_{j,k} f = \frac{(r_{j+1,k} f_{k-j})(H_{j,k-1} f) - (r_{j,k-1} f_{k-j})(H_{j+1,k} f)}{r_{j+1,k} f_{k-j} - r_{j,k-1} f_{k-j}}$$

and for $0 < k-j < n \le m$

[7] For the first part of this section (theorem 1 and its proof) one can suppose without loss of generality $i_j = j$ for all j. Actually, these notations will be needed only in the second part.

$$r_{j,k}\, f_n = \frac{(r_{j+1,k} f_{k-j})(r_{j,k-1} f_n) - (r_{j,k-1} f_{k-j})(r_{j+1,k} f_n)}{r_{j+1,k} f_{k-j} - r_{j,k-1} f_{k-j}}$$ 8)

Proof: For all $f: I \to \mathbb{R}$ and $x \notin \{x_0, \dots, x_m\}$ we have for $0 \le j < k \le m$

$$\begin{aligned} f(x) &= H_{j,k-1} f(x) + r_{j,k-1} f(x) \\ &= H_{j,k-1} f(x) + [x, x_{i_j}, \dots, x_{i_{k-1}} | f]\, g_{j,k}(x) \quad \text{and} \\ f(x) &= H_{j+1,k} f(x) + r_{j+1,k} f(x) \\ &= H_{j+1,k} f(x) + [x, x_{i_{j+1}}, \dots, x_{i_k} | f]\, g_{j+1,k+1}(x). \end{aligned}$$

Multiplying the first equation with $G_0 := g_{j,k+1} / g_{j,k}$ and the second with $G_1 := g_{j,k+1}/g_{j+1,k+1}$ and subtracting yields

$$\begin{aligned} f(x)(G_0 - G_1) &= G_0 H_{j,k-1} f - G_1 H_{j+1,k} f + \\ &+ g_{j,k+1} ([x, x_{i_j}, \dots, x_{i_{k-1}} | f] - [x, x_{i_{j+1}}, \dots, x_{i_k} | f]). \end{aligned}$$

Using the recurrence relation (10) for the divided differences it is easily seen that the difference enclosed in brackets may be replaced by

$$[x, x_{i_j}, \dots, x_{i_k} | f] \cdot \{[x, x_{i_j}, \dots, x_{i_{k-1}} | f_{k+1-j}] - [x, x_{i_{j+1}}, \dots, x_{i_k} | f_{k+1-j}]\}$$

where from (10) (or originally from an inspection of the proof of (10)) we know that the difference of divided differences in curved brackets is different from zero (as long as $x_{i_j} \neq x_{i_k}$). Therefore we get

8) For computation of $H_{0,m} f(x)$ according to theorem 1 in general $\frac{2}{3}m^3 + O(m^2)$ multiplications or divisions are needed. Computation of $H_{0,m} f(x)$ using NEWTON's formula (9) takes only $\frac{1}{3}m^3 + O(m^2)$ such operations. In case of polynomial interpolation the corresponding numbers are $m^2 + m$ and $\frac{1}{2}m^2 + \frac{3}{2}m$ respectively.

$$(11)\quad \frac{f(x)\cdot(G_0 - G_1)}{[x,x_{i_j},\dots,x_{i_{k-1}}|f_{k+1-j}] - [x,x_{i_{j+1}},\dots,x_{i_k}|f_{k+1-j}]} =$$

$$= \frac{G_0 H_{j,k-1}f - G_1 H_{j+1,k}f}{[x,x_{i_j},\dots,x_{i_{k-1}}|f_{k+1-j}] - [x,x_{i_{j+1}},\dots,x_{i_k}|f_{k+1-j}]} + r_{j,k}f,$$

where the factor of $f(x)$ is independent of f. Taking $f = g_{j,k+1}$ this factor is easily seen to be equal to one for all $x \notin \{x_0,\dots,x_m\}$. This fact, $g_{s,t+1}(x) = r_{s,t}\, f_{t+1-s}$, and

$$[x,x_{i_s},\dots,x_{i_t}|f] = r_{s,t}\, f/r_{s,t}\, f_{t+1-s} \text{ for } 0 \le s \le t \le m$$

together imply

$$G_0 - G_1 =$$

$$= r_{j,k}\, f_{k+1-j}\cdot\left(\frac{1}{r_{j,k-1}\, f_{k-j}} - \frac{1}{r_{j+1,k}\, f_{k-j}}\right) = \frac{r_{j,k-1}\, f_{k+1-j}}{r_{j,k-1}\, f_{k-j}} - \frac{r_{j+1,k}\, f_{k+1-j}}{r_{j+1,k}\, f_{k-j}},$$

or equivalently $(0 \le j < k \le m)$

$$r_{j,k}\, f_{k+1-j} = \frac{(r_{j+1,k}\, f_{k-j})(r_{j,k-1} f_{k+1-j}) - (r_{j,k-1} f_{k-j})(r_{j+1,k}\, f_{k+1-j})}{r_{j+1,k}\, f_{k-j} - r_{j,k-1}\, f_{k-j}}.$$

Using this in (11) we get, after some simple calculations, the first recurrence relation of the theorem and from this the second.

Now let $(f_0,\dots,f_{m+1})$ be an extended complete ČEBYŠEV-system of order $m+1$ over I, where I is a nonvoid real open interval, extended with respect to the differential operators (3) when m is replaced by $m+1$. Let $(x_0,\dots,x_m)^*$ be a system of possibly repeated knots which we assume to be identical with (4) and let $(x_{i_j},x_{i_{j+1}},\dots,x_{i_k})^*$ be an arbitrary sub-system of (4) whose possibly multiple knots are ordered in the natural order of $\mathbb{R}$. Let

$f: I \to \mathbb{R}$ denote an arbitrary real function defined on I such that $d^m f \in C(I)$. In the following we shall use the same notations as before where only a star will indicate that the knots may be repeated. Again the value of the interpolant $H^*_{j,k} f$ at a prescribed point $x \in I$ such that $x \notin \{x_0, \ldots, x_m\}$ can be computed by recurrence in most cases. Two different formulas are to be used depending exclusively on whether or not all knots are identical.

Theorem 2: For $0 \leq j < k \leq m$ there holds

$$H^*_{j,k} f = \begin{cases} \sum_{\ell=0}^{k-j} [x_{i_j}, x_{i_{j+1}}, \ldots, x_{i_{j+\ell}} | f]^* \cdot g^*_{j,j+\ell}(x) \\ \dfrac{(r^*_{j+1,k} f_{k-j})(H^*_{j,k-1} f) - (r^*_{j,k-1} f_{k-j})(H^*_{j+1,k} f)}{r^*_{j+1,k} f_{k-j} - r^*_{j,k-1} f_{k-j}} \end{cases}$$

provided $x_{i_j} = \ldots = x_{i_k}$ in the first case and $x_{i_j} \neq x_{i_k}$ in the second, where

$$r^*_{j,k} f_n := f_n(x) - H^*_{j,k} f_n$$

$$= \frac{(r^*_{j+1,k} f_{k-j})(r^*_{j,k-1} f_n) - (r^*_{j,k-1} f_{k-j})(r^*_{j+1,k} f_n)}{r^*_{j+1,k} f_{k-j} - r^*_{j,k-1} f_{k-j}}$$

when $x_{i_j} \neq x_{i_k}$ and $0 < k-j < n \leq m$.

Corollary: If $(f_0, \ldots, f_{m+1})$ denotes the system (6) with m replaced by m+1, then $H^*_{j,k} f$ can be computed totally by recurrence:

$$(12) \quad H^*_{j,k} f = H^*_{j,k-1} f + \delta^{k-j} f(x_{i_j}) \cdot r^*_{j,k-1} f_{k-j}$$

when $x_{i_j} = x_{i_{j+1}} = \ldots = x_{i_k}$, $0 \leq j < k \leq m$.

Proof: An inspection of the proof of theorem 1 shows that under the assumption of theorem 2 it remains valid also in the second case when $x_{i_j} \neq x_{i_k}$. In the first case, when $x_{i_j} = \dots = x_{i_k}$, the asserted formula reduces to the HERMITE-interpolation formula (9). We have only to establish the recurrence relation (12). It can be viewed as a generalization of TAYLOR's expansion formula.[9] For a proof by induction see [4], p. 395. Another proof proceeds by expanding the denominator and the numerator of

$$r^*_{j,k} f = \frac{V\binom{f_0, \dots, f_{k-j}, f}{x_{i_j}, \dots, x_{i_k}, x}^*}{V\binom{f_0, \dots, f_{k-j}}{x_{i_j}, \dots, x_{i_k}}^*}$$

along the (k-j+1)-st column, where $x_{i_j} = x_{i_{j+1}} = \dots$ $\dots = x_{i_k}$ and for $s \leq t$ $d^t f_s = w_t \cdot \delta_{s,t}$[10] leading to

$$r^*_{j,k} f = r^*_{j,k-1} f - \delta^{k-j} f(x_{i_j}) \cdot r^*_{j,k-1} f_{k-j}$$

which is equivalent with (12).

Remark: If the given ČEBYŠEV-system $(f_0, \dots, f_m)$ can be viewed as an extended complete ČEBYŠEV-system, extended with respect to two different systems of differential operators both of type (3) [compare example (ii) above] $(j=1,\dots,m)$

$$d^j := d_j d_{j-1} \dots d_1 \qquad \tilde{d}^j := \tilde{d}_j \tilde{d}_{j-1} \dots \tilde{d}_1$$

$$d_j f := D_j\left(\frac{f}{w_{j-1}}\right) \qquad \tilde{d}_j f = \tilde{D}_j\left(\frac{f}{\tilde{w}_{j-1}}\right)$$

$$D_j f(x) := \lim_{h \to 0} \frac{f(x+h) - f(x)}{v_j(x+h) - v_j(x)} \qquad \tilde{D}_j f(x) := \lim_{h \to 0} \frac{f(x+h) - f(x)}{\tilde{v}_j(x+h) - \tilde{v}_j(x)}$$

then the solutions $H^*_m f$, $\tilde{H}^*_m f$ resp. of the two generalized HERMITE-interpolation problems (8) with respect to the operators d^j, $\tilde{d}^j$ resp. are identical.

---

9) see [4], p.396 ff

10) $\delta_{s,t}$ KRONECKER delta

For a system $(f_0,\ldots,f_m)$ considered under example (ii) above, it is possible to compute the coefficients of the linear combination $\tilde{H}_m^* f$ of $(f_0,\ldots,f_m)$ that interpolates f in the classical HERMITEan sense (where in (8) instead of the operators (3) the operators (7) are prescribed) totally by recurrence, solving instead of the classical HERMITE-interpolation problem its generalized version (8), where in the definition of the operators d^j (see (3)) the functions w_k, v_j are the functions (2) basic for the construction of the ČEBYŠEV-system used. The same can be done by evaluating $\tilde{H}_m^* f$ at a prescribed point x using theorem 2 and its corollary. The equality $\tilde{H}_m^* f = H_m^* f$ is a consequence of the continuity of the functions $g_{j,k}^*$ and of the divided differences with simple knots of a sufficiently often differentiable function, differentiable in the generalized sense of the d^j and of the $\tilde{d}^j$, regarded as a function of the knots, when these are allowed to agree:

$$\lim\,[t_0,\ldots,t_k\,|\,f] = [x_0,\ldots,x_k\,|\,f]^* \quad (k=0,\ldots,m)$$

when $t_i \longrightarrow x_i$ $(i=0,\ldots,k)$ and all t_i are pairwise distinct. For a proof cf. [2], p.544.

References

[1] S. KARLIN and W. STUDDEN, "Tchebycheff Systems", Interscience, New York, (1966)

[2] MÜHLBACH, G., Newton- und Hermite-Interpolation mit Čebyšev-Systemen, ZAMM 54, 541-550, (1974)

[3] MÜHLBACH, G., A recurrence formula for generalized divided differences and some applications, Journal of Approximation Theory 9, No. 2, 165-172, (1973)

[4] MÜHLBACH, G., Asymptotische Fehlerdarstellungen bei der Approximation mit positiven linearen Operatoren, Rev. Roum. Math. Pures et Appl. Tome XVIII, No. 3, 391-406, (1973)

BEST SIMULTANEOUS APPROXIMATION IN THE L_1 AND L_2 NORMS

G.M. Phillips and B.N. Sahney*

§1. The problem of simultaneous Chebyshev approximation to two real-valued functions f_1 and f_2 defined on the interval [0,1] has been discussed by C.B. Dunham [4] and by J.B. Diaz and H.W. McLaughlin [1,2,3]. In what follows, we write $\|.\|_\infty$ to denote the supremum or Chebyshev norm. Diaz and McLaughlin [3] proposed the following definition.

Definition A If S is a non-empty family of real-valued functions on [0,1] and if there exists an $s^* \in S$ such that

$$\inf_{s\in S} \max\{\|f_1-s\|_\infty, \|f_2-s\|_\infty\} = \max\{\|f_1-s^*\|_\infty, \|f_2-s^*\|_\infty\}, \tag{1.1}$$

then s^* is said to be a best simultaneous approximation to f_1 and f_2.

They proved the following theorem.

Theorem A For any f_1 and f_2,

$$\inf_{s\in S} \left\| \left|\frac{f_1+f_2}{2} - s\right| + \tfrac{1}{2}|f_1-f_2| \right\|_\infty = \inf_{s\in S} \max\{\|f_1-s\|_\infty, \|f_2-s\|_\infty\}. \tag{1.2}$$

Thus the problem of approximating to f_1 and f_2 simultaneously is equivalent to the problem of approximating to $\frac{1}{2}(f_1+f_2)$ with the additive weight function $\frac{1}{2}|f_1-f_2|$.

On the other hand, W.H. Ling [8] has given the following definition.

---

* This research has been supported by the Scientific Research Council of Britain, the National Research Council of Canada and the Scientific Affairs Division of NATO.

Definition B If there exists an $s^* \in S$ such that

$$(1.3) \qquad \| |f_1-s^*| + |f_2-s^*| \|_\infty = \inf_{s\in S} \| |f_1-s| + |f_2-s| \|_\infty ,$$

then s^* is said to be a best simultaneous approximation to f_1 and f_2.

Using this definition, the following results are given by W.H. Ling [8].

Theorem B1 If

$$(1.4) \qquad \inf_{s\in S} \left\| \frac{f_1+f_2}{2} - s \right\|_\infty \geq \tfrac{1}{2}\|f_1-f_2\|_\infty$$

then

$$(1.5) \qquad \inf_{s\in S} \| |f_1-s| + |f_2-s| \|_\infty = 2 \inf_{s\in S} \left\| \frac{f_1+f_2}{2} - s \right\|_\infty .$$

In other words, under the condition (1.4), s^* is a best simultaneous approximation to f_1 and f_2 in the sense of Definition B if and only if s^* is a best approximation to $\tfrac{1}{2}(f_1+f_2)$ from S.

Theorem B2 If

$$(1.6) \qquad \inf_{s\in S} \left\| \frac{f_1+f_2}{2} - s \right\|_\infty < \tfrac{1}{2}\|f_1-f_2\|_\infty$$

then

$$(1.7) \qquad \inf_{s\in S} \| |f_1-s| + |f_2-s| \|_\infty = \|f_1-f_2\|_\infty .$$

Ling [8] deduces from this last result that if $\bar{S} \subset S$ is the set of best simultaneous approximations to f_1 and f_2, in the sense of Definition B, then

$$\bar{S} = \{s \in S \mid \left\| \frac{f_1+f_2}{2} - s \right\|_\infty < \tfrac{1}{2}\|f_1-f_2\|_\infty\}.$$

The object of this paper is to consider simultaneous approximations to two functions with respect to the L_1 and L_2 norms. We will write $\|.\|_p$ to denote the L_p norm.

The problem of best simultaneous approximation for two functions in abstract spaces has been discussed elsewhere [5,6] and a recent paper [7] has dealt with the problem for more than two functions.

§2. In this and the following section we consider simultaneous approximation in the L_1 norm.

Definition 1 If there exists an $s^* \in S$ such that

$$\inf_{s\in S} \int_0^1 \max\left\{|f_1(x)-s(x)|,\ |f_2(x)-s(x)|\right\}dx = \int_0^1 \max\left\{|f_1(x)-s^*(x)|,\ |f_2(x)-s^*(x)|\right\}dx\ , \tag{2.1}$$

then we say that s^* is a best simultaneous approximation to f_1 and f_2 in the L_1 norm.

We now state:

Theorem 1 For any f_1 and f_2 defined on $[0,1]$,

$$\inf_{s\in S} \left\|\frac{f_1+f_2}{2} - s\right\|_1 + \left\|\frac{f_1-f_2}{2}\right\|_1 = \inf_{s\in S} \int_0^1 \max\left\{|f_1(x)-s(x)|,\ |f_2(x)-s(x)|\right\}dx. \tag{2.2}$$

Thus s^* is the best simultaneous approximation to f_1 and f_2 in the L_1 norm if and only if it is the best approximation to their arithmetic mean.

To prove the theorem, we use the identity

$$|m+n| + |m-n| = 2 \max \{|m|, |n|\}, \tag{2.3}$$

where m and n are arbitrary real numbers. We replace m and n in (2.3) by $\frac{1}{2}(f_1(x)-s(x))$ and $\frac{1}{2}(f_2(x)-s(x))$ respectively, integrate over [0,1] and take the infimum of s over S. This completes the proof.

§3. We now explore the consequences of the following alternative definition.

Definition 2 If there exists an $s^* \in S$ such that

$$\inf_{s \in S} \left[\|f_1-s\|_1 + \|f_2-s\|_1 \right] = \|f_1-s^*\|_1 + \|f_2-s^*\|_1 \tag{3.1}$$

then s^* is said to be a best simultaneous approximation to f_1 and f_2 in the L_1 norm.

The following theorem is based on this definition.

Theorem 2 For any f_1 and f_2 defined on [0,1],

$$\begin{aligned} &\inf_{s \in S} \left[\|f_1-s\|_1 + \|f_2-s\|_1 \right] \\ &= 2 \inf_{s \in S} \int_0^1 \max\left\{ \left| \frac{f_1(x)+f_2(x)}{2} - s(x) \right|, \left| \frac{f_1(x)-f_2(x)}{2} \right| \right\} dx. \end{aligned} \tag{3.2}$$

To prove the theorem we replace m and n in (2.3) by $\frac{1}{2}(f_1(x)+f_2(x) - 2s(x))$ and $\frac{1}{2}(f_1(x)-f_2(x))$ respectively. Then (3.2) follows by integrating over [0,1] and taking the infimum. This completes the proof.

Corollary In the L_1 norm

(a) if for all x,

(3.3) $$\left|\frac{f_1(x)-f_2(x)}{2}\right| \le \left|\frac{f_1(x)+f_2(x)}{2} - s^*(x)\right|$$

then

(3.4) $$\inf_{s\in S}\left[\|f_1-s\|_1 + \|f_2-s\|_1\right] = 2\inf_{s\in S}\left\|\frac{f_1+f_2}{2} - s\right\|_1$$

and

(b) if for all x

(3.5) $$\left|\frac{f_1(x)-f_2(x)}{2}\right| > \left|\frac{f_1(x)+f_2(x)}{2} - s^*(x)\right|$$

then

(3.6) $$\inf_{s\in S}\left[\|f_1-s\|_1 + \|f_2-s\|_1\right] = \|f_1-f_2\|_1.$$

§4. We now consider simultaneous approximation in the L_2 norm.

Definition 3 If there exists an $s^* \in S$ such that

(4.1) $$\inf_{s\in S}\left[\|f_1-s\|_2^2 + \|f_2-s\|_2^2\right] = \|f_1-s^*\|_2^2 + \|f_2-s^*\|_2^2$$

then we say that s^* is a best simultaneous approximation to f_1 and f_2 in the L_2 norm.

The following theorem is based on Definition 3.

Theorem 3 For any f_1 and f_2,

(4.2) $$\inf_{s\in S}\left\|\frac{f_1+f_2}{2} - s\right\|_2^2 + \left\|\frac{f_1+f_2}{2}\right\|_2^2 = 2\inf_{s\in S}\left\{\left\|\frac{f_1-s}{2}\right\|_2^2 + \left\|\frac{f_2-s}{2}\right\|_2^2\right\}.$$

We observe that s* is a best simultaneous approximation to f_1 and f_2 in the L_2 norm if and only if it is a best approximation to $\frac{1}{2}(f_1+f_2)$. The proof of the theorem is obvious.

We note that, just as Theorem 1 is based on (2.3), we may regard Theorem 3 as being based on the Apollonius identity,

$$|\alpha+\beta|^2 + |\alpha-\beta|^2 = 2\{|\alpha|^2 + |\beta|^2\}$$

which holds for any real or complex α and β.

Acknowledgement It is a pleasure to thank A.S.B. Holland and J. Tzimbalario for helpful discussions related to this work.

REFERENCES

[1] J.B. Diaz and H.W. McLaughlin, Simultaneous Chebyshev approximation of a set of bounded complex-valued functions. J.Approximation Theory 6 (1969), pp.419-432.

[2] J.B. Diaz and H.W. McLaughlin, Simultaneous approximation of a set of bounded real functions. Math. Comp. 23 (1969), pp. 583-594.

[3] J.B. Diaz and H.W. McLaughlin, On simultaneous Chebyshev approximation and Chebyshev approximation with an additive weight function. J.Approximation Theory 6 (1972), pp. 68-71.

[4] C.B. Dunham, Simultaneous Chebyshev approximation of functions on an interval. Proc. Amer. Math. Soc. 18 (1967), pp. 472-477.

[5] D.S. Goel, A.S.B. Holland, C. Nasim and B.N. Sahney, On best simultaneous approximation in a normed linear space. To appear in Canadian Mathematical Bulletin.

[6] D.S. Goel, A.S.B. Holland, C. Nasim and B.N. Sahney, Characterisation of an element of best L^p-simultaneous approximation. To appear in S.R. Ramanujan memorial volume.

[7] A.S.B. Holland, B.N. Sahney and J. Tzimbalario, On best simultaneous approximation. To appear in J. Approximation Theory.

[8] W.H. Ling, On simultaneous approximation in the sum norms. To appear.

Comonotone Approximation and Piecewise Monotone Interpolation

Louis Raymon

Historically, the relationship between the theories of interpolation and approximation hasn't quite been what the uninformed would naively suspect. In 1912, Bernstein observed that the sequence of interpolation polynomials of the function $y = |x|$ on the triangular system of equally spaced points on $[-1,1]$ diverges for $0 < |x| < 1$. Earlier, Runge noted that the equally spaced interpolation to the function $f(x) = \frac{1}{1+x^2}$ on the interval $[-5,5]$ diverges on a subset of the interval, in spite of the fact that f is analytic throughout the interval. Bernstein and Faber independently discovered that for any triangular system of interpolation points on $[-1,1]$, there are continuous functions such that the polynomials of interpolation to f diverge at some points of $[-1,1]$. It would seem that requiring the polynomial to "approximate" with such great accuracy (i.e., to interpolate) on the points of interpolation in some way exhausts the "approximating energy" of the polynomials elsewhere on the interval, and they may approximate as badly as possible, i.e., they may diverge.

On the other hand, there is a broad theory of simultaneous approximation and interpolation. For example, there are the well known results that ensure the uniform convergence of the sequence of interpolating polynomials of functions f on a specific triangular systems of interpolation points (such as the zeros of the Chebychev polynomials) provided that f satisfies certain smoothness conditions. Paszkowski [14] was the first to prove that the Jackson estimates for the order of magnitude of the error in approximating a continuous function f on an interval by a polynomial of degree $\leq n$ are not increased if there is the added restriction that the polynomials also interpolate at a fixed number of points. Such a restriction on the approximation (i.e., one in which the restriction does not essentially alter the quality of the approximation being considered) may be termed an "efficient" restriction. Simultaneous approximation and interpolation may be thought of as restricted approximation, i.e., approximation with interpolatory constraints, or, on the other hand, as restricted interpolation - interpolation with approximation constraints.

For certain restrictions the relationship between problems of restricted approximation and problems of restricted interpolation is much closer than in the unrestricted case. This interdependence is well illustrated by the problems that will be discussed here. The problems are loosely stated as follows:

Question 1: Comonotone approximation: If a real valued function f is piecewise monotone on [a,b] - i.e., if f has only a finite number of local extrema on [a,b] - (and/or if the kth difference of f alternates in sign a finite number of times on [a,b]), how well can f be approximated by polynomials or other approximating functions that are "comonotone" with f (and/or with k^{th} difference copositive with the kth difference of f) on [a,b]?

Question 2: Piecewise monotone interpolation: Given points of interpolation $(x_0,y_0),\ldots,(x_k,y_k)$ with $x_0 < x_1 < \ldots < x_k$, $y_j \neq y_{j-1}$, $j = 1,\ldots,k$, what degree polynomial is necessary and/or sufficient to interpolate the given data subject to restrictions such a monotonicity on each of the intervals (x_{j-1},x_j)? What of such interpolation by functions other than ordinary polynomials?

In this discussion, we present a review of results related directly to these two problems. Most of these are recent. We shall not attempt the completeness of a thorough survey talk on monotone approximation and interpolation, which would include questions of best approximation, algorithms, approximation and interpolation by more general classes of functions and in norms more general than the sup norm in addition to the growing literature on simultaneous approximation and interpolation with norm preserving operators (SAIN).

The following result was first discovered in 1951.

Theorem 1: (Wolibner [18]) Let there be given points of interpolation $x_0 < x_1 < \ldots < x_n$, and real numbers y_i, with $y_{i-1} \neq y_i$, $i = 1,\ldots,n$. There is an algebraic polynomial p such that $p(x_i) = y_i$, $i = 0,\ldots,n$, and such that p is monotone on each of the intervals $[x_{i-1},x_i]$, $i = 1,\ldots,n$.

In order to prove Theorem 1, Wolibner found it necessary (or, at least, convenient) to first prove the following comonotone approximation theorem.

Theorem 2: (Wolibner [18]) Let there be given a continuous function f that is alternately nondecreasing and nonincreasing on the intervals $[x_0,x_1],[x_1,x_2],\ldots,[x_{n-1},x_n]$. Then f is the uniform limit of polynomials comonotone with f on $[x_0,x_n]$.

It is interesting to note that S.W. Young, who later proved Theorem 1 apparently unaware of Wolibner's proof, also proved Theorem 2 as a lemma [19]. It is not difficult to see that Theorem 1 implies Theorem 2, so that the two theorems are equivalent. A natural question to ask at this point was Question 2 above. Not too surprisingly at this point, this question turned out to be related to the question of the degree of approximation to a piecewise monotone function by comonotone polynomials. We briefly interrupt this discussion to present some formal definitions and notational conventions.

A function $f(x)$ on $[a,b]$ is said to be a piecewise monotone if $[a,b]$ may be partitioned into a finite number of subintervals on which f is alternately nondecreasing and nonincreasing. $f(x)$ and $g(x)$ are said to be comonotone on $[a,b]$ if they are piecewise monotone and are alternately nondecreasing and nonincreasing on the same subintervals. If f is piecewise monotone on $[a,b]$ we denote by $P_n^*(f)$ the set of all polynomials of degree $\leq n$ comonotone with f on $[a,b]$. The degree of comonotone approximation of f, $E_n^*(f)$, is defined by

$$E_n^*(f) = \inf_{p \in P_n^*} ||f-p||, \tag{1}$$

where $||\cdot||$ denotes the sup norm. If S is a set of comonotone functions, the degree of comonotone approximation to the set S is given by

$$E_n^*(S) = \sup_{f \in S} E_n^*(f) \tag{2}$$

A generalization of monotonicity and comonotonicity is obtained by considering the signs of the kth difference $\Delta^k(f)$ of f. If $[a,b]$ may be partitioned into a finite number of subintervals on which $\Delta^k f$ is alternately nonnegative and nonpositive, we denote by $P^*_{n,k}(f)$ the set of all polynomials p of degree $\leq n$ such that $\Delta^k f$ and $p^{(k)}$ are alternately nonpositive and nonnegative on the same subintervals, $k = 0,1,\ldots,$ $P^*_{n,1}(f)$ is the same as $P^*_n(f)$, i.e., a set of polynomials comonotone with f. $p^*_{n,0}(f)$ is a set of polynomials "copositive" with f. We extend the definition (1) above by

$$E^*_{n,k}(S) = \sup_{f \in S} E^*_{n,k}(f) .$$

If $\Delta^k f \geq 0$ throughout the interval [a,b] (or, if $\Delta^k f \leq 0$ throughout [a,b]) then $E^*_{n,k}(f)$ and $E^*_{n,k}(S)$ are called, respectively, the <u>degree of monotone approximation</u> to f and the <u>degree of monotone approximation</u> to S.

In this notation, Theorem 2 may be stated as follows: <u>If F is continuous and piecewise monotone on [a,b], then</u>

$$\lim_{n \to \infty} E^*_n(f) = 0.$$

<u>Monotone Approximation</u>

The first estimates of the degree of monotone approximation are due to O. Shisha [17]:

<u>Theorem 3:</u> (Shisha) Let $f \in C^k[a,b]$ and let $f^{(k)}(x) \geq 0$ on [a,b] and $f^{(p)} \in \text{Lip } 1$, $1 \leq k \leq p$. Then

$$E^*_{n,k}(f) \leq \frac{c_{p,k}}{(n-p)^{p-k+1}}$$

Roulier [15] obtained results that represented some improvement over Shisha's in certain cases where $k = p \geq 2$. Lorentz [4] streamlined Theorem 3, as follows.

Theorem 3': Let $f \varepsilon C^k[a,b]$ and let $f^{(k)}(x) \geq 0$ on $[a,b]$. Then

$$E^*_{n,k}(f) \leq \frac{2}{k!} (b-a)^k \; E_{n-k} (f^{(k)})$$

where $E_n(f)$ denotes the ordinary (unrestricted) degree of approximation to f.

For the case where $f \varepsilon C[a,b]$ is not assumed to be differentiable, but simply monotone, Roulier [15] obtained first estimates on $E^*_{n,1}(f)$ for $f \varepsilon C$. Using the Jackson Kernel, Lorentz and Zeller obtained a very satisfying result:

Theorem 4: (Lorentz and Zeller) If f is nondecreasing on $[a,b]$, then

$$E^*_{n,1}(f) = 0\,[\omega(f;1/n)],$$

where $\omega(f;\delta)$ is the modulus of continuity.

In view of Jackson's Theorem, Theorem 4 implies that, in a classwide sense, monotone approximation is an efficient restriction for the approximation of monotone functions. Lorentz and Zeller [6] have also shown, however, that there do exist particular monotone functions f such that

$$\limsup \frac{E^*_n(f)}{E_n(f)} = \infty \qquad (3)$$

Improvements in Theorem 3 for $k > 1$ analogous to those in Theorem 4 for $k = 1$ have been made, which fall somewhat short of being efficiency results:

<u>Theorem 5:</u> (Passow and Raymon [9]) If $\Delta^k f$ is ≥ 0 on $[a,b]$, Then for every $\varepsilon > 0$

$$E^*_{n,k}(f) = 0\ [\omega(f;1/n^{1-\varepsilon})].$$

If $f(x)$, in addition to being monotone, also satisfies certain smoothness conditions, the ordinary degree of approximation is improved. What of the degree of monotone approximation? Lorentz ([4] Theorem 12) and Roulier [15] found results for functions in C^1:

<u>Theorem 6:</u> (a) (Lorentz) If $f \varepsilon C^1$ $[a,b]$ and is monotone on this interval, then

$$E^*_{n,1}(f) = 0\ [\frac{1}{n}\ \omega(f';1/n)].$$

(b) (Roulier) If, in addition, $f' > 0$ on $[a,b]$,

$$E^*_{n,1}(f) = 0\ [E_n(f)].$$

(See, also, Zeller [20] and DeVore [1] for a further discussion of this problem.) For functions in C^m, R. DeVore [2] has recently extended Theorem 6(a). The proof is not trivial:

<u>Theorem 7:</u> If $f \varepsilon C^m[a,b]$ and f is monotone on this interval, then for $n > m+1$

$$E^*_{n,1}(f) = 0\ [\frac{1}{n^m}\ \omega(f^{(m)};\ 1/n)].$$

Comonotone Approximation

For f piecewise monotone on [a,b] it has been found, with the use of the Jackson Kernel, that "nearly comonotone" approximation is an efficient restriction on the approximation to f. A sequence of polynomials $\{p_n\}$ is said to approximate f nearly comonotonely if p_n is comonotone with f except on intervals I_{nj} centered at the turning points x_j of f, such that the lengths of these intervals approach 0 as $n \to \infty$.

Theorem 8: (Newman, Passow, Raymon [8]) If f is piecewise monotone on [a,b], there is a sequence $\{p_n\}$ of polynomials nearly comonotone with f such that

$$||f-p_n|| = 0\ [\omega(f;1/n)].$$

Originally, Theorem 8 was limited to functions f such that $|\ [f(x+h)-f(x)]/h\ | \geq \delta > 0$. DeVore [1] observed that this limitation is unnecessary. (See also Roulier [16]). Since kernels may be interpreted as weighted averages, it is not in general possible to obtain exact comonotone approximation results by the direct application of any kernel. Thus, in view of the well known representation theorems, comonotone approximation will not result from any positive linear operator. Using an argument that, essentially, combined the methods of Wolibner and Shisha, first estimates on the degree of comonotone approximation $E^*_{n,k}(f)$ for $f \varepsilon C^k$, $k \geq 2$, were obtained by Passow, Raymon and Roulier:

<u>Theorem 9:</u> If $f \varepsilon C^k[a,b]$, $k \geq 2$, is piecewise monotone with p turning points and if $f^{(k)} \varepsilon$ Lip 1 [a,b], Then

$$E^*_{n,1}(f) = 0\ (\frac{1}{n^{k-p-1}})$$

The following Theorem extends Theorem 4, with some loss of efficiency.

<u>Theorem 10:</u> (Passow and Raymon [9]) If f is piecewise monotone on [a,b], for any $\varepsilon > 0$

$$E^*_{n,1}(f) = 0\ [\omega(f;1/n^{1-\varepsilon})].$$

In a forthcoming PhD. dissertation, D. Myers improved the above estimate for a restricted class of functions. In particular, her results imply the following theorem:

<u>Theorem 11:</u> If L(x) is a piecewise linear continuous function on [a,b], Then

$$E^*_{n,1}(L) = 0\ (\frac{1}{n}).$$

Quantitative estimates on the degree of comonotone approximation $E^*_{n,k}$ for $k > 1$ have, for the most part, not yet been obtained. However, if a "nearly coconvex" sequence of polynomials is defined analogously to "nearly comonotone", a theorem analogous to Theorem 8 is obtained:

Theorem 12: (Myers) If f is piecewise convex on [a,b], there is a sequence $\{p_n\}$ of polynomials nearly coconvex with f such that

$$||f-p_n|| = 0\ [\omega(f;1/n)].$$

For k = 0, i.e., for "copositive" approximation, Passow and Raymon [10] have obtained the following efficiency theorem for "proper" piecewise monotone functions. A piecewise monotone function f is called proper if $f\ (x_i) \neq 0$ at relative maxima and minima x_i, $i = 1,\ldots,k$ of f.

Theorem 13: Let $f \in C[a,b]$ be a proper piecewise monotone function on this interval. Then

$$E^*_{n,0}(f) = 0[\omega(f;1/n)].$$

The Relation to The Degree of Piecewise Monotone Interpolation

Let $X = \{x_i\}_0^k$ where $0 = x_0 < x_1 < \ldots < x_k = 1$, and let $Y = \{y_i\}_0^k$ be real numbers such that $y_{i-1} \neq y_i$, $i = 1,\ldots,k$. A polynomial p(x) with the properties

(i) $p(x_i) = y_i$, $i = 0,\ldots,k$,

(ii) p(x) is monotone on (x_{i-1},x_i), $i = 1,\ldots,k$,

is said to interpolate Y at X piecewise monotonely; in case $y_i > y_{i-1}$ for all i, p(x) is said to interpolate monotonely. By Theorem 1, such a polynomial always exists. The smallest degree of a polynomial that interpolates Y at X (piecewise) monotonely is called the degree of (piecewise) monotone interpolation of Y with respect to X, and is denoted by N = N(X,Y). Let

$$\Delta = \Delta(Y) = \min_{1 \leq i \leq k} |y_i - y_{i-1}|;$$

$$M = M(X,Y) = \max_{1 \leq i \leq k} \left|\frac{y_i - y_{i-1}}{x_i - x_{i-1}}\right| .$$

Using Theorem 10 together with a convexity argument, the following result is obtained:

Theorem 14: (Passow and Raymon [11]) (a) For any $\varepsilon > 0$ there is a constant C, depending on ε and on the number of changes in monotonicity of the piecewise linear function joining the points of interpolation (x_i, y_i), i = 0,...,k of Y with respect to X, such that

$$N(X,Y) \leq C \left(\frac{M}{\Delta}\right)^{1+\varepsilon}.$$

(b) If the piecewise linear function in (a) is monotone, then

$$N(X,Y) \leq C\, M/\Delta.$$

Here the interconnection between Question 1 and Question 2 is readily seen. Theorem 14(b) is equivalent to Theorem

4. An improvement in the estimate in Theorem 10 or the extension of Theorem 11 to a classwide result (i.e., $E^*_{n,1}(S)$ where S is the set of all functions of a given modulus of continuity) would yield an improved estimate in Theorem 14(a), and conversely. When considering the questions of comonotone approximation and piecewise monotone interpolation by "Müntz polynomials," the two questions become inextricably entwined: We are unable to prove one theorem without the other. In these Theorems (15-18), we let $\Lambda = \{0 = \lambda_0, \lambda_1, \lambda_2, \ldots\}$ be a sequence of distinct nonnegative real numbers with the properties: (i) $\lim \lambda_i = \infty$, (ii) $\sum_1^\infty 1/\lambda_i$ diverges. Then $\sum_{i=0}^n a_i x^{\lambda_i}$ will be called a <u>Müntz polynomial with respect to Λ</u>.

<u>Theorem 15</u>: Let f be piecewise positive on [0,1]. Then for any Λ and for any $\varepsilon > 0$ there is a Müntz polynomial p with respect to Λ, copositive with f on [0,1], and such that $||f-p|| < \varepsilon$.

<u>Theorem 16</u>: Let f be piecewise monotone on [0,1]. Then for any Λ and for any $\varepsilon > 0$ there is a Müntz polynomial p with respect to Λ, comonotone with f on [0,1], and such that $||f-p|| < \varepsilon$.

<u>Theorem 17:</u> Let $X = \{0 = x_0 < x_1 < \ldots < x_m = 1\}$

$Y = \{y_0,\ldots,y_m\}$ with $y_{i-1} \neq y_i$, $i = 1,\ldots,m$. Then for any Λ there is a Müntz polynomial with respect to Λ that interpolates Y with respect to X piecewise monotonely.

Let $T15_k$ denote Theorem 15 where f has k changes of sign on on [0,1], k = 0,1,...,; let $T16_k$ denote Theorem 16 where f has k changes of monotonicity on [0,1]; let $T17_k$ denote Theorem 17 where the piecewise linear function joining the points $(x_0,y_0),\ldots,(x_m,y_m)$ has k changes of monotonicity on [0,1]. Theorems 15-17 then follow immediately and simultaneously from the following sequence of Lemmas:

<u>Lemma 1</u>: $T15_0$.

<u>Lemma 2</u>: $T15_k \Rightarrow T16_k$, $k = 0,1,\ldots$.

<u>Lemma 3</u>: $T16_k \Rightarrow T17_k$, $k = 0,1,\ldots$.

<u>Lemma 4</u>: $T17_k \Rightarrow T15_{k+1}$, $k = 0,1\ldots$.

These Theorems, as originally proved by Passow and Raymon [13] had an additional hypothesis that $1,2,\ldots k \in \Lambda$. In a private communication, D. Leviatan pointed out that this hypothesis was unnecessary. With regard to monotone approximation by Müntz polynomials with higher order differences, there is the following theorem.

<u>Theorem 18:</u> Let $f \in C[0,1]$ with $\Delta^k f > 0$ in [0,1]. Suppose $1,\ldots,k \in \Lambda$. Then for any $\varepsilon > 0$ there is a Müntz polynomial

p with respect to Λ with $p^{(k)}(x) \geq 0$ on [0,1] such that $||f-p|| < \varepsilon$.

Quantitative estimates on the degree of approximation have been obtained for Theorem 18 by Shisha [13], and for Theorem 16 by Myers [7].

An analysis of the results presented here reveals many gaps yet to be filled. Many of these, it would seem, will require new methods. Other questions of interest would involve simultaneous restrictions. For example, a comonotone coconvex approximation to a function is also a geometric or visual approximation to the function. The following is a theorem of this genre [3].

Theorem 19: Let f be a continuous piecewise monotone function with a finite number of zeros on [a,b]. Then for any $\varepsilon > 0$ there is a polynomial p copositive an comonotone with f such that $||f-p|| < \varepsilon$.

REFERENCES

[1] DeVore, R., Degree of Monotone approximation, in: Linear Operators and Approximation II, ISNM 25, Birkhäuser Verlag, Basel and Stuttgart, 1974, pp. 337-351.

[2] DeVore, R., Monotone approximation by polynomials, to appear.

[3] Hill, Passow, Raymon, Approximation with interpolatory constraints, Illinois J. Math., to appear.

[4] Lorentz, G.G., Monotone approximation. Proceedings of the 3rd Symposium on Inequalities, ed. O. Shisha, Academic Press, 1964, pp. 201-215.

[5] Lorentz, G.G. and K.L. Zeller, Degree of approximation by monotone polynomials I, J. Approximation Theory 1, (1968), pp. 501-504.

[6] Lorentz, G.G. and K.L. Zeller, Degree of approximation by monotone polynomials II, J. Approximation Theory 2, (1969), pp. 265-269.

[7] Myers, D., Comonotone and Coconvex Approximation, Thesis, Temple University, in progress.

[8] Newman, D.J., E. Passow and L. Raymon, Piecewise monotone polynomial approximation, Trans. Amer. Math. Soc., 172 (1972), pp. 465-472.

[9] Passow, E., and L. Raymon, Monotone and Comonotone Approximation, Proc. Amer. Math.Soc.,42(1974)pp.390-394.

[10] Passow, E. and L. Raymon, Copositive Polynomial Approximation, J. Approximation Theory,12 (1974), pp. 299-304.

[11] Passow, E. and L. Raymon, Degree of Piecewise Monotone Interpolation, Proc. Amer. Math. Soc., 48 (1975), pp. 409-412.

[12] Passow, E., L. Raymon and J. Roulier, Comonotone Polynomial Approximation, J. Approximation Theory 11, (1974), pp. 221-224.

[13] Passow. E., L. Raymon and O. Shisha, Piecewise Monotone Interpolation and Approximation with Müntz polynomials, Trans Amer. Math. Soc., to appear.

[14] Paszkowski, S., On Approximation with nodes, Rozprawy Mat. (1957).

[15] Roulier,J.,Monotone Approximation of certain classes of functions, J.Approximation Theory 1,(1968),pp.319-324.

[16] Roulier, J., Nearly Comonotone Approximation, Proc. Amer. Math. Soc., 47 (1975), pp.84-88.

[17] Shisha, O., Monotone Approximation, Pacific J. Math.,15 (1965), pp.667-671.

[18] Wolibner, W., Sur un polynome d' interpolation, Colloq. Math. 2, (1951), pp.136-137.

[19] Young, S.W., Piecewise monotone polynomial interpolation, Bull. Amer. Math. Soc., 73 (1967), pp.642-643.

[20] Zeller, K.L., Monotone Approximation, in Approximation Theory, ed. G.G. Lorentz, Academic Press, 1973, pp.523-525.

Angular Overconvergence for Rational Functions Converging Geometrically on $[0, +\infty)$

E. B. Saff[1] and R. S. Varga[2]

Dedicated to the memory of our teacher Prof. J. L. Walsh

§1. Introduction.

The classical results of Bernstein, Walsh, Gončar, and others concerning the overconvergence of rational functions are roughly of the following type (cf. [19]): It is assumed that

(i) $f(z)$ is defined (finite) on some compact set E in the complex plane $\mathbb{C}$;

(ii) $\{r_n(z)\}_{n=1}^{\infty}$ is a sequence of rational functions of respective degrees n which converge geometrically to f on E, i.e.,

$$\overline{\lim_{n\to\infty}}\{\|f-r_n\|_{L_\infty(E)}\}^{1/n} < 1;$$

and

(iii) the set of poles of the sequence $\{r_n(z)\}_{n=1}^{\infty}$ has no accumulation points on E.

It is then concluded that

(iv) the sequence $\{r_n(z)\}_{n=1}^{\infty}$ converges geometrically to an analytic extension of f on some open set in the plane containing E.

The aim of the present paper is to investigate the phenomenon of overconvergence in the case where E is a closed line segment $[a,b]$ and the hypothesis (iii) above

---

1) Research supported in part by the Air Force Office of Scientific Research under Grant AFOSR-74-2688, and by the University of South Florida Research Council.

2) Research supported in part by the Air Force Office of Scientific Research under Grant AFOSR-74-2729. and by the Energy Research and Development Administration (ERDA) under Grant E(11-1)-2075.

is weakened to allow accumulation points of poles at the endpoints of E, i.e., assumption (iii) is replaced by

(iii)' the set of poles of the sequence $\{r_n(z)\}_{n=1}^{\infty}$ has no accumulation points on the open subinterval (a,b) of E = [a,b].

Of course with the hypotheses (i), (ii), and (iii)', we must modify conclusion (iv) to read

(iv)' the sequence $\{r_n(z)\}_{n=1}^{\infty}$ converges geometrically to an analytic extension of f on some open set in the plane containing (a,b).

For the precise statements of such results on "angular overconvergence" it is sufficient to take $E = [0, +\infty)$, because any interval [a,b] can be mapped onto $[0, +\infty)$ by means of a bilinear transformation, and such bilinear transformations preserve rational functions of degree n. For example, one of the results which we prove asserts that if rational functions $r_n(z)$ of respective degrees n converge geometrically on $E = [0, +\infty)$, and the poles of the $r_n(z)$ lie outside an infinite sector of the form

$$\{z \in \mathbb{C}: |\arg z| < \phi_1\}, \quad 0 < \phi_1 \leq \pi,$$

then the $r_n(z)$ converge geometrically on some smaller infinite sector

$$\{z \in \mathbb{C}: |\arg z| < \phi_2\}, \quad 0 < \phi_2 < \phi_1.$$

It is important to note that a number of results have appeared in the literature ([8], [10], [11]) which give classes of functions f and examples of approximating rational functions $r_n(z)$ for which condition (ii) above is satisfied on $E=[0, +\infty)$. Furthermore, for some special sequences of approximating rational functions, the existence of pole-free open sets (in the plane) containing $(0, +\infty)$ follows from the results in [18], [12], [13], among others. Hence the main results of this paper, which we state in Section 2, have immediate applications. These applications will be discussed primarily in Section 3.

§2. Statements of Main Results.

We now introduce the necessary notation and state our main results. Their proofs will be given in [14].

For an arbitrary set A in the complex plane $\mathbb{C}$, we denote by $\|\cdot\|_A$ the sup norm on A, i.e.,

$$\|f\|_A := \sup \{|f(z)| : z \in A\}.$$

We use the symbol π_n to denote the set of all complex polynomials in the variable z having degree at most n, and let $\pi_{n,n}$ denote the set of all complex rational functions $r_n(z)$ of the form

$$r_n(z) = \frac{p_n(z)}{q_n(z)}, \text{ where } p_n \in \pi_n,\ q_n \in \pi_n,\ q_n \not\equiv 0.$$

The first three results which we state concern pole-free regions whose boundaries are tangent to the ray $E=[0,+\infty)$ at $x=+\infty$. It is convenient in this regard to introduce the set $\mathcal{H}$ which consists of all real nonnegative continuous functions h on $[0,+\infty)$ such that for x large, $h(x)>0$, and $h'(x)$ exists, is nonnegative, and satisfies

$$(2.1) \qquad \lim_{x\to+\infty} h'(x) = 0.$$

Corresponding to each $h \in \mathcal{H}$ we define generically the set $E_s(h)$, $0 \le s \le 1$, in the complex plane by

$$(2.2) \qquad E_s(h) := \{z = x + iy : x \ge 0 \text{ and } |y| \le sh(x)\}.$$

Notice that, by condition (2.1), the boundary of each set $E_s(h)$ defined in (2.2) makes an angle of zero with the positive real axis at $x=+\infty$.

Our first result is the following:

Theorem 2.1. *Assume that for a function f, defined and finite on $[0,+\infty)$, there exists a sequence of rational functions $\{r_n\}_{n=1}^{\infty}$, with $r_n \in \pi_{n,n}$ for all $n \ge 1$, and a real number $q>1$ such that*

$$(2.3)\quad \overline{\lim_{n\to\infty}}\ \{\|f-r_n\|_{[0,+\infty)}\}^{1/n} < \frac{1}{q} < 1.$$

Assume further that for some function $h \in \mathcal{H}$ the interior of the region $E_1(h)$ (defined in (2.2)) contains no poles of the $r_n(z)$ for all n sufficiently large. Then for every d satisfying the inequality

$$(2.4)\quad 0 < d < \frac{\sqrt{q}-1}{\sqrt{q}+1} < 1,$$

there exists a bounded subset K_d of $E_d(h)$ and an analytic function $F(z)$ on $E_d(h) - K_d$ with $F(x) = f(x)$ for all real x in this set, such that $\{r_n(z)\}_{n=1}^{\infty}$ converges geometrically to $F(z)$ on $E_d(h) - K_d$. Moreover

$$(2.5)\quad \overline{\lim_{n\to\infty}}\{\|F-r_n\|_{E_d(h)-K_d}\}^{1/n} < \frac{1}{q}\cdot\left(\frac{1+d}{1-d}\right)^2 < 1.$$

The next result shows that in certain cases the conclusion of Theorem 2.1 can hold on the whole set $E_d(h)$, rather than on $E_d(h) - K_d$.

Corollary 2.2. Assume that for a continuous function $g(\not\equiv 0)$ on $[0,+\infty)$ there exists a sequence of polynomials $\{p_n\}_{n=1}^{\infty}$, with $p_n \in \pi_n$ for all $n \geq 1$, and a real number $q > 1$ such that

$$(2.6)\quad \overline{\lim_{n\to\infty}}\{\|\frac{1}{g} - \frac{1}{p_n}\|_{[0,+\infty)}\}^{1/n} \leq \frac{1}{q} < 1.$$

Then, as is known [7, Theorem 3], there exists an entire function $G(z)$ of finite order with $G(x) = g(x)$ for all $x \geq 0$. Next assume that for some function $h \in \mathcal{H}$, with $h(x) > 0$, for all $x > 0$, the interior of the region $E_1(h)$ (defined in (2.2)) contains no zeros of $p_n(z)$ for all n large. If d satisfies (2.4) and if G is nonzero on the vertical segment $\{z=iy : |y| \leq dh(0)\}$, then

$$(2.7)\quad \overline{\lim_{n\to\infty}}\ \{\|\frac{1}{G}-\frac{1}{p_n}\|_{E_d(h)}\}^{1/n} \le \frac{1}{q}(\frac{1+d}{1-d})^2 < 1.$$

As a concrete application of Corollary 2.2, we first recall from Meinardus and Varga [8] that

$$(2.8)\quad \lim_{n\to\infty}\{\|e^{-x}-\frac{1}{s_n(x)}\|_{[0,+\infty)}\}^{1/n} = \frac{1}{2},$$

where $s_n(z) = \sum_{k=0}^{n} z^k/k!$ denotes the familiar n-th partial sum of e^z. It is further known from Saff and Varga [12] that for

$$(2.9)\quad \hat{h}(x) := 2(x+1)^{1/2},\ x \ge 0,$$

the region

$$(2.10)\quad E_1(\hat{h}) = \{z = x+iy:\ x \ge 0,\ |y| \le 2(x+1)^{1/2}\}$$

contains no zeros of the $s_n(z)$ for all n. Note that $\hat{h} \in \mathcal{H}$, and that with $G(z) = e^z$ (so that G is nonzero at every finite point z), with $p_n \equiv s_n$ for all $n \ge 1$, and with $q = 2$, the hypotheses of Corollary 2.2 are all fulfilled. Thus for any d satisfying $0 < d < (\sqrt{2}-1)/(\sqrt{2}+1)$, we have from (2.7) that

$$(2.11)\quad \overline{\lim_{n\to\infty}}\ \{\|e^{-z}-\frac{1}{s_n(z)}\|_{E_d(\hat{h})}\}^{1/n} \le \frac{1}{2}(\frac{1+d}{1-d})^2 < 1,$$

which is effectively the result of [11, Theorem 4.1]. We remark that for any $d > 0$ the set

$$(2.12)\quad E_d(\hat{h}) = \{z = x+iy:\ x \ge 0,\ |y| \le 2d(x+1)^{1/2}\}$$

is an unbounded parabolic region truncated at the origin.

As a consequence of Corollary 2.2 and of the results in [12], similar overconvergence results in unbounded parabolic regions also hold for each column of the Padé table for e^{-z}, i.e., for the Padé approximants $\{R_{\nu,n}(z)\}_{n=1}^{\infty}$ where the degree, ν, of the numerator is fixed.

Applications of Corollary 2.2 can in fact be made to a certain class of entire functions which contains the above example, and this will be described in the next section.

From Corollary 2.2 it is possible to deduce the following result which concerns geometric convergence on related unbounded sets whose widths grow more slowly at infinity.

<u>Corollary</u> 2.3. <u>With the hypotheses of Corollary</u> 2.2, <u>assume that</u> $c(x)$ <u>is a nonnegative continuous function on</u> $[0,+\infty)$ <u>with</u> $c(x) < h(x)$ <u>for all</u> $x > 0$, <u>such that</u>

$$(2.13)\qquad \lim_{x\to+\infty} \frac{c(x)}{h(x)} = 0,$$

<u>and let</u>

$$(2.14)\qquad \mathcal{C} := \{z = x+iy:\ x \geq 0,\ |y| \leq c(x)\}.$$

<u>If</u> G <u>is nonzero on the segment</u> $\{z = iy:\ |y| \leq c(0)\}$, <u>then</u>

$$(2.15)\qquad \overline{\lim_{n\to\infty}} \left\{\left\|\frac{1}{G} - \frac{1}{p_n}\right\|_{\mathcal{C}}\right\}^{1/n} = \overline{\lim_{n\to\infty}} \left\{\left\|\frac{1}{g} - \frac{1}{p_n}\right\|_{[0,+\infty)}\right\}^{1/n}.$$

The remaining results concern overconvergence on regions having a positive angle at infinity. In stating them it is convenient to introduce the sets $S(\theta,\mu)$ and $S(\theta)$ defined by

(2.16) $S(\theta,\mu) := \{z : |\arg z| < \theta,\ |z| > \mu\}$,

(2.17) $S(\theta) := \{z : |\arg z| < \theta\}$.

<u>Theorem</u> 2.4. <u>Assume that for a function</u> f, <u>defined and finite on</u> $[0,+\infty)$, <u>there exists a sequence of rational functions</u> $\{r_n\}_{n=1}^{\infty}$, <u>with</u> $r_n \in \pi_{n,n}$ <u>for all</u> $n \geq 1$, <u>and a real number</u> $q > 1$ <u>such that</u>

$$(2.18)\quad \overline{\lim_{n\to\infty}} \{\|f-r_n\|_{[0,+\infty)}\}^{1/n} < \frac{1}{q} < 1.$$

<u>Assume further that for some</u> θ_o <u>and</u> μ_o, <u>with</u> $0 < \theta_o \leq \pi$, $\mu_o > 0$, <u>the region</u> $S(\theta_o,\mu_o)$ <u>(defined in</u> (2.16)) <u>contains no poles of the</u> $r_n(z)$ <u>for all</u> n <u>large</u>. <u>Then for every</u> θ <u>satisfying the inequality</u>

$$(2.19)\quad 0 < \theta < 4\tan^{-1}\left\{\left(\frac{\sqrt{q}-1}{\sqrt{q}+1}\right)\cdot \tan\left(\frac{\theta_o}{4}\right)\right\},$$

<u>there exists a</u> $\mu = \mu(\theta) > 0$ <u>and an analytic function</u> $F(z)$ <u>on the closure</u> $\overline{S}(\theta,\mu)$ <u>with</u> $F(x) = f(x)$ <u>for all real</u> x <u>in this set, such that</u> $\{r_n(z)\}_{n=1}^{\infty}$ <u>converges geometrically to</u> $F(z)$ <u>on</u> $\overline{S}(\theta,\mu)$. <u>Moreover</u>

$$(2.20)\quad \overline{\lim_{n\to\infty}}\{\|F-r_n\|_{\overline{S}(\theta,\mu)}\}^{1/n} < \frac{1}{q}\cdot\left\{\frac{\sin[\frac{1}{4}(\theta_o+\theta)]}{\sin[\frac{1}{4}(\theta_o-\theta)]}\right\}^2 < 1.$$

It is interesting to note that while Theorem 2.1 cannot be deduced from Theorem 2.4, the former result can be considered as a limiting case of the latter. Indeed, for the situation of Theorem 2.1, we regard θ_o and θ as functions of x which tend to zero as $x\to+\infty$; specifically, we define θ_o and θ by the equations

$$\tan \theta_o = \frac{h(x)}{x}, \quad \tan \theta = \frac{dh(x)}{x}.$$

Then, on writing (2.19) in the equivalent form

$$\frac{\tan(\theta/4)}{\tan(\theta_o/4)} < \frac{\sqrt{q}-1}{\sqrt{q}+1},$$

and taking the limit as $x \to +\infty$, we derive the condition

$$\lim_{x\to+\infty} \frac{\tan(\theta/4)}{\tan(\theta_o/4)} = \lim_{x\to+\infty} \frac{\tan\theta}{\tan\theta_o} = d < \frac{\sqrt{q}-1}{\sqrt{q}+1},$$

which is the same as inequality (2.4) of Theorem 2.1.

Using Theorem 2.4 we can deduce the following analogs of Corollaries 2.2 and 2.3:

Corollary 2.5. *Let the functions* g, G, *and the sequence of polynomials* $\{p_n\}_{n=1}^{\infty}$ *be as in Corollary* 2.2 (*so that, in particular, inequality* (2.6) *holds*). *Assume further that no zeros of* p_n *lie in the infinite sector* $S(\theta_o)$ (*defined in* (2.17)), $0 < \theta_o \le \pi$, *for all* n *sufficiently large, and that* $g(0) \neq 0$. *If* θ *satisfies* (2.19), *then on the closure* $\overline{S}(\theta)$,

$$\overline{\lim_{n\to\infty}} \left\{\left\|\frac{1}{G} - \frac{1}{p_n}\right\|_{\overline{S}(\theta)}\right\}^{1/n} \le \frac{1}{q}\left\{\frac{\sin[\frac{1}{4}(\theta_o+\theta)]}{\sin[\frac{1}{4}(\theta_o-\theta)]}\right\}^2 < 1. \tag{2.21}$$

Corollary 2.6. *Let the functions* g, G, *and the sequence of polynomials* $\{p_n\}_{n=1}^{\infty}$ *be as in Corollary* 2.2 (*so that, in particular, inequality* (2.6) *holds*). *Assume that no zeros of* p_n *lie in* $S(\theta_o)$, $0 < \theta_o \le \pi$, *for all* n *sufficiently large, and that* $g(0) \neq 0$. *Then for any nonnegative continuous function* $c(x)$ *on* $[0, +\infty)$ *such that* $c(x) = o(x)$ *as* $x \to +\infty$ *and such that* (i) $c(0) = 0$ *if* $\theta_o = \pi/2$, (ii) $c(x) < x\tan(\theta_o)$ *for* $x > 0$ *if* $0 < \theta_o < \pi/2$, *we have*

$$(2.22)\quad \overline{\lim_{n\to\infty}}\left\{\left\|\frac{1}{G}-\frac{1}{p_n}\right\|_{\mathcal{G}}\right\}^{1/n} = \overline{\lim_{n\to\infty}}\left\{\left\|\frac{1}{g}-\frac{1}{p_n}\right\|_{[0,+\infty)}\right\}^{1/n},$$

where the region $\mathcal{G}$ *is defined as in* (2.14).

If, in Corollary 2.5, we weaken the hypothesis by replacing the reciprocals of polynomials, $1/p_n$, by arbitrary rational functions $r_n \in \pi_{n,n}$ whose poles omit a full sector, then we obtain the following less specific conclusion:

Theorem 2.7. *Assume that for a function* f, *defined and finite on* $[0,+\infty)$, *there exists a sequence of rational functions* $\{r_n\}_{n=1}^{\infty}$, *with* $r_n \in \pi_{n,n}$ *for all* $n \geq 1$, *and a real number* $q > 1$ *such that inequality* (2.18) *holds. Suppose further that the infinite sector* $S(\theta_o)$ *(defined in* (2.17)), $0 < \theta_o \leq \pi$, *contains no poles of the* $r_n(z)$ *for all* n *large. Then there exists a* θ, $0 < \theta < \theta_o$, *and a function* $F(z)$ *analytic on the sector* $S(\theta)$, *continuous on* $\overline{S}(\theta)$, *with* $F(x) = f(x)$ *for all* $x \geq 0$, *such that* $\{r_n(z)\}_{n=1}^{\infty}$ *converges geometrically to* $F(z)$ *on* $\overline{S}(\theta)$.

Theorem 2.7 has an important application to the problem (raised at the International Conference on Approximation Theory, Maryland, 1970) of finding a sequence of rational functions which converges geometrically to e^{-z} in an infinite sector. It is well-known that the sequence $1/s_n(z)$, $s_n(z) = \sum_{o}^{n} z^k/k!$, does not have this property because no infinite sector is devoid of zeros of $s_n(z)$ for all n large (cf. [3] or [15]). However, it is shown by the authors in [11] and [13], that certain sequences of Padé approximants of e^{-z} converge geometrically on $[0,+\infty)$

to e^{-x}, and furthermore have all their poles outside some infinite sector $\{z: |\arg z| < \theta_0\}$. Hence, by Theorem 2.7, such a sequence must converge geometrically to e^{-z} on some infinite sector $\{z: |\arg z| < \theta\}$, $0 < \theta < \theta_0$. The precise details of this application shall be reserved for a later occasion.

The last result of this section concerns rational functions which converge faster than geometrically on $[0, +\infty)$, i.e.,

$$(2.23) \quad \lim_{n\to\infty} \{\|f-r_n\|_{[0,+\infty)}\}^{1/n} = 0.$$

Corollary 2.8. If in Theorem 2.7, the assumption of inequality (2.18) is replaced by (2.23), then the sequence $\{r_n(z)\}_{n=1}^{\infty}$ converges faster than geometrically on every closed sector $\overline{S}(\theta)$, $0 < \theta < \theta_0$, i.e.,

$$(2.24) \quad \lim_{n\to\infty} \{\|F-r_n\|_{\overline{S}(\theta)}\}^{1/n} = 0.$$

§3. Some Applications.

In order to apply results such as Corollaries 2.2 and 2.3 we first need conditions on the entire function $G(z)$ which insure that there exists a sequence of polynomials p_n, with $p_n \in \pi_n$ for all $n \geq 1$, such that

$$(3.1)\quad \overline{\lim_{n\to\infty}} \{\|\frac{1}{G} - \frac{1}{p_n}\|_{[0,+\infty)}\}^{1/n} < 1.$$

Second, we need a specific result, like that of (2.9), which asserts that for an appropriate function $h \in \mathcal{H}$, the interior of the region $E_1(h)$ defined in (2.2) is free of zeros of the polynomials p_n in (3.1) for all n large. Results of both these types are already known for the case where the p_n are the n-th partial sums of the Maclaurin expansion for G. In order to state these results we remind the reader of some standard terminology.

If $g(z) = \sum_{k=0}^{\infty} a_k z^k$ is an entire function, we let $M_g(r) := \max\{|g(z)| : |z| = r\}$ denote its maximum modulus function, and let $\rho = \rho_g$ denote the order of g (for non-constant g), i.e., (cf. [2, p. 8], [16, p. 34])

$$(3.2)\quad \rho = \overline{\lim_{r\to+\infty}} \frac{\ln(\ln M_g(r))}{\ln r}.$$

Furthermore, an entire function $g(z)$ of order ρ, $0 < \rho < \infty$, is said to be of perfectly regular growth (cf. [16, p. 44]) if there exists a real $B > 0$ such that

$$(3.3)\quad 0 < B = \lim_{r\to+\infty} \frac{\ln M_g(r)}{r^\rho}.$$

We remark that if a non-constant entire function g satisfies a linear differential equation with rational function coefficients, then g is necessarily of perfectly regular growth (cf. [16, p. 108]).

We now state a result which gives sufficient conditions for geometric convergence on $[0, +\infty)$.

Theorem 3.1 (Meinardus and Varga [8]). *Let* $g(z) = \sum_{k=0}^{\infty} a_k z^k$ *be an entire function of perfectly regular growth* (ρ, B) *with real nonnegative coefficients* a_k. *Then*

$$(3.4) \quad \lim_{n\to\infty}\{\|\frac{1}{g} - \frac{1}{s_n}\|_{[0,+\infty)}\}^{1/n} = \frac{1}{2^{1/\rho}} < 1,$$

where $s_n(z) = \sum_{k=0}^{n} a_k z^k$ *denotes the* n-th *partial sum of the Maclaurin expansion for* g.

Concerning zero-free regions for the partial sums s_n we state a previously unpublished result from one of the authors' theses [18]. For related published results see [17].

Theorem 3.2. *Let* S *denote the set of all entire functions* $g(z) = \sum_{k=0}^{\infty} a_k z^k$ *for which*

(i) $a_o > 0$ *and* $a_k \geq 0$ *for all* $k \geq 1$;

(ii) *if* $a_m = 0$, *then* $a_{m+2j} = 0$ *for every* $j \geq 1$;

(iii) *if* $K := \{k: a_k > 0$ *and* $a_{k+2} > 0\}$ *is non-empty, then*

$$(3.5) \quad \inf_{k\in K} \{\frac{a_k}{(k+1)(k+2)a_{k+2}}\} > 0.$$

Then, for $g \in S$, *there exists a nondecreasing continuous function* h_g *defined on* $[0, +\infty)$ *with* $h_g(0) > 0$, *such that* $g(z)$ *and all its partial sums* $s_n(z) = \sum_{k=0}^{n} a_k z^k$, $n \geq 1$, *have no zeros in*

$$(3.6) \quad \{z = x+iy: x \geq 0 \text{ and } |y| \leq h_g(x)\}.$$

<u>Moreover, for each</u> $g \in S$, <u>the order</u> ρ_g <u>of</u> g <u>satisfies</u> $0 \le \rho_g \le 1$.

We remark that the set S of Theorem 3.2 contains many familiar elements. For example, $u(z) = e^z$, $v(z) = \cosh(\sqrt{z}) = \sum_0^\infty z^k/(2k)!$, the modified Bessel functions $J_n(iz)/(iz)^n$ for any $n \ge 0$, and the hypergeometric function ${}_1F_1(c;d;z)$ with $c > 0$, $d > 0$, are easily seen to be elements of S.

If $\mathcal{W}_g$ denotes the nonempty (from Theorem 3.2) collection of all positive nondecreasing continuous functions h_g on $[0, +\infty)$ for which $g(z)$ and all its partial sums $s_n(z)$, $n \ge 1$, have no zeros in the region defined by (3.6), then we define the (maximal) <u>width function</u> $H_g(x)$ by

$$(3.7) \quad H_g(x) := \sup\{h_g(x) : h_g \in \mathcal{W}_g\}, \text{ for each } x \ge 0.$$

The function $H_g(x)$ so defined is clearly nondecreasing on $[0, +\infty)$, and $g(z)$ and all its partial sums $s_n(z)$ have no zeros in the interior of the region defined by

$$(3.8) \quad \{z = x + iy : x \ge 0 \text{ and } |y| \le H_g(x)\}.$$

More precisely, it is easily seen that either $H_g(x) = +\infty$ for all x sufficiently large, or that H_g is a step function, i.e., there exists a denumerable set of points $\{z_j = x_j + iy_j\}_{j=1}^\infty$, with $0 \le x_j < x_{j+1}$ and $0 \le y_j \le y_{j+1}$ for all j, such that

$$H_g(x_j) = y_j \, ; \; H_g(x) = y_{j+1} \, , \text{ for } x_j < x \le x_{j+1} ; \; j \ge 1.$$

Here, each z_j is a zero of g or one of its partial sums s_n. Moreover, if g is of order $\rho_g > 0$, then a result of

Carlson [3] states that no proper sector, with vertex at the origin, can be devoid of zeros of the partial sums s_n, for all n large. Consequently, when $\rho_g > 0$, $H_g(x)$ is not only finite for all finite $x \geq 0$, but also $\lim_{j\to\infty} y_j/x_j = 0$.

The next corollary provides lower bounds for $H_g(x)$ for particular elements in $\mathcal{S}$.

Corollary 3.3. *Let $g(z) = \sum_0^\infty a_k z^k$ be an entire function such that $a_k > 0$ for all k and such that*

$$\text{(3.9)} \quad \inf_{k\geq 1} \left\{\frac{a_k}{k^2 a_{k+1}}\right\} > 0.$$

Then $g \in \mathcal{S}$ and its associated width function H_g of (3.7) satisfies, for some constant $c > 0$,

$$\text{(3.10)} \quad H_g(x) \geq cx^{1/2}, \quad \textit{for all } x \geq 0.$$

Proof. It is trivial to verify that $g(z) = \sum_0^\infty a_k z^k \in \mathcal{S}$. Furthermore, it follows from the hypotheses above that the entire function f defined by $f(z) := \sum_0^\infty a_k z^{2k}$ is also in $\mathcal{S}$. Thus, from Theorem 3.2, we can associate with f a continuous nondecreasing function h_f defined on $[0, +\infty)$, with $h_f(0) > 0$, such that f and all its partial sums $S_n(z)$ have no zeros in

$$\mathfrak{F} := \{z = x+iy: x \geq 0 \text{ and } |y| \leq h_f(x)\}.$$

But if $s_n(z)$ denotes the n-th partial sum of $g(z)$, then $s_n(z^2) = S_{2n}(z)$ for all $n = 1, 2, \cdots$, which allows us to relate the corresponding zeros of the partial sums of g with those of f. Thus, defining

$$\mathfrak{G} := \{z^2 : z \in \mathfrak{F}\},$$

then g and all its partial sums s_n have no zeros in $\mathcal{G}$. Now, since $h_f(0) > 0$ and h_f is nondecreasing on $[0, +\infty)$, then evidently

$$\mathcal{G} \supset \{z^2 : z = x+iy,\ x \geq 0 \text{ and } |y| \leq h_f(0)\}.$$

Thus, if H_g is the associated width function for g, the above inclusion implies that

$$H_g(t) \geq 2h_f(0)(t+h_f^2(0))^{1/2} \geq 2h_f(0)t^{1/2}, \text{ for all } t \geq 0,$$

which is the desired result of (3.10). ■

As previously noted, $u(z) = e^z$, of order $\rho_u = 1$, and $v(z) = \cosh(\sqrt{z})$, of order $\rho_v = 1/2$, are elements of the set S, and furthermore each is of perfectly regular growth. Moreover, for $u(z) = e^z$, the authors' result of (2.9) implies that

$$H_u(x) \geq 2(x+1)^{1/2}, \text{ for all } x \geq 0.$$

Also, applying Corollary 3.3 to $v(z) = \cosh(\sqrt{z})$ gives

$$H_v(x) \geq cx^{1/2}, \text{ for all } x \geq 0.$$

However, we believe that this last inequality can be improved. In fact, we conjecture more generally that, for any element $g \in S$ of perfectly regular growth, its associated width function satisfies

$$H_g(x) \geq cx^{(2-\rho_g)/2}, \text{ for all } x \geq 0.$$

As a consequence of Theorems 3.1 and 3.2, which apply to both e^z and $\cosh(\sqrt{z})$, we have the following application of Corollary 2.2.

Corollary 3.4. *For any $g \in S$ of order $\rho > 0$ which is of perfectly regular growth, let H_g be its associated non-*

decreasing width function of (3.7), *and let* $h \in \mathcal{H}$ *be any positive function for which* $h(x) \le H_g(x)$ *for all* $x \ge 0$. *Then for*

$$0 < d < (2^{1/2\rho}-1)/(2^{1/2\rho}+1),$$

we have

$$(3.11) \quad \overline{\lim_{n\to\infty}} \{\|\frac{1}{g} - \frac{1}{s_n}\|_{E_d(h)}\}^{1/n} \le \frac{1}{2^{1/\rho}} (\frac{1+d}{1-d})^2 < 1,$$

where the region $E_d(h)$ *is defined as in* (2.2), *and* $s_n(z)$ *denotes the* n-th *partial sum of* $g(z)$.

Proof. Because $g \in \mathcal{S}$ implies that the Maclaurin coefficients of g are all nonnegative, and because g is assumed to be of perfectly regular growth, then the conclusion (3.4) of Theorem 3.1 is valid. Next, by definition of $H_g(x)$ and the fact that $h(x) \le H_g(x)$ for all $x \ge 0$, it follows that g and all its partial sums s_n have no zeros in the interior of the region $E_1(h)$. Consequently, applying Corollary 2.2, with $q = 2^{1/\rho}$, gives the desired result of (3.11). ∎

We remark that the existence of a function $h \in \mathcal{H}$ satisfying the conditions of Corollary 3.4 is obvious. As a simple example, take h_g of Theorem 3.2 and set $h(x) \equiv h_g(0)$.

Concerning rational approximation to entire functions of order $\rho = 0$, it is shown in [7, Thm. 7] and in [4, Thm. 2] that if g is an entire function of order zero and satisfies certain growth and coefficient restrictions, then

$$(3.12) \quad \lim_{n\to\infty} \{\inf_{p\in\pi_n} \|\frac{1}{g} - \frac{1}{p}\|_{[0,+\infty)}\}^{1/n} = 0.$$

As an illustration of how our techniques apply to such situations, we present

Proposition 3.5. *Let* $g(z) = \sum_{k=0}^{\infty} z^k/a^{k^2}$, *where* $a \geq 2$, *and* *let* $s_n(z) = \sum_{k=0}^{n} z^k/a^{k^2}$. *Then, on every closed sector* $\overline{S}(\theta)$ *(defined in* (2.17)) *with* $0 < \theta < \pi$, *we have*

$$\lim_{n\to\infty} \left\{\left\|\frac{1}{g} - \frac{1}{s_n}\right\|_{\overline{S}(\theta)}\right\}^{1/n^2} = \frac{1}{\sqrt{a}} . \tag{3.13}$$

Of course, for the functions of Proposition 3.5, we see that the conclusion of (3.13) is far stronger, and implies the result of (3.12) as a special case.

The proof of Proposition 3.5 will be given in [14].

References

1. D. Aharonov and J. L. Walsh, "On the convergence of rational functions of best approximation to a meromorphic function", J. Math. Anal. Appl. 40(1972), 418-426.

2. Ralph P. Boas, Entire Functions, Academic Press, New York, 1954.

3. Fritz Carlson, "Sur les fonctions entières", Arkiv for Mathematik, Astronomi O. Fysik, Bd. 35A, No. 14.(1948), 18 pp.

4. P. Erdős and A. R. Reddy, "Rational Chebyshev approximation on the positive real axis", Notices of the Amer. Math. Soc. 21(1974), p. A-159.

5. A. A. Gončar, "On overconvergence of sequences of rational functions", Doklady Akademii Nauk SSSR 141 (1961), 1019-1022.

6. Günter Meinardus, Approximation of Functions: Theory and Numerical Methods. Springer Tracts in Natural Philosophy, Vol. 13, Springer-Verlag, New York, 1967.

7. G. Meinardus, A. R. Reddy, G. D. Taylor, and R. S. Varga, "Converse theorems and extensions in Chebyshev rational approximation to certain entire functions in $[0, +\infty)$", Trans. Amer. Math. Soc. 170(1972), 171-185.

8. Günter Meinardus and Richard S. Varga, "Chebyshev rational approximations to certain entire functions in $[0, +\infty)$", J. Approximation Theory 3(1970), 300-309.

9. G. Pólya and G. Szegö, Aufgaben und Lehrsätze aus der Analysis, Springer-Verlag, Berlin, 1964.

10. J. A. Roulier and G. D. Taylor, "Rational Chebyshev approximation on $[0, +\infty)$", J. Approximation Theory 11(1974), 208-215.

11. E. B. Saff and R. S. Varga, "Convergence of Padé approximants to e^{-z} on unbounded sets", J. Approximation Theory 13(1975), 470-488.

12. ___________, "Zero-free parabolic regions for sequences of polynomials", SIAM J. Math. Anal. (to appear).

13. ________, "On the zeros and poles of Padé approximants to e^z", Numer. Math. (to appear).

14. ________, "Geometric overconvergence of rational functions in unbounded regions", Pacific J. Math. (to appear).

15. G. Szegö, "Über eine Eigenschaft der Exponentialreihe", Berlin Math. Ges. Sitzunsber., 23(1924), 50-64.

16. Georges Valiron, Lectures on the General Theory of Integral Functions, Chelsea Publishing Co., New York, 1949.

17. Richard S. Varga, "Semi-infinite and infinite strips free of zeros", Rendiconti del Seminario Matematico Università e Politechnico di Tornio 11(1952), 289-296.

18. Richard Steven Varga, "Properties of a special set of entire functions and their respective partial sums", Ph.D. Thesis, Harvard University, 1954.

19. J.L. Walsh, Interpolation and Approximation by Rational Functions in the Complex Domain, Colloq. Publication Vol. 20, Amer. Math. Soc. Providence, R.I., fifth ed., 1969.

A GENERALIZATION OF MONOSPLINES AND PERFECT SPLINES

A. Sharma and J. Tzimbalario

1. <u>INTRODUCTION</u>. Let m be a natural number and let $r = -1,0,\cdots,m-1$. Then the class $\mathcal{S}_m^r = \{S(x)\}$ consisting of all functions with the following two properties is called the class of cardinal perfect splines in the sense of Glaeser [2]:

(1.1) $S(x) \in C^r(-\infty,\infty)$

(1.2) $S(x)$ restricted to $[\nu,\nu+1]$ represents a polynomial of degree m with the highest term $= (-1)^\nu x^m$.

Schoenberg and Cavaretta independently propose the problem of determining $S(x) \in \mathcal{S}_m^r$ which has the least Tchebycheff norm

$$\|S(x)\| = \sup_{x \in \mathbb{R}} |S(x)|.$$

The class M_m^r of cardinal monosplines of Schoenberg and Ziegler is similar to the class of perfect splines but differs from it in an essential way. $S(x) \in M_m^r$ $(r = -1,0,\cdots,m-2)$ if

(1.3) $$S(x) \in C^r(-\infty,\infty)$$

and

(1.4) $S(x)$ restricted to $[\nu,\nu+1]$ represents a polynomial of degree m with the highest degree term $= x^m$.

Our object in this note is to define a class of cardinal splines which we call t-perfect for a given real number t. We shall say that $S(x) \in \mathcal{S}^{s}_{m,t}$, for $s = -1,0,\cdots,m-1$, if the following two conditions are satisfied:

(1.5) $S(x) \in C^{s}(-\infty,\infty)$

(1.6) $S(x)$ restricted to $[\nu,\nu+1]$ represents a polynomial of degree m with the highest degree term $= t^{\nu}x^{m}$.

For $s = -1$, there are no continuity requirements between the polynomial components in successive intervals. We propose the following problem:

<u>Problem</u>. To determine $S(x) \in \mathcal{S}^{s}_{m,t}$ having the least t-norm $\|S\|_{t,\infty}$ where

(1.7) $$\|S(x)\|_{t,\infty} = \sup_{x \in \mathbb{R}} \left|\frac{S(x)}{t^{[x]}}\right| .$$

For $t = 1$, this reduces to the problem of cardinal monosplines and for $t = -1$ to the problem of cardinal perfect splines.

For $t = 1$, the solution of the problem is related to Bernoulli-Tchebycheff polynomials, while for $t = -1$, to Euler-Tchebycheff polynomial. For $t \neq \pm 1$, we need the Euler exponential polynomials.

In §2, we give a necessary condition for $F^{*}(x)$ to be an extremal solution of the problem. We study two special cases when $s = -1$ and $s = m-1$ in §3. In §4 we solve the problem when $\frac{m-2}{2} \leq s \leq m-2$ and $\operatorname{sign} t = (-)^{m-s}$. We devote §5 to solving the problem for $s = 0$ and

$\operatorname{sgn} t = (-)^m$, where our results of §4 do not apply.

2. A PROPERTY OF THE EXTREMAL SOLUTIONS. We shall need the following generalization of Lemma 4 of [6] and Lemma 3 of [1].

Theorem 1. *If* $F(x) \in \mathcal{S}^s_{m,t}$ *with finite t-norm* ρ, *then there exists another element* $F^*(x) \in \mathcal{S}^s_{m,t}$ *such that*

$$\|F^*\|_{t,\infty} \leq \rho$$

and

$$F^*(x+1) = tF^*(x) .$$

The proof follows the same argument as in [6] and is therefore left out.

Elsewhere in [7] we give a generalization of these results to L-splines. However the method used here seems to be of independent interest.

3. TWO SPECIAL CASES.

(I) When $s = -1$, there are no continuity requirements at the knots. Then in $[0,1]$ the optimal function $F(x)$ will coincide with the polynomial

$$t_m(x) = 2^{-2m+1} T_m(2x-1) = x^m + \cdots, \quad 0 < x < 1$$

and for any interval $(\nu, \nu+1)$,

$$F(x) = t^{\nu} t_m(x-\nu),$$

where $T_m(x) = \cos m\theta,\ x = \cos\theta$.

(II) When $t = 1$ and $s = m-1$ then the problem is void so it is enough to take $t \neq 1$ when $s = m-1$. We recall now that the exponential Euler polynomials $A_m(x;t)$ as given by Schoenberg [5] are:

$$A_m(x;t) = x^m + \binom{m}{1}a_1x^{m-1} + \cdots + a_m .$$

Also $A_m(x;t)$ is the unique monic polynomial satisfying

(3.1) $A_m^{(\nu)}(1;t) = tA_m^{(\nu)}(0;t),\ \nu = 0,1,\cdots,m-1,\ t \neq 1.$

Set

(3.2)
$$F^*(x) = A_m(x;t),\ 0 \leq x < 1,$$
$$F^*(x+1) = tF^*(x),\ x \notin [0,1).$$

Then it is easy to see that $F^*(x) \in \mathcal{S}_{m,t}^{m-1}$. It follows from a theorem of Schoenberg [5] that $F^*(x)$ is the unique t-perfect spline $\in \mathcal{S}_{m,t}^{m-1}$ satisfying

(3.3) $$F^*(x+1) = tF^*(x).$$

We shall now prove

<u>Theorem 2</u>. <u>If</u> $m \geq 1$, $s = m-1$, $t \neq 1$, <u>then</u> $F^*(x)$ <u>given by</u> (3.2) <u>is the unique spline</u> $\in \mathcal{S}_{m,t}^{m-1}$ <u>which minimizes the t-norm, that is</u>

$$\|S(x)\|_{t,\infty} = \sup_{x \in \mathbb{R}} \left|\frac{S(x)}{t^{[x]}}\right| .$$

<u>Proof</u> (a). We shall first prove it for $t < 0$. Suppose there exists $G(x) \in \mathcal{S}_{m,t}^{m-1}$ such that

(3.4) $$\|G\|_{t,\infty} \leq \|F^*\|_{t,\infty} .$$

Then $S(x) = F^*(x) - G(x)$ is a polynomial of degree $m-1$ on the whole real line since $S(x) \in C^{m-1}(\mathbb{R})$. By (3.3), $F^*(x)$ has an oscillatory behaviour in successive intervals and hence $S(x)$ must have infinitely many zeros. Hence $S(x) \equiv 0$ which proves the result.

(b). Assume $t > 1$. (If $0 < t < 1$, a similar argument can be used). We assume again that (3.4) holds. The coefficients a_k of the polynomial $A_m(x,t)$ are all positive, since

$$a_k = a_k(t) = \frac{\pi_k(t)}{(t-1)^k} .$$

where the polynomial $\pi_k(t)$ is the Euler-Frobenius polynomial whose zeros are all negative. It follows that $A_m(x;t) > 0$ for $0 < x < 1$. Hence $F^*(x) > 0$ for all x. If $\frac{F^*(x)}{t^{[x]}}$ attains its maximum ρ in $[0,1]$ at x_0, then since this function is periodic with period 1, we have

$$\frac{F^*(x_0-\nu)}{t^{[x_0-\nu]}} = \rho.$$

By (3.4), we have

$$2\rho \geq \frac{F^*(x_0-\nu) - G(x_0-\nu)}{t^{[x_0-\nu]}} \geq 0.$$

As $\nu \to \infty$, $t^{[x_0-\nu]} \to 0$ since $t > 1$. Hence $F^*(x_0-\nu) - G^*(x_0-\nu)$ approaches zero as $\nu \to \infty$, which is impossible since $F^*(x) - G^*(x)$ is a polynomial.

For $0 < t < 1$, we recall the relation

$$A_m\left(x;\frac{1}{t}\right) = (-1)^n A_n(1-x;t)$$

which reduces the case $0 < t < 1$ to the case $\frac{1}{t} > 1$ with x replaced by $1-x$.

This completes the proof of the theorem.

4. <u>RESULTS FOR</u> $\frac{n-2}{2} \leq s \leq n-2$. We shall first need a result that a subset of the sequence $\{A_m(x;t)\}$ forms a Tchebycheff set. More precisely we have

<u>Theorem 3</u>. <u>If</u> $t \neq 0,1$, $n \geq 2r-3$, <u>and</u> $\operatorname{sgn} t = (-1)^r$, <u>then the polynomials</u>

$$(4.1) \qquad A_n(x,t),\ A_{n-1}(x;t),\cdots,A_{n-r+2}(x;t)$$

form a Tchebycheff set on $[0,1]$.

We shall prove this theorem later. As a consequence of this theorem it follows that when $s \geq \frac{n-2}{2}$ and $\operatorname{sgn} t = (-1)^{n-s}$ there exists a unique polynomial $E_{n,s}(x)$ with

$$(4.2) \qquad E_{n,s}(x) = A_n(x;t)+C_{n-s-1}A_{n-1}(x;t)+\cdots+C_1A_{s+1}(x;t)$$

such that it has the minimum sup norm on $[0,1]$ among all polynomials of this form. The existence and uniqueness of this polynomial follows from the above theorem.

Corollary. There exist $n-s$ *points*

$$0 \leq x_1 < \cdots < x_{n-s} \leq 1$$

such that $E_{n,s}(x_\nu)$ *assumes the value* $\|E_{n,s}(x)\|_\infty$ *with alternating sign.*

We now define the spline $E^*_{n,s}(x)$ defined by

$$E^*_{n,s}(x) = E_{n,s}(x), \qquad 0 \leq x \leq 1,$$
$$E^*_{n,s}(x+1) = tE^*_n(x), \qquad \text{for all } x \notin [0,1].$$

Then it is clear that $E^*_{n,s}(x) \in \mathcal{S}^s_{n,t}$. We shall now prove

Theorem 4. Among all $F(x) \in \mathcal{S}^s_{n,t}$, *where* $s \geq \frac{n-2}{2}$ *and* $\operatorname{sgn} t = (-1)^{n-s}$, $t \neq 0,1$, $E^*_{n,s}(x)$ *is the unique t-perfect spline which minimizes the t-norm.*

The condition $s \geq \frac{n-2}{2}$ is not really necessary and is due to the method of proof adopted here. Elsewhere, by using a different method, we have proved a more general result for L-splines without this restriction.

For the proof of Theorems 3 and 4, we shall need some

lemmas.

<u>Lemma 1</u>. <u>Let</u> $n \geq r$ <u>and</u> $0 \leq x \leq 1$. <u>If</u> $\operatorname{sgn} t = (-1)^{k+1}$, <u>then</u> $\Pi_{n,r}(x;t)$ <u>as a polynomial in</u> x <u>has no zeros in</u> $[0,1]$ <u>where</u>

(4.3) $\Pi_{n,r}(x;t) =$

$$\begin{vmatrix}
1 & \binom{r}{1} & \cdots & \binom{r}{r-1} & 1-t & 0 & & \cdots \\
1 & \binom{r+1}{1} & \cdots & \binom{r+1}{r-1} & \binom{r+1}{r} & 1-t & & \cdots \\
\cdot & \cdot & \cdots & \cdot & \cdot & \cdot & \cdot & \cdot \\
1 & \binom{n-1}{1} & \cdots & \cdots & \cdots & \cdots & 1-t & 0 \\
1 & \binom{n}{1} & \cdots & \cdots & \cdots & \cdots & \binom{n}{n-1} & 1-t \\
x^{n-r+1} & \binom{n-r+1}{1}x^{n-r} & \cdots & \cdots & 1 & 0 & \cdots & 0 \\
\cdot & \cdot & \cdots & \cdot & \cdot & \cdot & \cdot & \cdot \\
x^{n-1} & \binom{n-1}{1}x^{n-2} & \cdots & \cdots & \cdots & \cdots & 1 & 0 \\
x^{n} & \binom{n}{1}x^{n-1} & \cdots & \cdots & \cdots & \cdots & \binom{n}{n-1}x & 1
\end{vmatrix}$$

The proof of this lemma is based on the following theorem for the proof of which we refer to [4]:

<u>Theorem 5</u>. [4] <u>If</u> $0 < x < 1$ <u>and</u> $n \geq r$, <u>the polynomial</u> $\Pi_{n,r}(x;t)$ <u>as a polynomial in</u> t <u>has real and simple zeros of</u> $\operatorname{sgn} (-1)^r$. <u>Furthermore the zeros of</u> $\Pi_{n,r}(x;t)$ <u>and</u> $\Pi_{n-1,r}(x;t)$ <u>interlace</u>.

The following relations are easy to verify from the definition of $\Pi_{n,r}(x;t)$

(4.4) $$\left.\begin{aligned} \Pi_{n,r}(0;t) &= \Pi_{n,r}(t) = 1 \\ \Pi_{n,r}(1;t) &= (-1)^{(n+1)(n-2r+1)}t^{n-2r+1} \end{aligned}\right\} r \leq n \leq 2r-1,$$

$$(4.5)\qquad \left.\begin{aligned} \Pi_{n,r}(0;t) &= \Pi_{n,r}(t) \\ \Pi_{n,r}(1;t) &= t\,\Pi_{n,r}(t) \end{aligned}\right\} \quad n \geq 2r-1 .$$

Using (4.4) and (4.5) and Theorem 5, we easily prove Lemma 1.

Let $H_r(a_n)$ denote a Hankel determinant so that

$$H_r(a_n) = \begin{vmatrix} a_n & a_{n-1} & \cdots & a_{n-r+1} \\ a_{n-1} & \cdots & \cdots & a_{n-r} \\ \cdots & \cdots & \cdots & \cdots \\ a_{n-r+1} & \cdots & \cdots & a_{n-2r+1} \end{vmatrix} .$$

We shall need the following identity of Lee:

<u>Lemma 2</u>. [4] <u>If</u> $n \geq 2r-1$, $r = 1,2,\cdots$, <u>then</u>

$$H_r\left(\frac{A_n(x;t)}{n!}\right) = C(n,r)\,\frac{\Pi_{n,r}(x;t)}{(1-t)^{n-r+1}} ,$$

<u>where</u>

$$C(n,r) = (-1)^{nr+[\frac{r}{2}]}\,\frac{1!2!\cdots r!}{n!(n-1)!\cdots(n-r+1)!} .$$

<u>Proof of Theorem 3</u>. The proof follows from Lemma 1 and 2 on using the method of Theorem 1.1 of Karlin ([3], p. 376 Lemma 5.3, p. 244).

<u>Proof of Theorem 4</u>. Suppose there exists $F(x) \in \mathcal{S}^s_{n,t}$ such that

$$\|F\|_{t,\infty} \leq \|E^*_{n,s}(x)\|_{t,\infty} .$$

Consider the difference $S(x) \equiv E^*_{n,s}(x) - F(x)$, which is a spline of degree $n-1$ and $\in C^s(-\infty,\infty)$. We assert that $S(x) \not\equiv 0$ in $[\nu-1,\nu]$ for any integer ν. For suppose there exists a ν such that

$$S(x) \equiv 0 , \quad \nu-1 \le x \le \nu,$$

$$S(x) \not\equiv 0 , \quad \nu \le x \le \nu+1 .$$

Since $S(x) \in C^{s}(-\infty,\infty)$, then $S(x) = \sum_{k=s+1}^{n-1} c_k (x-\nu)^k$ for $\nu \le x \le \nu+1$. Because there are $n-s$ points of equioscillation of $E^*_{n,s}(x)$ in $[\nu,\nu+1]$ by Corollary to Theorem 3, it follows that $S(x)$ has at least $n-s-1$ zeros in the open interval $(\nu,\nu+1)$. Counting the $s+1$ zeros of $S(x)$ at ν, it follows that $S(x)$ has at least n zeros in $[\nu,\nu+1]$. Hence $S(x) \equiv 0$ in $[\nu,\nu+1]$ contrary to our hypothesis. This proves our assertion.

We now use the upper and lower estimates on the zeros of cardinal splines obtained by Schoenberg and Ziegler [6]. Denoting the zeros of $S(x)$ in $[0,k]$ by $Z\{S(x);[0,k]\}$, we have then

$$Z\{S(x);\ [0,k]\} \ge k(n-s) - 1 ,$$

$$Z\{S(x);\ [0,k]\} \le (n-1) + (k-1)(n-s-1).$$

This is impossible for $k > s + 1$. This proves that $\|E^*_{n,s}(x)\| \ge \|F\|_{t,\infty}$, which completes the proof of Theorem 4.

5. <u>THE CASE $s = 0$</u>. This case is not a special case of Theorem 4, but here we can give an explicit formula for the extremal spline. More precisely we have

<u>Theorem 5</u>. <u>If</u> $|t| \neq 1$ <u>and</u> $\operatorname{sgn} t = (-1)^n$, <u>then there is a unique t-perfect spline</u> $S(x)$ <u>with minimum t-norm where</u>

$$S(x) = \frac{(1+\alpha)^n}{2^{2n-1}} \cos\left(n \text{ arc cos } \frac{2x-1+\alpha}{1+\alpha}\right), \quad 0 \le x \le 1,$$

$$S(x+1) = tS(x), \quad x \notin [0,1],$$

and

$$\alpha = \cot^2\left(\frac{1}{2n} \text{ arc cos } \frac{1}{2}\right) .$$

The proof of this theorem will depend on the following

Lemma 3. *Suppose* $\operatorname{sgn} t = (-1)^n$ *and* $|t| \neq 1$. *If* $P(x)$ *is a monic polynomial satisfying*

$$P(1) = tP(0)$$

and having least deviation from zero on $[0,1]$ *from the class of all such polynomials, then there are exactly* n *points of equioscillation.*

The proof of the lemma and Theorem 5 are given in detail in [7].

ACKNOWLEDGEMENT. We would like to thank Dr. A. Meir for several helpful suggestions and discussions.

REFERENCES

[1] A.S. Cavaretta, Jr., On cardinal perfect splines of least sup-norm on the real axis, J. Approximation Theory **8** (1973), pp. 285-303.

[2] G. Glaeser, Prolongement extremal de fonctions differentialles d'une variable, J. Approximation Theory **8** (1973), pp. 249-261.

[3] S. Karlin and W.J. Studden, Tchebycheff Systems, John Wiley and Sons Inc., 1966.

[4] S.L. Lee, A. Sharma, and J. Tzimbalario, A class of cardinal splines with Hermite type interpolation (to appear).

[5] I.J. Schoenberg, Cardinal Interpolation and Spline Functions IV. The Exponential Euler Splines, Linear Operators and Approximation (ISNM Vol. 20), Birkhäuser Verlag, Basel, 1972, pp. 382-404.

[6] I.J. Schoenberg and Z. Ziegler, On cardinal monosplines of least L_∞-norm on the real axis, J. Analyse Math. **23** (1970), pp. 409-436.

[7] A. Sharma and J. Tzimbalario, Cardinal t-perfect L-splines, SIAM Journal of Numerical Analysis (to appear).

ZERO PROPERTIES OF SPLINES

J. Tzimbalario

1. Introduction.

Let n and k be natural numbers. Set $x_1 < x_2 < \ldots < x_k$ as knots. Set also $x_0 = -\infty$, $x_{k+1} = +\infty$. Denote by

$$S(x) \equiv S_{n,k}(x) \equiv S_{n,k}(x,x_1,\ldots,x_k)$$

a spline function of degree n with the points $\{x_i\}_1^k$ as simple knots. In other words

(i) the restriction of $S(x)$ to any of the intervals (x_i,x_{i+1}) $(i = 0,1,\ldots,k)$ is a polynomial of degree at most n,

(ii) $S(x) \in C^{n-1}(\mathbb{R})$.

Let $\gamma \equiv \{\gamma_i\}_0^k$ be a real vector with k+1 non-zero components.

Definition. $S(x)$ is said to be a γ- spline if the coefficient of x^n in the polynomial component in (x_i,x_{i+1}) is $\gamma_i \neq 0$ $(i = 0,1,\ldots,k)$.

We denote the class of γ-splines of degree n with k knots by $P_{n,k}^{\gamma}$.

A particular case of γ-s is the case when $\gamma_i = t^i$ $(i = 0,1,\ldots,k)$ for some real fixed t.

This special class is denoted by $S_{n,k}^t$ and any element is called a t-perfect spline.

This generalisation of the well-known perfect splines was inspired by the recent study of the author and Professor A. Sharma [9], whom we would like to thank.

Defining

$$u_+ = \begin{cases} u \text{ if } u \geq 0 \\ 0 \text{ if } u < 0 \end{cases},$$

it is possible to represent every element from $P_{n,k}^{\gamma}$ in the form

$$(1) \qquad P(x) = \gamma_0 x^n + \sum_{\nu=0}^{n-1} a_\nu x^\nu + \sum_{i=1}^{k} \delta_i (x-x_i)_+^n ,$$

where $\delta_i = \gamma_i - \gamma_{i-1}$ $(i = 1,\ldots,k)$.
Without loss of generality we can assume that all δ_i are not zero.

Clearly there are n linear parameters i.e. the coefficients $\{a_\nu\}_0^{n-1}$ and k nonlinear parameters $\{x_i\}_1^k$.

2. General Properties

Before stating the results, it is essential to make precise the meaning of a multiple zero for a γ-spline. We shall use the criteria introduced by Schoenberg [10] and Johnson [4]:

If x is not a knot, x is a zero of multiplicity r if:

$$S(x) = \ldots = S^{(r-1)}(x) = 0$$

and

$$S^{(r)}(x) \neq 0.$$

We shall count zeros at knots as in the previous case if $r \le n-1$. Now if $S(x_i) = \ldots = S^{(n-1)}(x_i) = 0$, then x_i is said to be a zero of order

$$\begin{cases} \text{(i)} \quad n & \text{if } \gamma_{i-1}\gamma_i > 0, \\ \text{(ii)} \quad n+1 & \text{if } \gamma_{i-1}\gamma_i < 0. \end{cases}$$

The first theorem gives an upper bound for the number of zeros counting multiplicity.

Theorem 1 Let $S(x) \in P^{\gamma}_{n,k}$. Then $S(x)$ has at most $n + \tau(k)$ zeros counting multiplicities, where

$$\tau(\ell) \equiv \tfrac{1}{2}\sum_{i=1}^{\ell}[1 - \operatorname{sgn}(\gamma_{i-1}\gamma_i)] \qquad (\ell = 1,\ldots,k) \tag{2}$$

($\tau(\ell)$ is a counter for the number of changes of signs in the vector γ).

Corollary 1 If all γ_i are of the same sign, then $S(x)$ has at most n zeros, and if the signs are alternating, $S(x)$ has at most $n + k$ zeros.

Corollary 2 If $S(x)$ is a t-perfect spline, the number of the zeros is bounded from above by n if $t > 0$, and by $n + k$ if $t < 0$.

Proof of Theorem 1 The proof follows easily by induction. If $n = 1$, $S(x)$ is polygonal with k knots and with the slopes $\gamma_0,\gamma_1,\ldots,\gamma_k$. In this case the upper bound for the zeros is clearly $1 + \tau(k)$. For general n, repeated application of Rolle's theorem gives the result.

The next result gives a characterisation of the relative position of the zeros and the knots.

Theorem 2 Let $f(x)$ be a spline in $P^{\gamma}_{n,k}$ with zeros

$$-\infty < t_1 < t_2 < \ldots < t_{n+\tau(k)} < \infty.$$

Then

$$t_{\tau(i)} < x_i < t_{n+\tau(i)} \tag{3}$$

for i which satisfies $\gamma_{i-1}\gamma_i < 0$.

Proof Since, for $n = 1$, the spline is a polygonal line, the result is trivial. Let $n > 1$ and $\gamma_{i-1}\gamma_i < 0$. Suppose $t_{n+\tau(i)} \leq x_i$. In this case $f(x)$ on $(-\infty,x_i)$ is a γ-spline with i-1 knots and the highest coefficients $\gamma_0,\ldots,\gamma_{i-1}$. By Theorem 1 an upper bound for the number of zeros in $(-\infty,x_i)$ is $n + \tau(i-1)$. Then $\tau(i) = \tau(i-1)$, which means that $\gamma_{i-1}\gamma_i > 0$ in contradiction with our assumption.

Now suppose that $x_i \leq t_{\tau(i)}$. As before in (x_i,∞), $f(x)$ is a γ-spline with k-i knots and coefficients $\gamma_i,\gamma_{i+1},\ldots,\gamma_k$. Again by Theorem 1, $f(x)$ can have at most $n+\tau(k) - \tau(i)$ zeros. Since $x_i \leq t_{\tau(i)}$, we have in (x_i,∞) at least $n+\tau(k) - \tau(i-1)$ zeros, which again will contradict the fact that $\gamma_{i-1}\gamma_i < 0$.

3. Oscillatory Splines

Definition A spline $P^{\gamma}_{n,k}(x)$ is called an oscillatory γ-spline if it has $n+\tau(k) - 1$ relative extrema.

A t-perfect spline is oscillatory if it has n-1 relative extrema for $t > 0$ and $n+k - 1$ relative extrema for $t < 0$.

Remark It is interesting that the splines with all γ_i of the same sign have many properties similar to the polynomials.

In the rest of the paper we shall deal only with t-perfect splines, the most important case of γ-splines. The rest of the results are valid also for splines with all γ-s of the same sign or all signs alternating, and the proof is exactly the same, respectively. In order to study the case of general γ, the same method can be used, but the results and the proof are technically more complicated.

Theorem 3 Let e_i $(i = 1,\ldots,n+k-1)$ be real numbers so that

$$(4) \qquad (-1)^{i+n} \, e_i > o \qquad (i = 1,2,\ldots,n+k-1).$$

Then

(i) for any $t < 0$, there exists a unique t-perfect oscillatory spline $P(x) \in S^t_{n,k}$ having its relative extrema at the successive levels $y = e_i$ and with prescribed first left zero;

(ii) for any $t > 0$, there exists a unique t-perfect oscillatory spline $P(x) \in S^t_{n,k}$ having its relative extrema at the successive levels $y = e_i$, $(i = 1,2,\ldots,n-1)$, with prescribed knots x_i $(i = 1,\ldots,k)$, and with prescribed first left zero.

Proof. (i) If $t < 0$, the procedure developed by Cavaretta [1] for the case $t = -1$ can be adjusted (with only very minor modifications) in order to prove the result. We shall give here only a brief sketch of the proof. The proof is given by induction. Clearly the case n = 1, where the spline is a polygonal line, is trivial. Let g(x) be a t-perfect oscillatory spline of degree n-1 with the extrema $e_i = (-1)^{i+1}$ (i = 1,...,n+k-2) and with the zeros $t_1 < t_2 < \ldots < t_{n+k-1}$. By the induction assumption, t_1 can be prescribed and must be chosen at the right side of the prescribed zero a of the required spline.

The function

$$(5) \qquad P^*(x) = n \int_a^x g(s)ds$$

is a t-perfect oscillatory spline with the same knots as g(s) and with relative extrema at t_i (i = 1,...,n+k-1).

Any t-perfect spline has the representation:

$$(6) \qquad P(A,x) = P(x) = x^n + \sum_{i=0}^{n-1} a_i x^i + \sum_{i=1}^{k} \delta_i (x-x_i)_+^n$$

where $\delta_i = t^i - t^{i-1}$, and

$$(7) \qquad A = (a_0,\ldots,a_{n-1},x_1,\ldots,x_k)$$

is the set of parameters.

Using Cavaretta's method, the parameters can be continuously deformed such that $P^*(x)$ is transformed into the desired function.

(ii) If t > 0, the problem is simpler because of its linearity Here we shall give the proof in more detail in order to illustrate the deformation method. The notations will be parallel to those of Cavaretta [1].

Any spline will have still the same general representation (6) but this time with the last term in the right side completely determined and the parameters will be only the coefficients $a_i (i = 0,1,\ldots,n-1)$.

Let us denote here

$$A \equiv (a_0, a_1, \ldots, a_{n-1}).$$

Exactly as in the case of $t < 0$, we construct $P^*(x)$, a t-perfect oscillatory spline with the prescribed knots and the first left zero the prescribed one.

Let A^* be the corresponding set of coefficients for $P^*(x)$. Let Γ be the maximal arcwise connected set in $\mathbb{R}^n$ containing A^* such that for $A \in \Gamma$, $P'(A,t)$ has exactly n-1 distinct simple zeros, $t_i (i = 1,\ldots,n-1)$.
Clearly Γ is open.

By solving a certain system of differential equations, we shall construct a continuously differentiable arc $A(s)$, $s \in [0,1]$, such that $A(0) = A^*$ and $P(A(1),x)$ is the desired spline.

Let

$$A(s) = (a_0(s), \ldots, a_{n-1}(s)) \tag{8}$$

be a vector of functions belonging to $C^1[0,1]$ such that

$$A(0) = A^*$$

and

$$\sum_{i=0}^{n-1} \frac{da_i(s)}{ds} (t_j(s))^i = e_j - d_j^* \qquad (j = 0,1,\ldots,n-1) \tag{9}$$

where $t_0(A) \equiv a$, and e_0 is arbitrarily chosen such that $e_0 < e_1$, and

$$\begin{cases} t_j(s) = t_j(A(s)) & (j = 0,1,\ldots,n-1) \\ d_j^* = P(A^*, t_j(A^*)) & (j = 0,1,\ldots,n-1). \end{cases} \tag{10}$$

Obviously, if $A(s)$ is a solution of the above system, we get

$$\left.\frac{\partial P(A(s),x)}{\partial x}\right|_{t_j(s)} = 0 \qquad (j = 0,1,\ldots,n-1), \tag{11}$$

and

$$\frac{dt_0(s)}{ds} = 0,$$

and then,

$$(12)\qquad \begin{cases} \dfrac{dd_j(s)}{ds} = e_j - d_j^* & (j = 0,1,\ldots,n-1) \\ d_j(0) = d_j^* \end{cases}$$

where $d_j(s) = P(A(s),t_j(s))$ $(j = 0,1,\ldots,n-1)$.

Since the solution of (12) is

$$(13)\qquad d_j(s) = d_j^* + s(e_j - d_j^*) \qquad (j = 0,1,\ldots,n-1),$$

we have to prove only that (9) has a solution for all $s \in [0,1]$. The derivatives $\frac{da_j(s)}{ds}$ can be found explicitly in terms of $a_0,a_1,\ldots,a_{n-1}$ because the respective determinant is a Vandermonde.

By this way, we get

$$(14)\qquad \frac{da_j(s)}{ds} = f_j(a_0,a_1,\ldots,a_{n-1},s) \qquad (j = 0,1,\ldots,n-1)$$

where f_j are C^1 functions in all variables. Then the system can be solved in some interval near zero. The rest of the proof follows mutatis mutandi as the respective proof in Cavaretta [1].

4. Chebyshev t-splines

The problem of finding functions in different classes with least supremum norm in given intervals has always many applications. By the method which will be presented here, if $t = 1$ (in this case the knots are arbitrary because the 1-perfect splines are the polynomials) it is possible to get the well-known Chebyshev polynomials. Since for $t > 0$ and not 1, the knots can be prescribed, the problem will have more meaning for $t < 0$, and this will be our basic assumption here.

Let us deal with functions in $S^t_{n,k}$ with all the knots in (-1,1).

Definition A Chebyshev t-spline with k knots is a spline from $S^t_{n,k}$ which has minimal supremum norm on [-1,1], where the coefficients a_i are free to vary and also the knots in (-1,1).

Now we shall only state the next result which can be proved in exactly the same way as Theorem 2 in [1].

Theorem 4 Let $t < 0$ be a constant. There is a unique spline in $S^t_{n,k}$ which is a Chebyshev t-spline and this spline has exactly n+k+1 points of equioscillation.

Remark This result can be obtained also for the class $P^{\gamma}_{n,k}$ for γ-s alternating in signs.

5. The Fundamental Theorem of Algebra for t-perfect Splines

Let us consider the converse problem of Theorem 1 with respect to the t-perfect splines, i.e. the construction of splines with prescribed zeros.

Theorem 5 Let t_i (i = 1,...,n+k) be real numbers so that

$$t_1 < t_2 < \dots < t_{n+k}.$$

(i) If $t < 0$, there exists a unique t-perfect spline $P(x)$ with k knots such that

$$P(t_i) = 0, \qquad i = 1,2,\dots,n+k. \tag{19}$$

(ii) If $t > 0$, and if $z_1 < z_2 < \dots < z_k$ is a set of k real numbers, then there exists a unique t-perfect spline $P(x)$ with the knots z_i (i = 1,...,k) and such that

$$P(t_i) = 0, \qquad i = 1,\dots,n. \tag{20}$$

Proof We shall give here only the outline of the proof, because it is similar to that of Theorem 3.

By Theorem 3, there is a t-perfect oscillatory spline $P(x)$ of degree $n+1$ with k knots (which are prescribed if $t > 0$).

This spline has $n + k$ extrema if $t > 0$.

The spline $P'(x)/(n+1)$ has $n+k$ zeros if $t < 0$ and n zeros if $t > 0$.

By a continuous deformation, we can get the desired spline.

Remark The case of multiple zeros can be obtained again by some continuity arguments.

REFERENCES

1. Cavaretta, A.S.,"Oscillatory and Zero Properties for Perfect Splines and Monosplines", Journal d'Analyse (1973).

2. Coddington, E.A. and Levinson, N.,"Theory of Ordinary Differential Equations", McGraw-Hill, New York 1955.

3. Fitzgerald, G.H. and Schumaker, L.L.,"A Differential Equation Approach to Interpolation at External Points", Journal d'Analyse (1969).

4. Johnson, R.S.,"On Monosplines of Least Deviation", T.A.M.S. 96 (1960) pp.458-477.

5. Karlin, S.,"Total Positivity", Stanford Univ. Press, 1969.

6. Karlin, S. and Michelli, C.,"The Fundamental Theorem of Algebra for Monosplines Satisfying Boundary Conditions", Israel Journal of Math. 11 (1972).

7. Karlin, S. and Schumaker,L.L., "The Fundamental Theorem of Algebra for Monosplines" , Journal d'Analyse 1969.

8. Michelli, C., "The Fundamental Theorem of Algebra for Monosplines with Multiplicities", IBM Research Report 1971.

9. Sharma, A. and Tzimbalario, J.,"t-perfect L Splines", (to appear).

10. Schoenberg, I.J. and Whitney, A.,"On Polya Frequency Functions III", T.A.M.S. 1953.

A REVIEW OF THE REMEZ EXCHANGE ALGORITHM FOR APPROXIMATION WITH LINEAR RESTRICTIONS

Bruce L. Chalmers

§1. Introduction and examples. This paper demonstrates a Remez exchange algorithm applicable to approximation of real-valued continuous functions of a real variable by polynomials of degree smaller than n with various linear restrictions. As special cases are included the notion of restricted derivatives approximation (examples of which are monotone and convex approximation and restricted range approximation) and the notion of approximation with restrictions at poised Birkhoff data (examples of which are bounded coefficients approximation, ϵ-interpolator approximation, and polynomial approximation with restrictions at Hermite and "Ferguson-Atkinson-Sharma" data and pyramid matrix data). Furthermore, the exchange procedure is completely simplified in all the cases of approximation with restrictions at poised Birkhoff data. Also, results are obtained in the cases of general linear restrictions where the Haar condition prevails. In the other cases (e.g., monotone approximation) the exchange in general requires essentially a matrix inversion, although insight into the exchange is provided and partial alternation results are obtained which lead to simplifications.

In the following is described in detail the setting and background of our investigation.

Let E denote a compact subset of the real line and let CE denote the real space of real-valued continuous functions on E with the usual supremum norm $\|\cdot\|_E$ or $\|\cdot\|$. Let V^n be an n-dimensional subspace of CE. For some fixed index space A, let $\{L^\alpha\}_{\alpha\in A}$ be a compact space (usual norm) of linear functionals defined on V^n, such that, for each p in V^n, $L^\alpha p$ is a continuous function on A, where A inherits the norm topology of $\{L^\alpha\}_{\alpha\in A}$. Now set

$$V_0^n = \{p \in V^n : \ell_\alpha \leqq L^\alpha p \leqq u_\alpha,\ \alpha \in A\} ,$$

where ℓ_α and u_α are extended-real-valued functions on A and $\ell_\alpha < +\infty$, $u_\alpha > -\infty$, ℓ_α and u_α are finite on closed sets relative to which they are continuous, and $\ell_\alpha \leqq u_\alpha$. We assume, moreover, that $\ell_\alpha = u_\alpha$ implies α is an isolated point of A.

Let e_x represent point evaluation at x in E. Now consider any fixed f in $CE - V_0^n$ with the restriction that if $L^\alpha = e_x$ for some α in A and some x in E, then $\ell_\alpha \leqq f(x) \leqq u_\alpha$. We are concerned then with approximating f by elements of V_0^n.

We showed in [2] that, if V_0^n is not empty and V^n satisfies a generalized Haar condition, then there exists a unique best approximation p in V_0^n to f.

For the treatment in this paper we will need an additional hypothesis on V^n which is satisfied by all the examples in question (namely, nearly Haar).

Definition A. If any $k^{\underline{th}}$ order subset of $\{N_i\}_{i=1}^{n+1}$ has rank k , we will say $\{N_i\}_{i=1}^{n+1}$ has uniform rank k , i.e., $\Sigma_{j=1}^{k}\alpha_j N_{i_j} = 0$ implies all $\alpha_j = 0$. We will also say the (n+1)-tuple $(N_1, N_2, \ldots, N_{n+1})$ has uniform rank k .

Definition B. V^n is called nearly Haar on $\Omega = \{L^\alpha\}_{\alpha\in A} \cup \{e_x\}_{x\in E}$ provided

> the set of n-tuples $(R_1, R_2, \ldots, R_n) \in \Omega^n$, where the R_i are linearly dependent, forms a closed nowhere dense subset of Ω^n (or alternatively the set of (n+1)-tuples $(R_1, R_2, \ldots, R_{n+1}) \in \Omega^{n+1}$ which do not have uniform rank n forms a closed nowhere dense subset of Ω^{n+1}) .

Definition C. V^n is called Haar (on Ω) if any distinct n elements in Ω are linearly independent, i.e. any set of n+1 distinct elements in Ω has uniform rank n .

We now give several examples, including all those mentioned above. Existence and uniqueness of best approximation in all these examples was shown in [2] as a consequence of the fact that a generalized Haar condition held. We emphasize that the above stated Haar condition implies the generalized Haar condition and therefore uniqueness of best approximation holds from V_0^n whenever V^n is Haar (on Ω) . For bibliographical references to previous work on these examples, see [2]. In the sequel P_{n-1} will represent the space of real polynomials of degree no larger than n-1, i.e.

$P_{n-1} = \{c_0 + c_1 x + \cdots + c_{n-1}x^{n-1}\}$.

Example 1. Monotone Approximation and Restricted Range Approximation are special cases of Restricted Derivatives Approximation (R.D.A.) described as follows: Let $\{k_i\}_{i=0}^q$ be a fixed set of $q+1$ integers satisfying $0 \leq k_0 < k_1 < \cdots < k_q \leq n-1$. Consider approximation in $C[a,b]$ by the set $N = \{p \in P_{n-1} : \ell_i(x) \leq p^{(k_i)}(x) \leq u_i(x),\ i = 0, 1,\ldots,q\}$, where $\ell_i(x) < +\infty$ and $u_i(x) > -\infty$ are finite on closed sets relative to which they are continuous, and $\ell_i(x) < u_i(x)$ on $[a,b]$, $i = 0,1, \ldots,q$. Also assume that if $k_0 = 0$ then $\ell_0(x) \leq f(x) \leq u_0(x)$, and if $k_i > 0$, then $\ell_i(x)$ and $u_i(x)$ are either differentiable or identically infinite on $[a,b]$.

Restricted derivatives approximation fits into our scheme as follows. First let $E = [a,b]$ and let $A = \cup_{i=0}^q \{(x,i) : x \in E\}$ endowed with the obvious topology. Let $V_n = P_{n-1}$. Then for each $\alpha = (x,i) \in A$, we have $L^\alpha p = p^{(k_i)}(x)$. Next we set $\ell_\alpha = \ell_i(x)$, $u_\alpha = u_i(x)$ for all $\alpha = (x,i)$ in A , and we find that $N = V_0^n = \{p \in V^n : \ell_\alpha \leq L^\alpha p \leq u_\alpha,\ \alpha \in A\}$.

One can check that the remaining conditions hold. Surely $\{L^\alpha = e_x^{k_i}$, where $e_x^{k_i}$ denotes point evaluation at x of the k_ith derivative$\}$ is a compact set in the dual of V^n, $L^\alpha p = p^{(k_i)}(x)$ is a continuous function on A for each p in V^n , and if $L^\alpha = e_x$ for some α in A and some x in E , then $\ell_\alpha \leq f(x) \leq u_\alpha$.

The fact that V^n is nearly Haar follows

immediately from well-known properties of $(n-1)^{\underline{st}}$ degree algebraic polynomials (see [4]).

Example 2. Let E be any compact subset of the real line and let $V^n \subset CE$. Further consider s elements $L^1, L^2, \ldots, L^s$ in the dual of V^n such that for each choice of r functionals $L^{j_1}, L^{j_2}, \ldots, L^{j_r}$, $\{L^{j_1}, L^{j_2}, \ldots, L^{j_r}, e_{x_{r+1}}, e_{x_{r+2}}, \ldots, e_{x_n}\}$ is an independent set for every choice of $n-r$ points $\{x_{r+1}, x_{r+2}, \ldots, x_n\}$, $r \leq s$, provided $e_{x_j} \neq L^{j_i}$, $i = 1,2,\ldots,r; j = r+1,\ldots,n$. Let $V_0^n = \{p \in V^n : \ell_i \leq L^i p \leq u_i,\ i = 1,2,\ldots,s\}$.

This example fits into our scheme where $A = \{1,2,\ldots,s\}$ with the discrete topology. V^n is Haar on Ω by definition.

As special cases of Example 2, we have the three specific examples of Bounded Coefficients Approximation, ϵ-Interpolator Approximation, and Polynomial Approximation with Interpolation.

§2. Characterization of Best Approximation.

Definition. For p in V_0^n a set $\mathcal{S} = \{L^\alpha\}_{\alpha \in I_1} \cup \{e_x\}_{x \in I_3}$ of symbols representing distinct elements in the dual of V^n (where $I_1 \subset A$, $I_3 \subset E$) is called an extremal set for f and p if

(i) $L^\alpha p = u_\alpha$ (or ℓ_α), $\alpha \in I_1$,

(ii) $|(f-p)(x)| = \|f-p\|$, $x \in I_3$,

(iii) $e_x \notin \{L^\alpha\}_{\alpha \in I_1}$ if $|(f-p)(x)| = \|f-p\|$.

Definition. Let $\mathcal{S}' = \{L^\alpha\}_{\alpha \in I_1} \cup \{e_x\}_{x \in I_3}$ be the extremal set for f and p, maximal with respect to I_1 and I_3.

Define a "signature" function σ on $\mathcal{S}'$ by

$$\sigma(e_x) = 1 \quad \text{if} \quad e_x p = f(x) - \|f-p\|,$$
$$\sigma(e_x) = -1 \quad \text{if} \quad e_x p = \|f-p\| - f(x),$$
$$\sigma(L^\alpha) = 1 \quad \text{if} \quad L^\alpha p = \ell_\alpha,$$
$$\sigma(L^\alpha) = -1 \quad \text{if} \quad L^\alpha p = u_\alpha.$$

That is, σ is 1 at lower extrema and -1 at upper extrema. Then define $\mathcal{S}^\sigma = \{\sigma(R)R : R \in \mathcal{S}'\}$.

The following characterization theorem is a restatement of Theorem 3.8.5 in [5] for the present setting (and holds, it should be emphasized, in the absence of any Haar conditions).

Theorem 1. p is a best approximation to f if and only if 0 is in the convex hull of some k ($\leq n+1$) members of $\mathcal{S}^\sigma$; i.e., $0 = \sum_{i=1}^{k} \lambda_i N_i$ where $N_i \in \mathcal{S}^\sigma$, $\lambda_i > 0$, $i = 1,2,\ldots,k \leq n+1$.

Examples. Examples of this theorem are found in many of the works on constrained approximation (e.g., in the case of monotone approximation see [8, p. 5]). See also [3, p. 73].

Definition. Following Laurent [5], we extend the domain of f to $\{\pm L^\alpha\}_{\alpha \in A}$, and introduce an auxiliary distance d of f from V_0^n by $d = \inf_{q \in V_0^n} \rho(f-q)$ where $\rho(h) = \sup\{|N(h)|\}$, with N ranging over $\{\pm e_x\}_{x \in E} \cup \{\pm L^\alpha\}_{\alpha \in A}$; i.e., $\rho(h) = \max\{\|h\|_E, \sup_{\alpha \in A} |(\pm L^\alpha)h|\}$.

Notation. $N = \pm e_x$ will indicate that $|N(f-p)| = \|f-p\|$. $N = \pm L^\alpha$ will indicate that $N \in \mathcal{S}^\sigma$ but $N \neq \pm e_x$.

In the case where $k = n+1$ and $\{N_i\}_{i=1}^{n+1}$ has uniform rank n, then in Theorem 1, the quantity d is related to the linear combination $\ell = \sum_{i=1}^{n+1} \lambda_i N_i$ (assuming $\sum_{N_i=\pm e_x} \lambda_i = 1$) via the following restatement of Theorem 3.9.3 in [5]:

Theorem 2. $\ell(f) = d = \|f-p\|$, where p is the best approximation to f, and p and d are determined by the $n+1$ linear relations $N_i p = N_i f - d$ if $N_i = \pm e_x$, $N_i p = N_i f$ if $N_i = \pm L^\alpha$.

§3. Convergence of a Remez Exchange Algorithm. Remez Algorithm: Suppose at the νth stage we have the extremal set (of uniform rank n) $s_\nu^\sigma = \{N_i^\nu\}_{i=1}^{n+1} = \{\pm L_i^\nu\}_{i=1}^{k_\nu} \cup \{\pm e_{x_i^\nu}\}_{i=k_\nu+1}^{n+1}$ for f and p_ν, the best approximation to f from $V_\nu^n = \{p \in V^n : \ell_i^\nu \leq L_i^\nu p \leq u_i^\nu,\ i = 1,2,\ldots,k_\nu\}$ with respect to $\|\ \|_{E^\nu}$ where $E^\nu = \{x_i^\nu\}_{i=k_\nu+1}^{n+1}$. Let $d_\nu = \|f-p_\nu\|_{E^\nu}$.

In order to construct the $(\nu+1)^{\text{st}}$ stage, let N_*^ν be a functional yielding the maximum deviation of $f-p_\nu$ on $\{e_x\}_{x\in E} \cup \{\pm L^\alpha\}_{\alpha\in A}$; i.e., N_*^ν yields $\max\{\max_{N=\pm e_x} N(f-p_\nu)-d_\nu,\ \max_{N=\pm L^\alpha} N(f-p_\nu)\} = e_\nu$; then exchange one of $\{N_i^\nu\}_{i=1}^{n+1}$ for N_*^ν via the Exchange procedure (see below) to obtain the set $s_{\nu+1}^\sigma = \{N_i^{\nu+1}\}_{i=1}^{n+1} = \{\pm L_i^{\nu+1}\}_{i=1}^{k_{\nu+1}} \cup \{\pm e_{x_i^{\nu+1}}\}_{i=k_{\nu+1}+1}^{n+1}$. Next calculate $p_{\nu+1}$ (and therefore also $d_{\nu+1}$) according to the prescription of Theorem 2.2.

Note 1. By Theorem 2.2 we have that in the above algorithm $p_{\nu+1}$ is the best approximation to

f from $V^n_{\nu+1} = \{p \in V^n : \ell^{\nu+1}_i \leq L^{\nu+1}_i p \leq u^{\nu+1}_i,\ i = 1, 2, \ldots, k_{\nu+1}\}$ with respect to $\| \|_{E^{\nu+1}}$ where $E^{\nu+1} = \{x^{\nu+1}_i\}^{n+1}_{i=k_{\nu+1}+1}$. (Also $d_{\nu+1} = \|f-p_{\nu+1}\|_{E^{\nu+1}}$.)

Note 2. For the 1st stage of the algorithm, let $k_1 = 0$ and $E^1 = \{x^1_i\}^{n+1}_{i=1}$ be a set where p_1 does not interpolate f. Then $d_1 > 0$.

Exchange Procedure (see Laurent [5,p.147]). Suppose (1) $0 = \Sigma^{n+1}_{i=1} \lambda_i N_i$, where $\lambda_i > 0$ and $\{N_i\}^{n+1}_{i=1}$ has uniform rank n. Given N_*, we can find an N_j to replace by N_* so that $0 = \Sigma_{i \neq j} \lambda^*_i N_i + \lambda^*_j N_*$ where $\lambda^*_i \geq 0$ and $\lambda^*_j > 0$ as follows:

$$N_* - N_{n+1} = \Sigma^n_{i=1} \alpha'_i N_i \text{ , whence}$$

(2) $N_* = \Sigma^{n+1}_{i=1} \alpha'_i N_i$ with at least one $\alpha'_i > 0$ (namely α'_{n+1}). Then multiply (2) by $\lambda > 0$ so that

(3) $\lambda N_* = \Sigma^{n+1}_{i=1} \alpha_i N_i$ where $\lambda_i \geq \alpha_i$ with equality holding for some i. Then subtracting (3) from (1) yields the result.

Note 3. Computationally the above exchange involves an $n \times n$ matrix inversion to determine $\{\alpha'_i\}^n_{i=1}$ and a linear search to determine $\lambda^{-1} = \max_i \alpha'_i / \lambda_i$.

Note 4. If V^n is Haar on Ω then $\lambda_i = \alpha_i$ for exactly one i.

Note 5. The Haar condition imposed in ([5, p. 140]) in the discussion of constrained approximation precludes both L^α and $-L^\alpha$ from being restraining functionals, since $L^\alpha p = 0$

implies $(-L^{\alpha})p = 0$. Thus the imposition of both an upper and lower restriction on the same functional L^{α}, i.e. $\ell_{\alpha} \leqq L^{\alpha}p$ and $-u_{\alpha} \leqq (-L^{\alpha})p$ is not allowable. This theory therefore does not yield as an example the Remez algorithm for ordinary restricted (both above and below) range approximation, obtained in [13]. ([5, p. 149, Proposition 3.9.6] does not hold if both L^{α} and $-L^{\alpha}$ occur because then $\lambda L^{\alpha} + \lambda(-L^{\alpha}) = 0$. The method of proof in [5] does extend, however, to upper and lower restrictions on L^{α} if L^{α} is isolated in the dual.)

The methods of [5] can, however, be combined with the methods of [13] to obtain a Remez algorithm in the general setting with upper and lower restrictions as we see in Theorem 1.

<u>Theorem 1</u>. If V^n is Haar on Ω , then the sequence $\{p_{\nu}\}$ generated by the Remez Algorithm converges to the best approximation p .

<u>Notation</u>. If ω belongs to the dual of V^n , then $\|\omega\| = \sup|\omega(h)|$, where h ranges over those functions in V^n of norm 1 .

<u>Definition</u>. If Ω_r is a finite subset of Ω such that $\sup_{\omega_1 \in \Omega_r} \inf_{\omega_2 \in \Omega_r} \|\omega_1 - \omega_2\| < r$, then Ω_r is called a discretization of Ω with mesh size r . If any n arbitrary distinct elements of Ω_r are linearly independent over V^n , then Ω_r is called non-singular.

<u>Theorem 2</u>. If P_{n-1} is nearly Haar on Ω then the sequence p_{ν}^{r} generated by the Remez Algorithm on any non-singular discretization Ω_r

converges to the best approximation p^r. Furthermore, if the sequence r_m tends to zero, p^{r_m} converges uniformly to the best approximation p.

It is easily seen that the result in [6, Theorem 4] proved for convex approximation goes over to the case of Restricted Derivatives Approximation (R.D.A.).

Theorem 3. In the case of R.D.A. there exist constants C and D independent of m such that

$$\|f-p\| - \|f-p^{r_m}\| \leq C \cdot r_m^2$$

and

$$\|f-p^{r_m}\| - \|f-p\| \leq \omega(f;r_m) + D \cdot r_m,$$

where $\omega(f;\sigma)$ is the modulus of continuity function of f.

Theorem 4. In the case of one-sided R.D.A. (i.e., $N = \{p \in P_{n-1} : c_i(x) \leq \epsilon_i p^{(k_i)}(x), \epsilon_i = \pm 1$, $i = 0,1,\ldots,q\})$, the $\lambda_i^{\nu_m}$ occurring in the Remez Algorithm are uniformly bounded independently of the discretization Ω_{r_m}.

§4. Simplifying the Exchange. Suppose $\{R_m\}_{m=1}^{n+1}$ is an arbitrary subset of $\Omega = \{L^\alpha\}_{\alpha \in A} \cup \{e_x\}_{x \in E}$ of uniform rank n, and we have a priori knowledge about the signs of the coefficients α_m in

$$(1) \qquad 0 = \Sigma_{m=1}^{n+1} \alpha_m R_m, \text{ where } \Sigma|\alpha_m| > 0 .$$

Then the Exchange Procedure can be simplified, since the problem is to replace some $R_{m_0}^\nu$ by R_*^ν so that the signs of the coefficients in (1) are preserved; i.e., if $R_m^{\nu+1} = R_m^{\nu}$ $(m \neq m_0)$ and $R_{m_0}^{\nu+1} = R_*^{\nu}$, then, in the dependency $0 = \sum_{i=1}^{n+1} \alpha_m^{\nu+1} R_m^{\nu+1}$, sign $(\alpha_m^{\nu+1})$ = sign (α_m^{ν}), $m = 1,2,\ldots,n+1$.

If, in particular, the signs of the coefficients in (1) are known for every choice of $\{R_m\}_{m=1}^{n+1} \subset \Omega$, then the matrix inversion and linear search of the Exchange Procedure can be eliminated.

Notation. We will sometimes refer to sign α_m in (1) as sign R_m.

Definition. Fix $\{x_1,x_2,\ldots,x_k\} \subset [a,b]$. Let $\mathcal{L}_i^j$ denote the linear functional on P_{n-1} defined by $\mathcal{L}_i^j p = p^{(j)}(x_i)$. Following Schoenberg, let $\mathcal{E} = (e_{ij})_{i=1,2,\ldots,k}^{j=0,1,\ldots,n-1}$ be an m-incidence matrix, i.e. each e_{ij} is 0 or 1 and $\sum_{(i,j)} e_{ij} = m$. If $\mathcal{E}$ is an m-incidence matrix, let $m_j = \sum_{i=1}^k e_{ij}$, $j = 0,1,\ldots,n-1$, and $M_j = \sum_{p=0}^j m_p$, $j = 0,1,\ldots,n-1$. Then $\mathcal{E}$ is said to satisfy the Pólya conditions if $M_j \geq j+1$ for $j = 0,1,\ldots,m-1$, and the strong Pólya conditions if $M_j \geq j+2$ for $j = 0,1,\ldots,m-2$. The n-incidence matrix $\mathcal{E}$ is said to be poised if the set of n linear functionals $\{\mathcal{L}_i^j : e_{ij} = 1\}$ is linearly independent on P_{n-1}.

Definition. (See [4, p. 26]). Let the m-incidence matrix $\mathcal{E}$ have k rows. Let f_i be the column index of the first one which appears in row i. $\mathcal{E}$ is called a pyramid matrix if, for each i, $e_{ij} = 1$ implies $e_{ij'} = 1$ for $f_i \leq j' \leq j$, and there is some value $i (1 \leq i \leq k)$ so that $f_1 \geq f_2 \geq \cdots \geq f_i$ and $f_i \leq f_{i+1} \leq \cdots \leq f_k$.

As examples, if the transpose matrix $\mathcal{E}^t$ is

$$\begin{pmatrix} 0 & 0 & 1 & 1 & 0 \\ 0 & 0 & 1 & 0 & 0 \\ 0 & 1 & 1 & 0 & 0 \\ 1 & 1 & 0 & 0 & 0 \\ 0 & 0 & 0 & 0 & 1 \\ 0 & 0 & 0 & 0 & 0 \\ 0 & 0 & 0 & 0 & 0 \\ 0 & 0 & 0 & 0 & 0 \end{pmatrix}$$

(henceforth in the examples we will often suppress the printing of the last zero rows of $\mathcal{E}^t$), or

$$\begin{pmatrix} 0 & 0 & 0 & 1 & 0 & 0 & 0 \\ 0 & 0 & 1 & 0 & 1 & 0 & 0 \\ 0 & 1 & 0 & 0 & 0 & 1 & 0 \\ 1 & 0 & 0 & 0 & 0 & 0 & 1 \end{pmatrix},$$

then $\mathcal{E}$ is a pyramid matrix, while if $\mathcal{E}^t$ is

$$\begin{pmatrix} 1 & 0 & 1 \\ 0 & 1 & 0 \\ 0 & 0 & 0 \end{pmatrix},$$

then $\mathcal{E}$ is not a pyramid matrix.

We determine completely the signs of the coefficients α_i in the dependency $0 = \Sigma_{m=1}^{n+1}\alpha_m R_m$, where $\{R_m\}_{m=1}^{n+1}$ is any set of derivative evaluations $(e_{x_i}^j)$ whose $(n+1)$-incidence matrix is a pyramid matrix satisfying the <u>strong</u> Pólya conditions. This then yields a simple alternation scheme completely simplifying the Exchange

Procedure in a large class of linearly restricted approximation problems where V^n is Haar on Ω including, for example, Bounded Coefficients Approximation.

<u>Notation</u>. The situation of the preceding paragraph will be referred to in the sequel as the pyramid situation, and we will say $R = (R_1, R_2, \ldots, R_{n+1}) \in PM_n$. We will also assume $R \in PM_n$ is ordered so that if $R_u = e^j_{x_i}$ then $R_{u+1} = e^\ell_{x_i}$ or $e^\ell_{x_{i+1}}$ where $j \geqq \ell$ if $i \leqq i_o$ and otherwise $\ell \geqq j$.

<u>Theorem 1</u>. Consider the general pyramid situation $(R \in PM_n)$. Let $R_k = e^j_{x_i}$ and $R_{k+1} = e^r_{x_s}$ and $\sigma = \text{sign}\,(R_k)\,\text{sign}\,(R_{k+1})$. If $r < j$, then $\sigma = (-1)^{j-r+1}$, while if $r \geqq j$, then $\sigma = -1$.

<u>Corollary 1</u>. Theorem 1 provides the alternation pattern for Bounded Coefficients Approximation, thereby completely simplifying the Exchange Procedure in this case.

<u>Example</u>. Recall that $V^n_0 = \{p = \Sigma^{n-1}_{i=0} a_i x^i \in P_{n-1}[a,b] : c_i \leqq a_i \leqq b_i \ (i = 0,1,\ldots,n-1)\}$, where c_i and b_i are given extended real numbers, and $c_0 \leqq f(0) \leqq b_0$. Then $V^n_0 = \{p \in P_{n-1}[a,b] : \ell_i \leqq e^i_0 p \leqq u_i \ (i = 0,1,\ldots,n-1)\}$ where $\ell_i = i!c_i$ and $u_i = i!b_i$. Suppose at stage ν the $n+1$ functionals forming the extremal set are

$$\overset{+}{e}{}^{15}_0 \ \overset{+}{e}{}^{12}_0 \ \overset{+}{e}{}^{11}_0 \ \overset{+}{e}{}^{10}_0 \ \overset{-}{e}{}^{6}_0 \ \overset{-}{e}{}^{5}_0 \ \overset{-}{e}_{x_1} \ \overset{+}{e}_{x_2} \cdots \overset{-}{e}_{x_{k-1}} \ \overset{+}{e}_{x_k}$$

where $x_i < x_{i+1}$ and we write $\overset{\pm}{R}$ for $\pm R$. Thus if N^ν_*

$= \overset{+}{e}_x$, where $x_1 < x < x_2$, N_*^ν replaces $e_{x_2}^+$ while if $N_*^\nu = \bar{e}_x$, N_*^ν replaces $\bar{e}_{x_1}$. If $N_*^\nu = \bar{e}_x$, where $0 < x < x_1$ then N_*^ν replaces $e_{x_1}^-$ while if $N_*^\nu = \overset{+}{e}_x$, N_*^ν replaces e_0^{-5} . If $N_*^\nu = e_0^{-7}$ then N_*^ν replaces e_0^{+10} , while if $N_*^\nu = e_0^{+7}$ then N_*^ν replaces e_0^{-6} . If $N_*^\nu = \overset{+}{e}_x$ where $x_{n+1} < x$ then N_*^ν replaces $\overset{+}{e}_{x_{n+1}}$, while if $N_*^\nu = \bar{e}_x$ then N_*^ν replaces e_0^{+15} .

Note. Theorem 1 also gives the alternation scheme for Bounded Coefficients Approximation on $[a,0]$.

Example.

$$\overset{+}{e}_{x_1} \quad \bar{e}_{x_2} \cdots \overset{+}{e}_{x_{k-1}} \bar{e}_{x_k} e_0^{+5} \, e_0^{-6} \, e_0^{+10} \, e_0^{-11} \, e_0^{+12} \, e_0^{-15} .$$

Definition. $A \geqq B$ will mean that for $a \in A$ either $a \geqq b$ for all $b \in B$ or $a \leqq b$ for all $b \in B$.

Corollary 2. Theorem 1 provides the alternation pattern for approximation from $V_0^n = \{p \in P_{n-1}[a,b] : \ell_i(x) \leqq p^{(i)}(x) \leqq u_i(x)$ on E_i , where $[a,b] \subset E_0$ and $E_i \geqq \cup_{j=0}^{i-1} E_j$ for $i = 0,1, \ldots, n-1\}$.

As examples we have restricted range with bounded coefficients approximation and approximation by polynomials which are required to be monotone (or convex) in an interval outside $[a,b]$.

Corollary 3. The alternation scheme of Theorem 1 also holds if we replace each $R_m = e_{x_i}^j$ by $S_m = \int_{-\infty}^{\infty} (\quad)^{(j)} d\mu_{ji}$ where μ_{ji} is a positive measure such that the convex hull of its support c_{ji} contains x_i and such that none of the c_{ji} intersect.

<u>Definition</u>. $\{M_i\}_{i=1}^m = \{e_{x_i}^{j_i}\}_{i=1}^m$ is called a set of poised Birkhoff data if $M = \{M_i\}_{i=1}^m \cup \{e_{x_\mu}\}_{\mu=m+1}^n$ is linearly independent in the dual of P_{n-1} for all choices of distinct $e_{x_\mu} \notin \{M_i\}_{i=1}^m$. We will say that P_{n-1} is Haar on $[a,b]$ relative to $\{M_i\}_{i=1}^m$.

Consider now approximation problems where $V^n = P_{n-1}[a,b]$ and included in $\Omega = \{M_i\}_{i=1}^m \cup \{e_x\}_{x\in[a,b]}$ is a set $\{M_i\}_{i=1}^m$ of poised Birkhoff interpolatory data. That is, $V_0^n = \{p \in V^n : M_i p = m_i \ (i = 1,\ldots,m)\}$, and $M_i f = m_i$ if $M_i = e_x$. We assume without loss of generality that all $m_i = 0$, for we can approximate $f-p_0$ where $M_i p_0 = m_i (i = 1,\ldots,m)$. Thus we approximate from $V_0^{n-m} = P_{n-1} \cap \{p : M_i p = 0 \ (i = 1,\ldots,n)\}$.

<u>Theorem 2</u>. Suppose P_{n-1} is Haar on $[a,b]$ relative to $\{M_i\}_{i=1}^m$. Then we approximate from $V_0^{n-m} = P_{n-1} \cap \{p : M_i p = 0 \ (i = 1,\ldots,m)\}$. If $R = (e_{x_1}, e_{x_2}, \ldots, e_{x_{n-m+1}})$ is an $(n-m+1)$-tuple of extremals where $x_1 < x_2 < \cdots < x_{n-m+1}$, then alternation occurs as follows: sign (e_{x_i}) sign $(e_{x_{i+1}}) = (-1)^{q+1}$ where q is the number of Hermite data in $\{M_i\}_{i=1}^m$ at all x where $x_i < x < x_{i+1}$.

<u>Corollary 4</u>. Theorem 2 provides the

alternation pattern for approximation with interpolation at poised Birkhoff data, thereby completely simplifying the Exchange Procedure.

Example. An example is Polynomial Approximation with Interpolation (at strictly Hermite data). (See [7].)

Theorems 1 and 2 are special cases of the following result.

Theorem 3. Let V_0^n be as in Corollary 2 and let $\{M_i^j\}_{i=1}^{m_j}$ be poised Birkhoff data on each E_j, $j = 0,1,\ldots,n-1$. Let $\Sigma m_j = m$. Then the Remez Algorithm can be applied to $V_0^{n-m} = V_0^n \cap \{p: M_i^j p = 0\}$ with the following alternation scheme. Let $R \in PM_{n-m}$. Let $R_k = e_{x_i}^t$ and $R_{k+1} = e_{x_s}^r$ and $\sigma = \text{sign}(R_k)\ \text{sign}(R_{k+1})$.

(i) If $r < t$, then $\sigma = (-1)^{t-r+1}$

(ii) If $r = t$, then $\sigma = (-1)^{q+1}$ where q is the number of Hermite data in $\{M_i^j\}$ at all x where $x_i < x < x_s$.

(iii) If $r > t$, then $\sigma = -1$.

Example. As a further example we have restricted range approximation with Hermite interpolation.

Finally we wish to consider the case where V^n is not Haar on Ω, and investigate what alternation properties hold and what simplifications can be made in the Exchange Procedure. We will confine our attention to Restricted Derivatives Approximation (R.D.A.).

From Theorem 2.1 we know that p is a best approximation to f if and only if

$$(2) \qquad 0 = \Sigma_{i=1}^{r}\lambda_i\sigma_i e_{x_i} + \Sigma_{i=r+1}^{r+s+1}\lambda_i\sigma_i e_{x_i}^{j_i}$$

for some $\mathfrak{J} = \{e_{x_i}^{k_i}\}_{i=1}^{r+s+1} \subset \mathfrak{S}'$ where $j_i > 0$, $\sigma_i =$ sign $e_{x_i}^{k_i}$, all $\lambda_i > 0$ and $r+s \leq n+1$.

Theorem 4. In the case of R.D.A., suppose that if $x_i \in (a,b)$ then $e_{x_i}^{j_i}$ and $e_{x_i}^{j_i+1}$ do not both belong simultaneously to $\mathfrak{J}$. Let t equal the number of $e_{x_i}^{j_i}$ where $x_i \in (a,b)$. Then if $r+s+t = n+1$, we have alternation occurring among the e_{x_i} in (2); i.e., if $x_1 < x_2 < \cdots < x_r$ then $\sigma_i\sigma_{i+1} = -1$, $i = 1,2,\ldots,r-1$. In fact, the alternation pattern among $\{e_{x_i}^{j_i}\}_{x_i = a \text{ or } b} \cup \{e_{x_i}\}_{i=1}^{r}$ is that of Theorem 1 (the pyramid case).

Example. In [6, Example 7] we have that in the case of Monotone Approximation where $f(x) = x^7$, on $[-1,1]$ and $n = 7$, the extremal set is $\{\bar{e}_{-1}, \overset{+}{e}_{-.8}, \bar{e}_{.8}, \overset{+}{e}_{1}, e_{-.4}^{+1}, e_{+.4}^{+1}\}$.

Example. In [6, Example 2] we have that in the case of Convex Approximation where $f(x) = |x|$ on $[-1,1]$ and $n = 6$, the extremal set is $\{\bar{e}_{-1}, \overset{+}{e}_{-.4}, \bar{e}_{0}, \overset{+}{e}_{.4}, \bar{e}_{1}, e_{-1}^{+2}, e_{1}^{+2}\}$.

Theorem 5. In if (2) there are only t $e_{x_i}^{j_i}$ such that $x_i \in (x_1, x_r)$ where $t = 0$ or 1, then

(i) if $t = 0$, then $r+s = n+1$ and the alternation pattern coincides with that in Theorem 1 (i.e., the pyramid situation).

(ii) if $t = 1$, then $n \leqq r+s \leqq n+1$.

If $n = r+s$, the alternation pattern is again that of Theorem 1 among all the functionals excluding $e_{x_i}^{j_i}$ which itself appears with a positive coefficient in (2).

If $n+1 = r+s$, the alternation pattern is the same as in (ii) with exactly one exception.

<u>Examples</u>. Theorems 4 and 5 reveal the alternation patterns for all the twelve examples in [6]. For example in [6, Example 5] we have that in the case of Convex Approximation where $f(x) = \ell n\, 2.1 - \ell n(x+1.1)$ on $[-1,1]$ and $n = 6$, the extremal set is $\{\overset{+}{e}_{-1}, \bar{e}_{-.9}, \overset{+}{e}_{-.6}, \overset{+2}{e_0}, \overset{+}{e}_{.6}, \bar{e}_1, \overset{+2}{e_1}\}$ and in accordance with Theorem 5, if either $\overset{+}{e}_{-.6}$ or $\overset{+}{e}_{.6}$ is suppressed, the remaining functionals (ignoring $\overset{+2}{e_0}$) alternate as in the pyramid case. ($\pm e_0^3$ would replace either $\overset{+}{e}_{-.6}$ or $\overset{+}{e}_{.6}$ in the Exchange Procedure.)

<u>Theorem 6</u>. For any approximation problem discussed in this paper, if there is exactly one extremal (n+1)-tuple $N = (N_1, N_2, \ldots, N_{n+1})$ with distinct entries for f and p the best approximation, then in the Remez algorithm, $N^\nu = (N_1^\nu, N_2^\nu, \ldots, N_{n+1}^\nu)$ converges to N. Thus there exists K such that $\nu \geqq K$ implies that, in the Exchange

Procedure, N_*^ν replaces that $N_{i_0}^\nu$ yielding $\min_{1 \leq i \leq n+1} \|N_i^\nu - N_*^\nu\|$.

Proofs and further details and results will appear elsewhere.

REFERENCES

[1] C. Carasso, Convergence de l'algorithms de Remes, J. Approx. Theory 11(1974), 149-158.

[2] B. L. Chalmers, A unified approach to uniform real approximation by polynomials with linear restrictions, Trans. Amer. Math. Soc. 166 (1972), 309-316.

[3] E. W. Cheney, Introduction to Approximation Theory, McGraw-Hill, N.Y., 1966.

[4] David Ferguson, The question of uniqueness for G. C. Birkhoff interpolation problems, J. Approx. Theory 2(1969), 1-28.

[5] P. J. Laurent, Approximation et Optimisation, Hermann, Paris, 1972.

[6] J. T. Lewis, Approximation with convex restraints, Univ. of Rhode Island, Tech. Rep. #11, August 1970.

[7] H. L. Loeb, D. G. Moursund, L. L. Schumaker, G. D. Taylor, Uniform generalized weight function polynomial approximation with interpolation, SIAM J. Numer. Anal. 6(1969), 284-293.

[8] G. G. Lorentz and K. L. Zeller, Monotone approximation by algebraic polynomials, Trans. Amer. Math. Soc. 149(1970), 1-18.

[9] R. A. Lorentz, Uniqueness of best approximation by monotone polynomials, J. Approx. Theory 4(1971), 401-418.

[10] J. A. Roulier, Polynomials of best approximation which are monotone, J. Approx. Theory 9(1973), 212-217.

[11] J. A. Roulier and G. D. Taylor, Uniform approximation by polynomials having bounded coefficients, Abh. Math. Sem. Univ. Hamburg 36(1971) 126-135.

[12] O. Shisha, Monotone approximation, Pacific J. Math. 15(1965), 667-671.

[13] G. D. Taylor and M. J. Winter, Calculation of best restricted approximations, SIAM J. Numer. Anal. 7(1970), 248-255.

An Algorithm for Constrained, Non-linear, Tchebycheff Approximations

GARY A. GISLASON

1. Introduction

In this paper we consider the problem of obtaining certain non-linear Tchebycheff approximations which satisfy Hermite interpolation conditions. The setting is C[0,1] where

$$||f|| = \max_{0 \le x \le 1} |f(x)| \quad \text{for} \quad f \in C[0,1].$$

2. Basic Results

Let Ω be an open subset of real n-space, R^n, where n is a positive integer and for each $A = (a_1, \ldots, a_n)$ in R^n we define

$$||A|| = \max_{1 \le i \le n} |a_i|.$$

To each $A \in \Omega$ we assign a continuous, real-valued function $F(A,x)$, $x \in [0, 1]$, such that the partial derivatives

$$\frac{\partial F(A,x)}{\partial a_i} \qquad i = 1, \ldots, n$$

and

$$\frac{\partial^j F(A,x)}{\partial x^j} \qquad j = 1, \ldots, m$$

exist and are continuous in A and x, where m is to be specified later. Set

$$W(A) = \{\sum_{i=1}^{n} c_i \frac{\partial F(A,x)}{\partial a_i} : c_1, \ldots, c_n \text{real}\}$$

and denote by $d(A)$ the dimension of $W(A)$. Finally, set

$$V = \{F(A,x) : A \in \Omega\}.$$

Now suppose we are given points $0 \leq y_1 < \ldots < y_k \leq 1$ and positive integers $m_1, \ldots, m_k$ where $m = \max m_i$. Let $V(g)$ denote the set of functions $F(A,x) \in V$ which satisfy (1):

$$(1)\quad F^{(j)}(A,y_i) = g^{(j)}(y_i) \qquad \begin{array}{l} i = 1, \ldots, k \\ j = 0, 1, \ldots, m_i - 1. \end{array}$$

<u>Definition 1.</u> $F(A^*,x) \in V$ is a best approximation to g if and only if $F(A^*,x) \in V(g)$ and

$$||g - F(A^*)|| = \inf\{||g - F(A)|| : F(A,x) \in V(g)\}.$$

We impose the following additional conditions on the family V:

(2) For each $A \in \Omega$ the mixed partial derivatives

$$\frac{\partial^{j+1} F(A,x)}{\partial a_i \partial x^j} \qquad \begin{array}{l} i = 1, \ldots, n \\ j = 1, \ldots, m \end{array}$$

exist and are continuous in A and x.

(3) For each $A \in \Omega$ the vector space $W(A)$ is an extended Haar subspace of $C^m[0,1]$ of dimension $d(A)$ and order $m + 1$.

(4) If $A, A' \in \Omega$ and $F(A,x) \neq F(A',x)$ then $F(A,x) - F(A',x)$ can have at most $d(A) - 1$ zeros counting multiplicities up to order $m + 1$.

The following are a few examples of such approximating families:

<u>Example 1.</u> Let $\Omega = R^{n+1}$. For each $A = (a_0, a_1, \ldots, a_n)$ in Ω define

$$F(A,x) = \sum_{i=0}^{n} a_i x^i \quad \text{for} \quad x \in [0,1].$$

Then V is the space of polynomials of degree at most n defined on $[0,1]$, and $d(A) = n + 1$.

Example 2. Let m and n be positive integers. For each A in R^{m+n+2} write $A = (a_0, a_1, \ldots, a_m, b_0, b_1, \ldots, b_n)$. Let

$$\Omega = \{A \in R^{m+n+2}: \sum_{j=0}^{n} b_j x^j > 0,\ x \in [0,1]\}.$$

For each $A \in \Omega$ set

$$F(A,x) = \sum_{i=0}^{m} a_i x^i \Big/ \sum_{j=0}^{n} b_j x^j.$$

Then $V = R_n^m[0,1]$. If μ and ν denote, respectively, the degrees of $\Sigma a_i x^i$ and $\Sigma b_j x^j$, then $d(A) = m+n+1 - \min(m-\mu, n-\nu)$.

Example 3. Let n be a positive integer and set

$$\Omega = \{A \in R^{2n} : a_{n+i} \neq a_{n+j} \text{ for } i \neq j \text{ and } 1 \leq i, j \leq n\}.$$

For each $A \in \Omega$ let

$$F(A,x) = \sum_{i=1}^{n} a_i e^{a_{n+i} x} \quad \text{for} \quad x \in [0,1].$$

The V is the classical exponential family. In this case, $d(A) = n + k$ where k denotes the number of non-zero a_i for $1 \leq i \leq n$.

The following theorem, which appears in [4], characterizes best approximations and extends a result in [6]. See also [5].

Theorem 1. $F(A^*,x)$ is the unique best approximation to $g(x)$ if and only if there exist

$$q = d(A^*) - \sum_{i=1}^{k} m_i + 1$$

points $0 \leq x_1 < \ldots < x_q \leq 1$ such that

$$g(x_{r+1}) - F(A^*,x_{r+1}) = (-1)^{1+k(r)}$$
$$\times (g(x_r) - F(A^*,x_r)),\ r = 1, \ldots, q-1,$$

and

$$|g(x_1) - F(A^*,x_1)| = ||g-F(A^*)||.$$

Here, $k(r)$ denotes the sum of the multiplicities m_i for which $x_r < y_i < x_{r+1}$.

The next result is an appropriate generalization of a theorem of de La Vallée-Poussin, [7], and is needed in the sequel.

Theorem 2. If $F(A',x) \varepsilon V(g)$ and

$$\operatorname{sgn}(g(x'_{r+1}) - F(A',x'_{r+1})) = (-1)^{1+k'(r)} \times \operatorname{sgn}(g(x'_r) - F(A',x'_r)),\ r = 1, \ldots, q-1,$$

for some set of $q = d(A') - \Sigma m_i + 1$ points $0 \leq x'_1 < \ldots$ $\ldots < x'_q \leq 1$ distinct from the y_i then

$$||g - F(A)|| \geq \min_{1 \leq r \leq q} |g(x'_r) - F(A',x'_r)|$$

for all $F(A,x) \varepsilon V(g)$. Here, $k'(r)$ denotes the sum of the multiplicities m_i for which $x'_r < y_i < x'_{r+1}$.

Definition 2. $F(A',x) \varepsilon V$ is said to be normal if

$$d(A') = \max_{A \varepsilon \Omega} d(A).$$

We point out that the Strong Uniqueness Theorem, [4], which is used later in the convergence proof, requires that $F(A^*,x)$ be normal. Hence, that stipulation is made now. In addition, it is assumed for the remainder of the discussion that $F(A^*,x)$ is the best approximation to the function

$$g \varepsilon C^{m-1}[0,1],\ ||g - F(A^*)|| > 0,\ q = d(A^*) - \Sigma m_i + 1 \geq 1,$$

and $\sigma(x) = \operatorname{sgn}(g(x) - F(A^*,x))$ for $x \varepsilon [0,1]$. We turn now to the problem of determining the best approximation, $F(A^*,x)$. To this end we let

$$X = \{x \in [0,1] : |g(x) - F(A^*,x)| = ||g - F(A^*)||\},$$
$$Q = \{(u_1, \ldots, u_q) : 0 \leq u_1 < \ldots < u_q \leq 1 \text{ and}$$
$$u_r \neq y_i \text{ for } 1 \leq r \leq q \text{ and } 1 \leq i \leq k\},$$
$$k_r(u) = \sum_{i \in I} m_i,\ I = \{i : u_r < y_i < u_{r+1}\}, \text{ for } 1 \leq r \leq q-1,$$
$$C = \{u \in Q : u_r \in X \text{ for } 1 \leq r \leq q \text{ and}$$
$$\sigma(u_{r+1}) = (-1)^{1+k_r(u)}\sigma(u_r) \quad \text{for } 1 \leq r \leq q-1\} \text{ and}$$
$$Q(\delta) = \{u \in Q : \operatorname{dist}(u,C) \leq \delta\}.$$

Here, $u \in Q$ implies $\operatorname{dist}(u,C) = \inf\{||u-v|| : v \in C\}$ and $||u|| = \max |u_r|$. In addition, we will have need to refer to the system

$$\begin{aligned} g(u_{r+1}) - F(A,u_{r+1}) &= (-1)^{1+k_r(u)} \\ &\times (g(u_r) - F(A,u_r)),\ 1 \leq r \leq q-1, \\ g(u_1) - F(A,u_1) &= \lambda \end{aligned} \tag{5}$$

where $F(A,x) \in V(g)$, λ is real, and $u \in Q$.

Relying in part on the extended local Haar condition, (3), one is able to prove the existence of positive numbers γ^*, δ^*, and ε^* for which the following assertions are valid.

Lemma 1. $Q(\delta^*)$ is closed (hence compact) and if $u \in Q(\delta^*)$ then $u_r \neq y_i$ for $1 \leq r \leq q$ and $1 \leq i \leq k$.

Lemma 2. If $u \in Q(\delta^*)$ then there exists $A(u) \in \Omega$ and $\lambda(u)$ real such that $A(u)$, $\lambda(u)$, and u satisfy (5), $F(A(u),x)$ is in $V(g)$, and $||A(u) - A^*|| \leq \varepsilon^*$.

Lemma 3. Suppose there exists $F(A,x) \in V(g)$ and $u \in Q$ for which

$$\begin{aligned} \operatorname{sgn}(g(u_{r+1}) - F(A,u_{r+1})) &= (-1)^{1+k_r(u)} \\ &\times \operatorname{sgn}(g(u_r) - F(A,u_r)) \quad \text{for } 1 \leq r \leq q-1, \end{aligned}$$

and $\min |g(u_r) - F(A,u_r)| \geq \gamma^*$. Then $u \in Q(\delta^*)$.

The proofs of the above results are straightforward generalizations of the corresponding results in [1] for the case of no interpolation and are therefore omitted.

Now, for each $u \in Q$ let $\delta(u) = |\lambda(u)|$ if there exists $A(u) \in \Omega$ such that $A(u)$, $\lambda(u)$, and u satisfy (5). Otherwise, set $\delta(u) = 0$. Observe that by Theorem 2, $\delta(u) \leq ||g-F(A^*)||$. By Theorem 1 it follows that

$$\max_{u \in Q} \delta(u) = ||g - F(A^*)||.$$

In the algorithm which follows we take the approach of maximizing the functions $\delta(u)$ as a means of determining $F(A^*,x)$. The method is described in [2] for the case of approximation by n-parameter families.

Now, if $2 \leq i \leq q-1$ we define $\delta_u^{\,i}(x) = \delta(u_1, \ldots, u_{i-1}, x, u_{i+1}, \ldots, u_q)$ for $u_{i-1} < x < u_{i+1}$. Let $\delta_u^{\,1}(x) = \delta(x, u_2, \ldots, u_q)$ if $0 \leq x < u_2$ and $\delta_u^{\,1}(x) = \delta(u_2, \ldots, u_q, x)$ in case $u_q < x \leq 1$. Finally, set $\delta_u^{\,q}(x) = \delta(u_1, \ldots, u_{q-1}, x)$ for $u_{q-1} < x \leq 1$ and $\delta_u^{\,q}(x) = \delta(x, u_1, \ldots, u_{q-1})$ for $0 \leq x < u_1$.

Algorithm 1. Suppose $A(u)$, $\lambda(u)$, and u satisfy (5) with $||A(u) - A^*|| \leq \varepsilon^*$ and $|\lambda(u)| \geq \gamma^*$. Let $\bar{x} \in [0,1]$ and $\bar{m} \in \{1, \ldots, q\}$ be such that

$$\delta_u^{\,\bar{m}}(\bar{x}) = \max_{1 \leq i \leq q} \max_{0 \leq x \leq 1} \delta_u^{\,i}(u).$$

If $2 \leq \bar{m} \leq q-1$ construct $\bar{u}$ from u and $\bar{x}$ be setting $\bar{u}_i = u_i$ for $1 \leq i \leq q$ and $i \neq \bar{m}$, and $\bar{u}_{\bar{m}} = \bar{x}$. If $\bar{m} = 1$ there are two possibilities to consider. In case $0 \leq \bar{x} < u_2$ let $\bar{u}_1 =$

$\bar{x}$ and $\bar{u}_i = u_i$ for $2 \leq i \leq q$. If $u_q < \bar{x} \leq 1$ set $\bar{u}_i = u_{i+1}$ for $1 \leq i \leq q-1$ and $\bar{u}_q = \bar{x}$.

In case $\bar{m} = q$ there are also two possibilities to consider. If $0 \leq \bar{x} < u_1$ set $\bar{u}_1 = \bar{x}$ and $\bar{u}_i = u_{i-1}$ for $2 \leq i \leq q$. If $u_{q-1} < \bar{x} \leq 1$ let $\bar{u}_i = u_i$ for $1 \leq i \leq q-1$ and $\bar{u}_q = \bar{x}$. Finally, denote $\bar{u}$ by u and begin again.

A second algorithm is discussed in [2] and [3] in which, essentially, one maximizes cyclically the functions $\delta_u{}^i(x)$, updating the q-tuple u after each maximization. This method, which is better suited to actual computation than Algorithm 1, can be adapted to the present setting and shown to converge. One obvious feature of Algorithm 1 and the variation described above is their simplicity when compared, for example, with the Remez Algorithm, [1]. Convergence, however, may not be as rapid as for the Remez Algorithm, which is known to converge quadratically under certain conditions.

We now establish an estimate of the rate of convergence of Algorithm 1.

<u>Theorem 3.</u> Let $A^{(o)}$, $u^{(o)}$, and $\lambda^{(o)}$ satisfy (5) with $||A^{(o)} - A^*|| \leq \varepsilon^*$ and $|\lambda^{(o)}| \leq \gamma^*$. If $\{u^{(\nu)}\}_{\nu=1}^{\infty}$ is generated by Algorithm 1 and $\{A^{(\nu)}, \lambda^{(\nu)}\}_{\nu=1}^{\infty}$ are the corresponding solutions to (5) for which $||A^{(\nu)} - A^*|| \leq \varepsilon^*$ then

$$||F(A^*) - F(A^{(\nu)})|| \leq K\theta^{\nu}$$

where K and θ are constants independent of ν and $0 \leq \theta < 1$.

<u>Proof:</u> First, let A, u, and λ satisfy (5) with $||A-A^*|| \leq \varepsilon^*$ and $\delta(u) \leq \gamma^*$. Suppose $\bar{u}$ is generated from u according to

Algorithm 1. Denote by $\bar{A}$ and $\bar{\lambda}$ the corresponding solutions to (5) for which $||\bar{A} - A^*|| \leq \varepsilon^*$. Let $x' \in [0,1]$ satisfy the relation $|g(x') - F(A,x')| = ||g - F(A)||$. If x' is not already one of the coordinates of u, then there is a q-tuple $u' \in Q$ such that $u'_i = x'$ for some i and exactly one of the relations $u'_j = u_{j-1}$, $u'_j = u_j$, $u'_j = u_{j+1}$ holds for all $j \neq i$, and such that

$$\operatorname{sgn}(g(u'_{r+1}) - F(A,u'_{r+1})) = (-1)^{1+k_r(u')} \times \operatorname{sgn}(g(u'_r) - F(A,u'_r)), \quad 1 \leq r \leq q-1.$$

The verification is easy but rather tedious and is omitted. By construction of u' we have

$$\min_{1\leq r\leq q} |g(u'_r) - F(A,u'_r)| \geq \min_{1\leq r\leq q} |g(u_r) - F(A,u_r)| = \delta(u) \geq \gamma^*.$$

By Lemma 3, $u' \in Q(\delta^*)$. Hence, by Lemma 2 there exists a vector $A' \in \Omega$ and a real number λ' such that $||A' - A^*|| \leq \varepsilon^*$ and A', u', and λ' satisfy (5). Applying the reasoning of [1] to the sign properties of extended Haar systems we write

$$\delta(u') = |\lambda'| = \sum_{r=1}^{q} \theta_r \, |g(u'_r) - F(A,u'_r)|$$

where $\theta_r \geq 1-\theta > 0$, $\Sigma\, \theta_r = 1$, and θ is independent of u'. Now, by definition of $\delta(\bar{u})$ we have $\delta(\bar{u}) \geq \delta(u')$. Hence, we obtain the following inequalities:

$$\begin{aligned}\delta(\bar{u}) - \delta(u) &\geq \delta(u') - \delta(u)\\ &\geq \sum_{r=1}^{q} \theta_r \, |g(u'_r) - F(A,u'_r)| - \delta(u)\\ &\geq \sum_{r=1}^{q} \theta_r [|g(u'_r) - F(A,u'_r)| - \delta(u)]\end{aligned} \tag{6}$$

$$\geq (1 - \theta)\ [||g - F(A)|| - \delta(u)]$$
$$\geq (1 - \theta)\ [||g - F(A^*)|| - \delta(u)].$$

From (6) it follows that

$$||g - F(A^*)|| - \delta(\bar{u})$$
$$= ||g - F(A^*)|| - \delta(u) - (\delta(\bar{u}) - \delta(u))$$
$$\leq \theta(||g - F(A^*)|| - \delta(u)).$$

In terms of the quantities $A^{(\nu)}$, $u^{(\nu)}$, and $\lambda^{(\nu)}$ the above inequality implies

$$(7)\quad \begin{aligned} ||g - F(A^*)|| - \delta(u^{(\nu)}) &\leq \theta(||g - F(A^*)|| - \delta(u^{(\nu-1)})) \\ &\leq \theta^{\nu}(||g - F(A^*)|| - \delta(u^{(0)})). \end{aligned}$$

From inequalities (6) and (7) we obtain

$$\begin{aligned} ||g - F(A^{(\nu)})|| - ||g - F(A^*)|| &\leq ||g - F(A^{(\nu)})|| - \delta(u^{(\nu)}) \\ &\leq \frac{1}{1-\theta}(\delta(u^{(\nu+1)}) - \delta(u^{(\nu)})) \\ (8)\qquad &\leq \frac{1}{1-\theta}(||g - F(A^*)|| - \delta(u^{(\nu)})) \\ &\leq M\theta^{\nu}. \end{aligned}$$

We remark that the constant M is independent of ν. Now, from the Strong Uniqueness Theorem of [4], there exists a constant $\alpha > 0$ such that if $F(A,x) \in V(g)$, then $||g - F(A)|| \geq ||g - F(A^*)|| + \alpha\, ||F(A^*) - F(A)||$. Hence, from the above inequality, with $A = A^{(\nu)}$, and inequality (8) we obtain

$$||F(A^*) - F(A^{(\nu)})|| \leq K\theta^{\nu}. \quad \blacksquare$$

BIBLIOGRAPHY

[1] R. B. Barrar, H. L. Loeb, "On the Remez Algorithm for Non-linear Families," *Numer. Math.* 15(1970), pp. 382-391.

[2]. P. C. Curtis, Jr., "n-Parameter Families and Best Approximation," *Pacific J. Math.* 9(1959), pp. 1013-1027.

[3]. P. C. Curtis, W. L. Frank, "An Algorithm for the Determination of the Polynomial of Best Approximation to a Function Defined on a Finite Point Set," *J. Assoc. Comp. Mach.* 6(1959), pp. 395-404.

[4]. G. A. Gislason, H. L. Loeb, "Non-linear Tchebycheff Approximation with Constraints," *J. Approx. Th.* 6(1972), pp. 291-300.

[5]. H. L. Loeb, D. G. Moursund, L. L. Schumaker, G. D. Taylor, "Uniform Generalized Weight Function Polynomial Approximation with Interpolation," *SIAM J. Num. Analysis.* 6(1969), pp. 284-293.

[6]. A. L. Perrie, "Uniform Rational Approximation with Osculatory Interpolation," Doctoral Dissertation, University of Oregon, Eugene, 1969.

[7] C. J. de La Vallée Poussin, "Lecons sur l'approximations des functions d'une variable reelle," Paris, Gauthier-Villars, 1952.

TRANSFINITE INTERPOLATION AND APPLICATIONS TO ENGINEERING PROBLEMS

CHARLES A. HALL

ABSTRACT

Interpolation schemes which yield interpolants that match a given function on a nondenumerable set of points were termed transfinite by W. J. Gordon and the author. Such schemes afford the analyst the ability to construct interpolants that pass through a specified network of curves where more classical interpolation schemes are restricted to a finite number of points of interpolation. Transfinite schemes based on blending function theory are discussed as well as their application to various engineering problems related to finite difference and finite element analyses and process control computing.

1.0 INTRODUCTION

Multivariate and vector-valued interpolation problems occur frequently in engineering applications and require formulae which are easily constructed and which can be efficiently evaluated. In the following sections, four such problems are described which were successfully solved using essentially the same interpolation formula even though the problems are fundamentally distinct.

Interpolation schemes which yield interpolants that match a given function on a nondenumerable set of points were termed transfinite by W. J. Gordon and the author [10]. As background for the latter sections, we now give some basic formulae along with an indication of their approximation theoretic properties.

Consider given a function $f(x,y)$ defined on $[0,1]\times[0,1]$ and two univariate schemes. The first,

$$P_x : f \to P_x[f] \tag{1}$$

which treats y as a parameter, and the second

$$P_y : f \to P_y[f] \tag{2}$$

which treats x as a parameter. Assuming that P_x and P_y are projectors (idempotent linear transformations), two bivariate interpolation schemes are formed by compounding the univariate projectors: the *tensor product* scheme

$$P_x P_y : f \to P_x[P_y[f]] \tag{3}$$

and the *Boolean sum* scheme

$$P_x \oplus P_y : f \to P_x[f]+P_y[f]-P_x P_y[f] \tag{4}$$

Gordon [9] has shown (under the commutativity assumption $P_x P_y = P_y P_x$) that $P_x \oplus P_y$ is *algebraically better* than $P_x P_y$ in the sense that it is the maximal element and $P_x P_y$ is the minimal element in the lattice of all projectors which are combinations of P_x and P_y under the binary operations of Boolean sum and operator multiplication (cf. Figure 1).

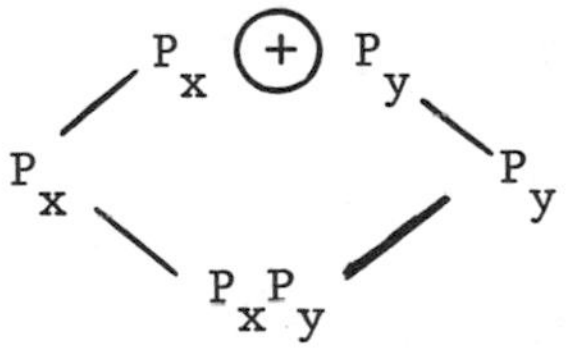

FIGURE 1: Distributive lattice with partial ordering $A \leq B$ if and only if $A(B) = A$.

The remainders also form a lattice (cf. Figure 2) which suggests that the error in the Boolean sum scheme is "smaller" in some sense than for the tensor product scheme. That is, *analytically* $P_x \oplus P_y$ is better than $P_x P_y$. For specific choices of the projectors P_x and P_y this has been shown to be the case and explicit error bounds have been derived [3,5,9,10]. (See also the Appendix.)

$$(I-P_xP_y) = (I-P_x) \oplus (I-P_y)$$

$$I-P_x \qquad\qquad I-P_y$$

$$I-(P_x \oplus P_y) = (I-P_x)(I-P_y)$$

FIGURE 2: Lattice of remainders.

As observed in [9], the approximant $P_x \oplus P_y[f]$ possesses all the interpolatory properties of both $P_x[f]$ and $P_y[f]$, while $P_xP_y[f]$ possesses only those properties common to $P_x[f]$ and $P_y[f]$. Let us be more specific and define

$$P_x[f] \equiv \sum_{i=0}^{N} f(x_i,y)\phi_i(x) \tag{5}$$

where $\phi_i(x)$ is any function that satisfies the cardinality conditions

$$\phi_i(x_j) = \delta_{ij} \quad \text{(Kronecker delta)}, \tag{6}$$

$0 = x_0 < x_1 < \ldots < x_N = 1$. Similarly,

$$P_y[f] \equiv \sum_{j=0}^{M} f(x,y_j)\psi_j(y) \tag{7}$$

where

$$\psi_j(y_k) = \delta_{kj} \tag{8}$$

and $0 = y_0 < y_1 < \ldots < y_M = 1$. From (6) and (8) we observe that by construction

$$\begin{aligned} P_x[f](x_i,y) &= f(x_i,y), \ 0\leq y\leq 1, \ 1 = 0,\ldots,N \\ P_y[f](x,y_j) &= f(x,y_j), \ 0\leq x\leq 1, \ j = 0,\ldots,M \end{aligned} \tag{9}$$

and the tensor product interpolant satisfies

$$\begin{aligned} &P_xP_y[f](x_i,y_j) = f(x_i,y_j), \\ &i = 0,\ldots,N; \ j = 0,\ldots,M \end{aligned} \tag{10}$$

while the ("maximal") Boolean sum interpolant satisfies

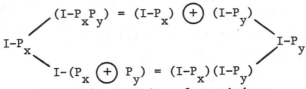

$$P_x \oplus P_y[f](x_i,y) = f(x_i,y), \quad 0 \le y \le 1, \; i = 0,\ldots,N$$

$$P_x \oplus P_y[f](x,y_j) = f(x,y_j) \quad 0 \le x \le 1, \; j = 0,\ldots,M. \tag{11}$$

The functions $\{\phi_i(x)\}_{i=0}^N$ (and $\{\psi_j(y)\}_{j=0}^M$) are termed "blending functions" since they blend the curves $\{f(x_i,y)\}_{i=0}^N$ into a surface $P_x[f]$, [8]. Such surfaces as $P_x[f]$, $P_y[f]$ and $P_x \oplus P_y[f]$ are referred to as <u>blended surfaces</u> and the schemes associated with the projectors P_x, P_y and $P_x \oplus P_y$ are <u>transfinite</u> in the sense that they match f on more than a finite number of points.

In [10,11] <u>vector-valued transfinite</u> interpolation was proposed as a means of constructing domain transformations or natural coordinatizations of various domains. At this point we simply note that the appropriate formulae (E.g. eq. (16).) follow from (5), (7) and (4) by interpreting f as a vector-valued function. (See [11] for details.)

With this very terse summary of formulae we now discuss four engineering applications of transfinite interpolation: surface interpolation, construction of finite elements, finite difference models for curved domains, and efficient storage of tabular data.

2.0 SURFACE INTERPOLATION

The classic application of bivariate tensor product and transfinite (blending) interpolation is the mathematical representation of smooth surfaces. Work in this area was spearheaded by the Mathematics Department of General Motors Research Laboratories over the past 15 years, [2,7,8]. Automobile exterior surfaces are extremely "smooth" and an explicit mathematical representation to be used in computer-aided design necessitates constructable surfaces of interpolation with continuous curvature as well as continuous gradient.

Typically, the given data consists of a set of points

$$((x_i, y_j, f(x_i, y_j): \quad 0 \leq i \leq p_1, \ 0 \leq j \leq p_2) \tag{12}$$

and due to their regularity, the tensor product scheme (3) can easily be implemented with $\phi_i(x)$ and $\psi_j(y)$ being cubic splines, $N = p_1$ and $M = p_2$. The resulting surface, passes through the given data, is of continuity class C^2 and is termed the <u>bicubic spline interpolant</u> of the given data, [2,7]. Alternatively, we can choose $N \ll p_1$, $M \ll p_2$, construct a network of <u>cubic</u> splines $\{f(x_i, y)\}_{i=0}^{N}$ $\{f(x, y_j)\}_{j=0}^{M}$ and blend these together using cubic spline blending functions $\phi_i(x)$ and $\psi_j(y)$. The resulting cubic spline blended [8] (transfinite) surface is also of continuity class C^2, but in general its construction is dependent on far fewer data than the bicubic spline. The stencils given in Figure 3 are indicative of the data used to construct the two surfaces and in

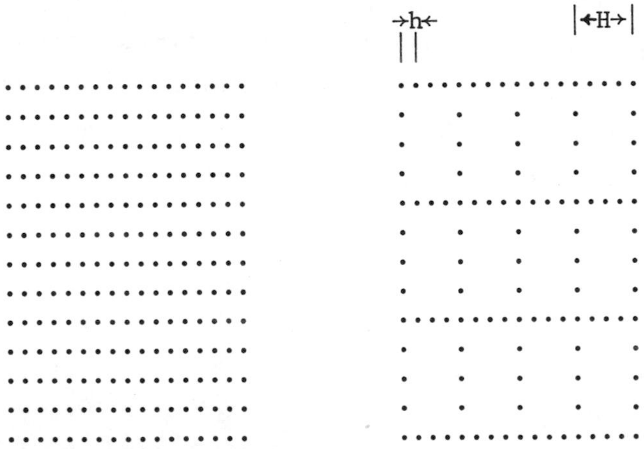

FIGURE 3: Data at 221 points determine the bicubic spline surface while data at 113 points determine the cubic spline blended surface of comparable accuracy.

this connection the Boolean sum interpolant is transfinite in the sense that the network of cubic splines are matched exactly. Gordon [9] proved that choosing $h = H^2$ in Figure 3 then asymptotically (as $H \to 0$) the bicubic spline $P_x P_y[f]$ and spline blended interpolant $P_x \oplus P_y[f]$ both converge to a function $f \varepsilon C^{(4,4)}$ at the same rate, $O(h^4)$. (Cf. (32) and also [3] for explicit error bounds.)

3.0 CONSTRUCTION OF FINITE ELEMENTS

The transfinite formulae of the generic form (4) play a central role in the construction of various standard finite element stencils and the evaluation of their convergence properties [10]. In (4) choose the blending functions to be the Lagrange cardinal basis functions

$$\phi_i(x) = \prod_{k \neq i} \frac{(x-x_k)}{(x_i-x_k)},$$
$$\psi_j(y) = \prod_{k \neq j} \frac{(y-y_k)}{(y_j-y_k)}. \tag{13}$$

First, consider the case M = N = 1 and (4) which reduces to the bilinearly blended transfinite formula

$$\begin{aligned} P_x \oplus P_y[f] &= (1-x)f(0,y) + x\, f(1,y) \\ &+ (1-y)f(x,0) + y\, f(x,1) \\ &- (1-x)(1-y)f(0,0) - (1-x)\, y\, f(0,1) \\ &- x(1-y)f(1,0) - xy\, f(1,1). \end{aligned} \tag{14}$$

Note that $P_x \oplus P_y[f]$ matches f on the boundary of the unit square and if we choose the boundary curves appropriately, (14) reduces to the interpolation formula associated with various finite element stencils as indicated in Figure 4.

Similarly, for the case M = N = 2, the biquadratically blended formula becomes

$$P_x \oplus P_y[f] = \phi_0(x)f(0,y) + \phi_1(x)f(\tfrac{1}{2},y) + \phi_2(x)f(1,y) + \psi_0(y)f(x,0) + \psi_1(y)f(x,\tfrac{1}{2}) + \psi_2(y)f(x,1) - \sum_{i=0}^{2}\sum_{j=0}^{2} f(x_i,y_j)\phi_i(x)\psi_j(y) \quad (15)$$

where $\phi_0(x) = 2(\frac{1}{2}-x)(1-x)$, $\phi_1(x) = 4x(1-x)$, $\phi_2(x) = 2x(x-\frac{1}{2})$ and $\psi_j(y) = \phi_j(x)$. Appropriate choices of the network curves produce the formulae associated with the finite element stencils in Figure 5.

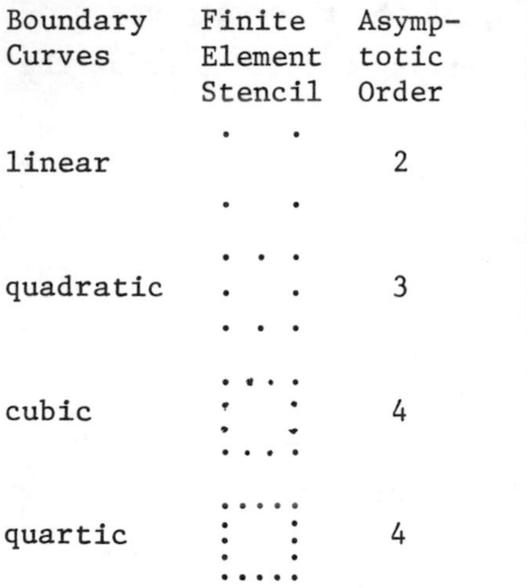
Boundary Curves
Finite Element Stencil
Asymptotic Order
linear
2
quadratic
3
cubic
4
quartic
4

FIGURE 4: Finite element stencils - bilinear blending.

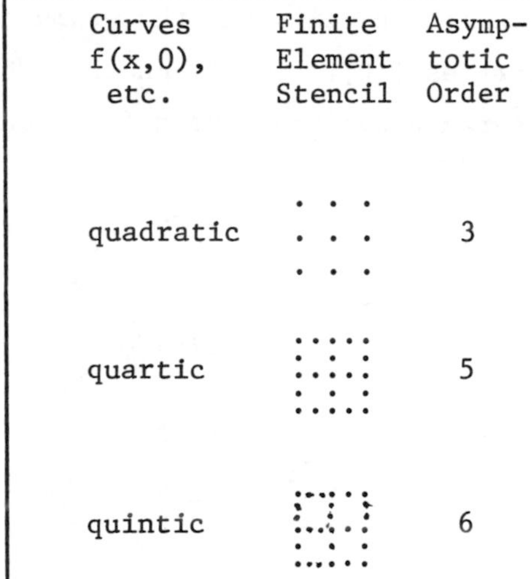
Curves f(x,0), etc.
Finite Element Stencil
Asymptotic Order
quadratic
3
quartic
5
quintic
6

FIGURE 5: Finite element stencils - biquadratic blending.

In Gordon and Hall [10], details of such constructions are given, along with error analyses. In [5] the authors generalize such formulae to include L-spline blending functions and a network of L-spline curves. (See also the Appendix.) Birkhoff, Cavendish and Gordon [1] propose the use of piecewise Hermite and cubic spline boundary functions in conjunction with the transfinite formula (14) to develop finite element stencils that allow flexibility in abutting different size elements. Such schemes, studied further in [4,6], are valuable in locally refining a finite element mesh or in "clustering" degrees of freedom in regions of anticipated erratic behavior of the solution.

So far, our domain has been the unit square $\mathcal{S}$: $[0,1]\times[0,1]$, however, in practice the element domain is $[0,h_x]\times[0,h_y]$ or, more generally, a domain $\mathcal{E}$ with curved boundaries. The most commonly used domain transformations are also transfinite formulae of the generic form (14) where now the "curves" are in fact "vectors". For example, if $\mathcal{E}$ is the domain in Figure 6 with boundary segments parametrized as indicated, then $\overline{T}: \mathcal{S} \rightarrow \mathcal{E}$ where

$$\begin{aligned}\overline{T}: \begin{bmatrix} x(s,t) \\ y(s,t) \end{bmatrix} &= (1-s)\begin{bmatrix} x(0,t) \\ y(0,t) \end{bmatrix} + s\begin{bmatrix} x(1,t) \\ y(1,t) \end{bmatrix} \\ &+ (1-t)\begin{bmatrix} x(s,0) \\ y(s,0) \end{bmatrix} + t\begin{bmatrix} x(s,1) \\ y(s,1) \end{bmatrix} \\ &- (1-s)(1-t)\begin{bmatrix} x(0,0) \\ y(0,0) \end{bmatrix} - (1-s)t\begin{bmatrix} x(0,1) \\ y(0,1) \end{bmatrix} \\ &- s(1-t)\begin{bmatrix} x(1,0) \\ y(1,0) \end{bmatrix} - st\begin{bmatrix} x(1,1) \\ y(1,1) \end{bmatrix}.\end{aligned} \tag{16}$$

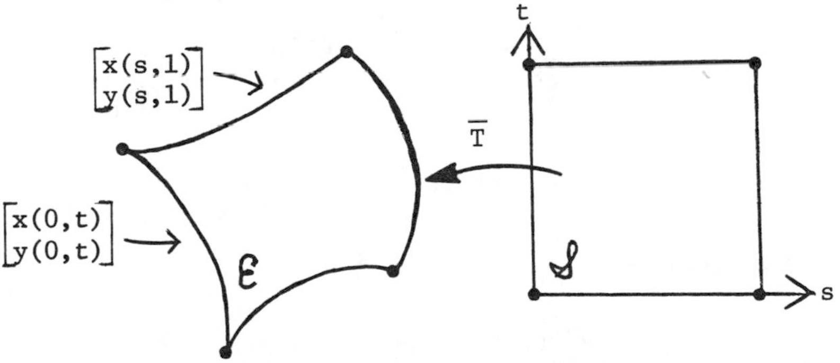

FIGURE 6: Domain Transformation.

If the boundary curves, e.g. $\begin{bmatrix} x(s,0) \\ y(s,0) \end{bmatrix}$ are parametrized as polynomials of the <u>same</u> form as used to approximate the dependent variable f (by, say u) and also the <u>same</u> blending functions are used then we obtain the <u>isoparametric</u> formulae of Zienkiewicz [21], (cf. Figure 7). Similarly, <u>sub</u>- and <u>super</u>-<u>parametric</u> elements can be

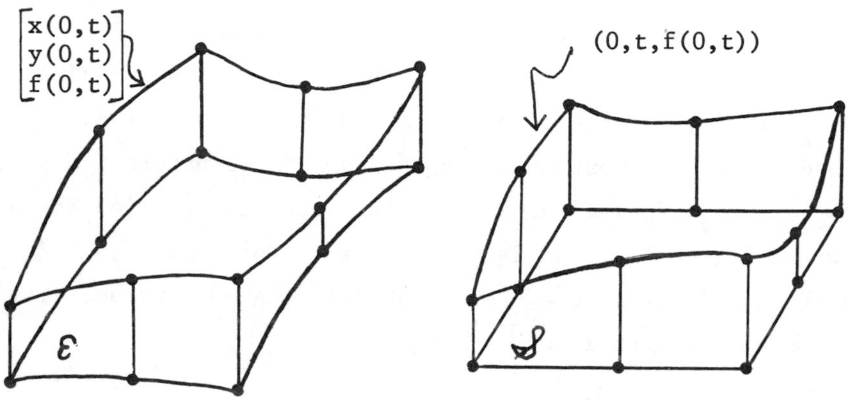

FIGURE 7: An isoparametric element generated from the transfinite formula:

$$\begin{aligned}\begin{bmatrix}x(s,t)\\y(s,t)\\u(s,t)\end{bmatrix} &= (1-s)\begin{bmatrix}x(0,t)\\y(0,t)\\f(0,t)\end{bmatrix} + s\begin{bmatrix}x(1,t)\\y(1,t)\\f(1,t)\end{bmatrix}\\ &+ (1-t)\begin{bmatrix}x(s,0)\\y(s,0)\\f(s,0)\end{bmatrix} + t\begin{bmatrix}x(s,1)\\y(s,1)\\f(s,1)\end{bmatrix}\\ &- (1-s)(1-t)\begin{bmatrix}x(0,0)\\y(0,0)\\f(0,0)\end{bmatrix} - (1-s)t\begin{bmatrix}x(0,1)\\y(0,1)\\f(0,1)\end{bmatrix}\\ &- s(1-t)\begin{bmatrix}x(1,0)\\y(1,0)\\f(1,0)\end{bmatrix} - st\begin{bmatrix}x(1,1)\\y(1,1)\\f(1,1)\end{bmatrix},\end{aligned} \tag{17}$$

where for example,

$$\begin{aligned}\begin{bmatrix}x(0,t)\\y(0,t)\\f(0,t)\end{bmatrix} &= (\tfrac{1}{2}-t)(1-t)\begin{bmatrix}x(0,0)\\y(0,0)\\f(0,0)\end{bmatrix}\\ &+ 4t(1-t)\begin{bmatrix}x(0,\tfrac{1}{2})\\y(0,\tfrac{1}{2})\\f(0,\tfrac{1}{2})\end{bmatrix} + 2t(t-\tfrac{1}{2})\begin{bmatrix}x(0,1)\\y(0,1)\\f(0,1)\end{bmatrix}.\end{aligned} \tag{18}$$

derived by making different choices. So for example, the element in Figure 8 is obtained from parametrizing the curves $\begin{bmatrix} x(0,t) \\ y(0,t) \end{bmatrix}$, etc. by quadratics and linear blending these curves together with linear boundary curves f(0,t), etc.

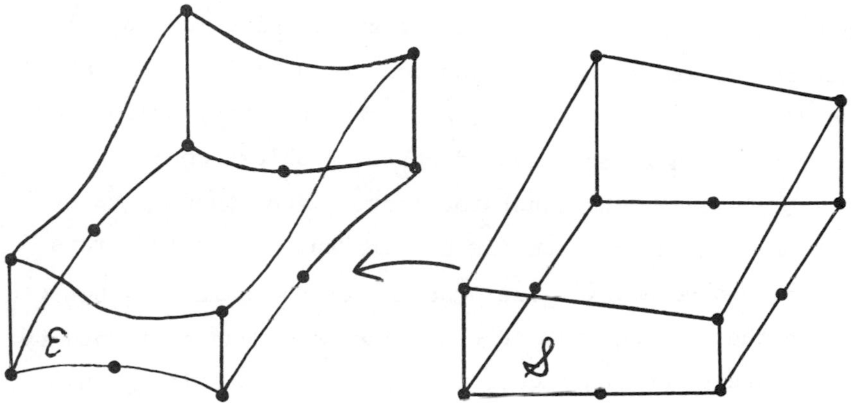

FIGURE 8: A sub-parametric element generated from the linear blending (transfinite) scheme.

4.0 FINITE DIFFERENCE MODELS FOR CURVED DOMAINS

In engineering applications, one is often faced with the problem of analyzing physical phenomena such as heat transfer in structures which are rather oddly shaped. Conventional finite difference programs approximate curved interfaces and boundaries by polygonal arcs and translate interface and boundary conditions to such polygonal approximations. Alternatively, we can exploit vector-valued transfinite interpolation and map the given problem domain one-to-one into a square parameter domain, [5,12]. This is equivalent to the introduction of a generalized curvilinear coordinate system on the original problem domain which preserves the integrity of the boundary as constant coordinate lines. As in [12], the given boundary value problem is then cast in terms of the induced curvilinear coordinate system and a finite difference model formulated and solved with respect to this more "natural" coordinate system. In this respect, our approach is akin to the classical handling of axially symmetric problems in terms of natural cylindrical or polar coordinates. See Morse and Feshback [14, p. 655] for a catalogue of classical curvilinear coordinate systems and Gordon and Hall [11] for natural coordinate systems which preserve the integrity of the boundary for a much wider class of curved problem domains.

Utilizing such transformations, given boundary value problems with curved domains can be "automatically" transformed into boundary value problems of the same type but with domain the unit square. For domains which can be mapped by the chosen transformations, the program discretizing the differential operator can then be treated as a "black-box" with little or no guidance or interference from the user, [12].

Consider the curved domain $\mathcal{R}$ given in Figure 9, and suppose we seek to solve Laplace's equation $\nabla^2 u = 0$ subject to Dirichlet boundary conditions so that the solution is $u = 5-x+y$. We proceed as follows:

(i) Four boundary points are designated as images of the corners of the unit square, S under the mapping $\overline{T}$ in (16) where now we are treating the entire domain R as an element.

(ii) The four boundary segments $\overline{F}(0,t) = \begin{bmatrix} x(0,t) \\ y(0,t) \end{bmatrix}$, etc. connecting these points are input as indicated in Figure 9.

(iii) The vector valued linearly blended map (16) is now completely specified and the graph of constant s and t coordinate lines are displayed in Figure 9 with one line emanating from the "fifth" corner.

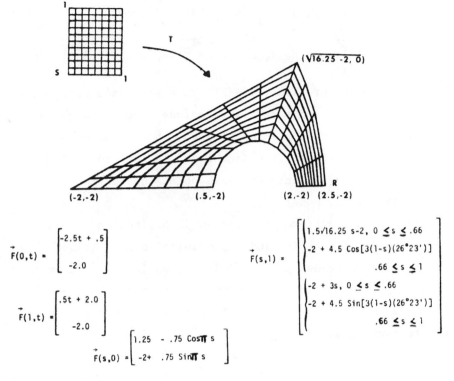

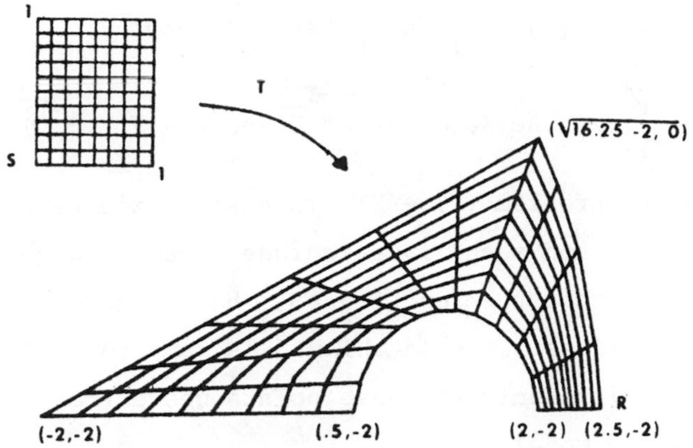

$$\vec{F}(0,t) = \begin{bmatrix} -2.5t + .5 \\ -2.0 \end{bmatrix} \qquad \vec{F}(1,t) = \begin{bmatrix} .5t + 2.0 \\ -2.0 \end{bmatrix}$$

$$\vec{F}(s,0) = \begin{bmatrix} 1.25 - .75 \cos \pi s \\ -2 + .75 \sin \pi s \end{bmatrix}$$

$$\vec{F}(s,1) = \begin{bmatrix} \begin{cases} 1.5\sqrt{16.25}\, s - 2, & 0 \le s \le .66 \\ -2 + 4.5 \cos[3(1-s)(26°23')] & .66 \le s \le 1 \end{cases} \\ \begin{cases} -2 + 3s, & 0 \le s \le .66 \\ -2 + 4.5 \sin[3(1-s)(26°23')] & .66 \le s \le 1 \end{cases} \end{bmatrix}$$

FIGURE 9: The given domain R in the x-y plane.

(iv) The boundary value problem is cast in its variational formulation: Find u such that

$$F[u] = \min_{w \in W_2^1} F[w] \tag{19}$$

where

$$F[w] = \tfrac{1}{2}\int \{w_x^{\,2} + w_y^{\,2}\}dxdy \tag{20}$$

and W_2^1 is the Sobolev space of functions with square integrable generalized derivatives which are zero on the boundary of **R** . The "change of coordinates" (16) can be effected to yield

$$F[w] = \frac{1}{2}\int\{(w_s w_t)B(w_s w_t)^T\}dsdt \tag{21}$$

where $J = \begin{bmatrix} x_s x_t \\ y_s y_t \end{bmatrix}$ is the Jacobian of $\overline{T}$ and $B = J^{-1}(J^{-1})^T|\det J|$.

The minimization problem (19)-(21) is discretized using finite differences [12], and mesh lines include lines through extra corner points so that the functions B_{ij} have discontinuities only along mesh lines. The Jacobian J is computed and inverted (numerically) only at those points needed for the finite difference model.

The finite difference solution is plotted in the parameter domain (Figure 10) and the original domain (Figure 11). The interested reader is referred to [10] for more illuminating examples and discussion of this technique. Suffice to say here that the key ingredient is the vector-valued transfinite mapping (16).

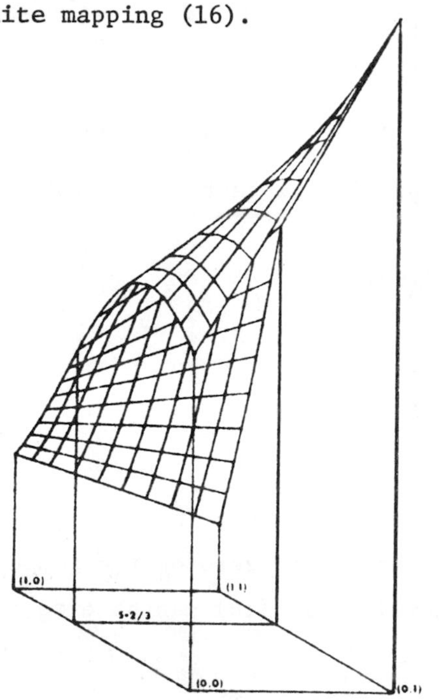

FIGURE 10: Finite difference solution over the parameter domain S.

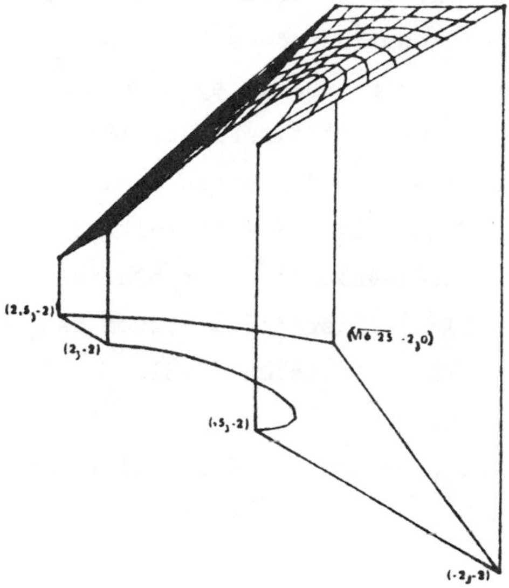

FIGURE 11: Finite difference solution plotted over original domain R. Solving the transformed problem introduced a discretization error 0.1%.

5.0 STORAGE OF TABULAR DATA

In a real time environment or simulation it is often desirable to have the capability of retreiving good approximate values of tabulated data, e.g. thermodynamic properties of steam and water are necessary in the operation of nuclear and fossil fired power plants. In [13], it was determined that such thermodynamic properties (enthalpy, specific volume, etc.) can best be considered as surfaces constructed over a domain in the temperature-pressure plane (cf. Figure 12) and their approximation reduces to a surface fitting problem not totally unlike those considered in §2. However, there are three important distinctions: (1) the "surface" or tabular data is fixed, (2) the domain is curved and (3) the size of computer (25-30K) involved in process control limits the amount of information that can be stored and efficiently retrieved.

Transfinite interpolation was used in [13] to construct domain transformations and to interpolate steam and water properties. Note that if enthalpy is considered as the dependent variable and pressure-temperature as the independent variables,then the domain is as illustrated in Figure 13; similarly, when pressure is chosen as the dependent variable, the domain is as illustrated in Figure 14. Now for either of these domains the transfinite (vector-valued) interpolation scheme $\overline{T}$ in (16) induces a curvilinear coordinate system as indicated in Figures 13 and 14. The associated "table look-up" problems are then respectively:

(1) Given a temperature and pressure, find enthalpy, <u>or</u> (2) Given a temperature and enthalpy, find pressure. In both cases we must

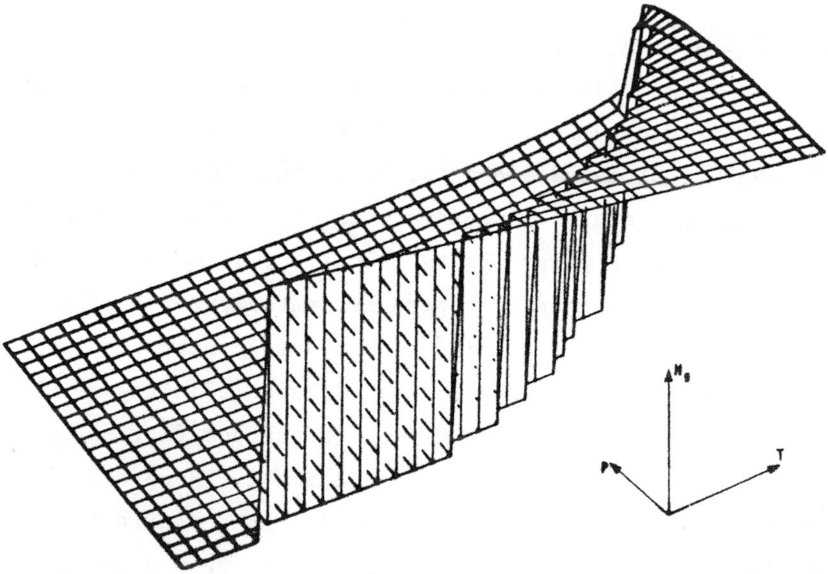

FIGURE 12: Enthalpy of superheated steam versus temperature and pressure.

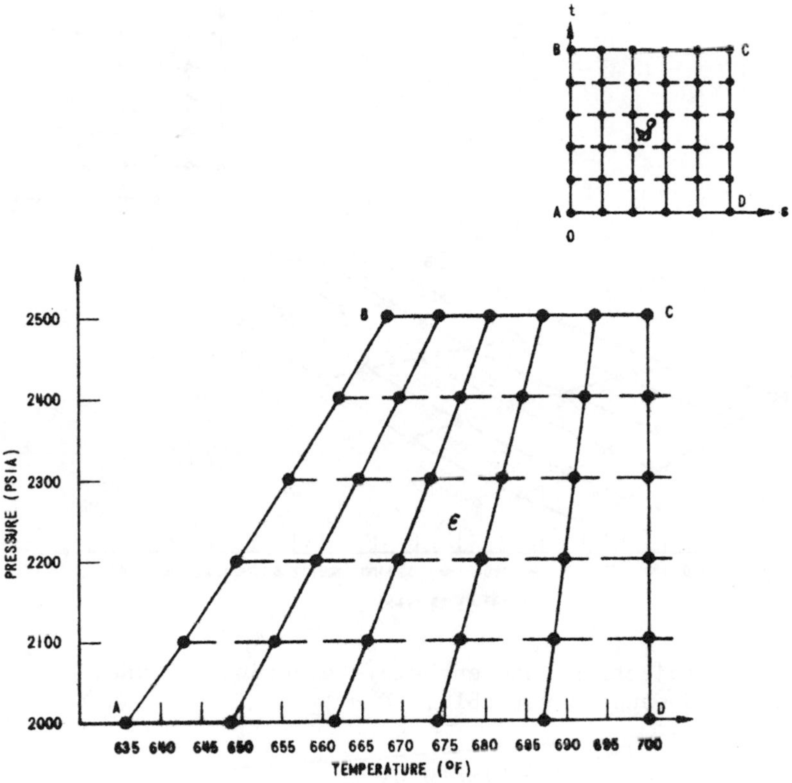

FIGURE 13: Projection into temperature-pressure plane: enthalpy the dependent variable. The saturation line AB is curved.

in fact invert the map $\overline{T}$ and then evaluate, for example, a linear blended surface (14) at the appropriate (s,t) value.

The first scheme was chosen since the inversion of $\overline{T}$ is linear while in general (and in particular for Figure 14) this inversion results in having to solve a system of non-linear equations. For the coordinatization in Figure 13, the inverse is

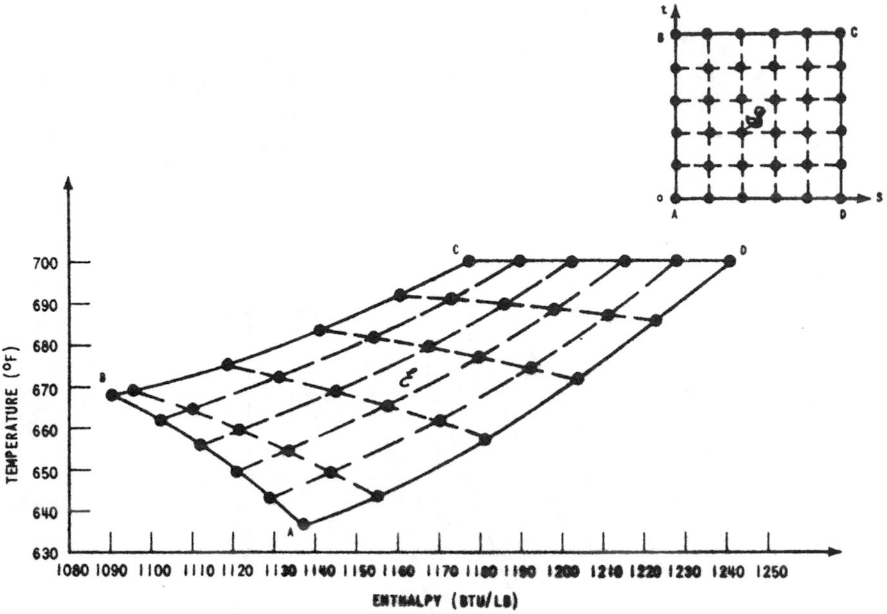

FIGURE 14: Projection into enthalpy-temperature plane: pressure the dependent variable.

$$
\begin{aligned}
t &= \frac{P(0,t)-P(0,0)}{P(1,1)-P(0,0)}\ , \\
s &= \frac{T(s,t)-T(0,t)}{T(1,t)-T(0,t)}\ .
\end{aligned} \tag{22}
$$

Given a temperature T and pressure P, the special form of the domain implies that $P(0,t) = P$ from which the first equation is (22) yields the t value associated with P and T. Given t, the second equation in (22) yields the s value associated with P and $T \equiv T(s,t)$. We now have the (s,t) coordinates corresponding to the given (P,T) coordinates. Enthalpy is then given by the linear blended formula (14).

If a wide range of pressures and temperatures are desired, then the domain can be subdivided into several elements and the above procedure applied to each element. The fact that the "surface" or tabular data is _fixed_ allows one to _a priori_ determine a configuration and number of elements which yield the desired accuracy and remain within the storage requirements of the process control computer. In [13], for a narrow [wide] range of temperatures and pressures, roughly $\frac{2}{5}\left[\frac{1}{100}\right]$ the tensor product table data was used in the linear blended scheme. Also in the tensor product scheme, _ad hoc_ strategies were necessary near the curved saturation line. In short, the transfinite interpolation scheme has been implemented by Boris Mutafelija of the Pressurized Water Reactor Group at Westinghouse Electric Corporation and performs quite satisfactorily.

REFERENCES

1. G. Birkhoff, J. C. Cavendish and W. J. Gordon, "Multivariate Approximation by Locally Blended Univariate Interpolants," _Proc. Nat. Acad. Sci._, 71, 3423-3425, (1974).

2. G. Birkhoff and C. DeBoor, "Piecewise Polynomial Interpolation and Approximation," in _Approximation Functions_, H. Garabedian, ed., 164-190 Amsterdam, Elsevier, 1965.

3. R. E. Carlson and C. A. Hall, "Error Bounds for Bicubic Spline Interpolation," _J. Approx. Theory_, 7, 41-47, (1973).

4. J. C. Cavendish, "Local Mesh Refinement Using Rectangular Blended Finite Elements," to appear: _J. Comp. Physics_.

5. J. C. Cavendish, W. J. Gordon and C. A. Hall, "Ritz-Galerkin Approximations in Blending Function Spaces," To appear: _Numer. Math_.

6. J. C. Cavendish, W. J. Gordon and C. A. Hall, "Substructured Macro Elements Based on Locally Blended Interpolation," In preparation.

7. C. DeBoor, "Bicubic Spline Interpolation," _J. Math. and Phys._, 41, 212-218, (1962).

8. W. J. Gordon, "Spline-Blended Surface Interpolation Through Curve Networks," J. Math. Mech., 10, 931-952, (1968).

9. W. J. Gordon, Distributive Lattices and the Approximation of Multivariate Functions," in Approximation with Special Emphasis on Spline Functions, 223-277, I. J. Schoenberg, ed., New York, Academic Press, 1969.

10. W. J. Gordon and C. A. Hall, "Transfinite Element Methods: Blending-Function Interpolation Over Curved Element Domains," Numer. Math. 21, 109-129, (1973).

11. W. J. Gordon and C. A. Hall, "Construction of Curvilinear Coordinate Systems and Applications to Mesh Generation," Int. J. Numer. Meth. Engg., 7, 461-477, (1973).

12. C. A. Hall, R. W. Luczak and A. G. Serdy, "Numerical Solution of Steady State Heat Flow Problems Over Curved Domains," To appear: TOMS.

13. C. A. Hall and B. A. Mutafelija, "Transfinite Interpolation of Steam Tables," J. Comp. Physics, 18, 79-91, (1975).

14. P. Morse and H. Feshbach, Methods of Theoretical Physics, Part I, McGraw Hill, New York, 1953.

15. M. Schultz and R. S. Varga, "L-splines," Numer. Math., 10, 345-369, (1967).

16. M. Schultz, "L^∞-multivariate Approximation Theory," SIAM J. Numer. Analysis, 6, 161-183, (1969).

17. M. Schultz, "L^2-multivariate Approximation Theory," SIAM J. Numer. Analysis, 6, 184-209, (1969).

18. M. Schultz, Spline Analysis, Prentice-Hall, Inc., Englewood Cliffs, New Jersey, 1973.

19. B. K. Swartz and R. S. Varga, "Error Bounds For Spline and L-spline Interpolation," J. Approx. Theory, 6, 6-149, (1972).

20. R. S. Varga, Functional Analysis and Approximation Theory in Numerical Analysis, SIAM, 1972.

21. O. C. Zienkiewicz, The Finite Element Method in Engineering Science, McGraw Hill, London, 1971.

APPENDIX: ASYMPTOTIC CONVERGENCE RESULTS

L-splines have been given considerable attention in the approximation theory literature [15-20]. One natural extension for multivariate approximation is tensor products of univariate L-splines. Another extension, exploited by Cavendish, Gordon and the author [5] is provided by blending function theory in which Boolean sums of univariate L-spline approximation operators are considered. Let

$$\pi_x : 0 = x_0 < x_1 < \ldots < x_{N+1} = 1 \tag{23}$$

be a partition of [0,1] with mesh gauge and mesh ratio

$$h_x \equiv \max_i |x_i - x_{i-1}|,$$
$$\beta_x \equiv h_x / \min_i |x_i - x_{i-1}| \tag{24}$$

respectively. Consider the nth order differential operator L_x defined by

$$L_x[u](x) \equiv \sum_{j=0}^{n} c_j(x) D^j u(x), \quad n \geq 1 \tag{25}$$

where $c_j(x) \in C^j[0,1]$, $0 \leq j \leq n$ and $c_n(x) \geq \delta > 0$ for $x \in [0,1]$. The L-spline space, $S(L_x, \pi_x, z_x)$ associated with the operator L_x, partition π_x and integer $z_x (1 \leq z_x \leq n)$, is defined to be the collection of all real-valued functions $w(x)$ defined on [0,1] such that

$$L_x^* L_x[w](x) = 0, \quad x \in (0,1) - \{x_i\}_{i=1}^{N}$$
$$D^k w(x_i-) = D^k w(x_i+),$$
$$0 \leq k \leq 2n-1-z_x, \quad 0 < i < N+1$$

where $L_x^*[v](x) \equiv \sum_{j=0}^{n} (-1)^j D^j \{c_j(x) v(x)\}$ is the formal adjoint of L_x.

From Varga [20] and Swartz and Varga [19] we have $S(L_x, \pi_x z_x) \subset W_\infty^{2n-z_x}[0,1]$, dim $S(L_x, \pi_x, z_x) = 2n + z_x N$ and given $g \varepsilon W_2^{2n}[0,1]$, there exists a unique element $s \varepsilon S(L_x,\pi_x,z_x)$ such that

$$
\begin{aligned}
&\text{(i) } D^j[g-s](x_i) = 0, \\
&\qquad 0 \le j \le z_x-1,\ 0 < i < N+1 \\
&\text{(ii) } D^j[g-s](0) = D^j[g-s](1) = 0, \\
&\qquad 0 \le j \le n-1 \\
&\text{(iii) } ||D^j[g-s]||_{L_q[0,1]} \\
&\qquad \le K_x h_x^{2n-j-1/2+1/q} ||g||_{W_2^{2n}[0,1]}, \\
&\qquad 0 \le j \le 2n-1,\ 2 \le q \le \infty,
\end{aligned}
\tag{26}
$$

where K_x is a constant and β_x is bounded as $h_x \to 0$.

Schultz [17, Theorem 4.6] establishes that for a rectangular domain $\mathcal{S}$, if P_x and P_y are <u>orthogonal projectors</u> of $L_2(\mathcal{S})$ onto $S(L_x,\pi_x,z_x)$ and $S(L_y,\pi_y,z_y)$ respectively, then the <u>tensor product</u> projector $P_x P_y$ satisfies <u>for</u> $k = 0$, $0 \le \ell \le 2m-1$ <u>or</u> $\ell = 0$, $0 \le k \le 2n-1$

$$
\begin{aligned}
&||(f-P_xP_y[f])^{(k,\ell)}||_{L_2(\mathcal{S})} \\
&\le ||(f^{(0,\ell)} - P_x[f^{(0,\ell)}])^{(k,0)}||_{L_2(\mathcal{S})} \\
&+ ||(f^{(k,0)} - P_y[f^{(k,0)}])^{(0,\ell)}||_{L_2(\mathcal{S})} \\
&= 0(h_x^{2n-k} + h_y^{2m-\ell})
\end{aligned}
$$

for sufficiently smooth functions f. (cf. Figure 2).

In [5], the following result is proven which establishes convergence of a very general class of transfinite L-spline interpolants.

<u>THEOREM</u> (Cavendish, Gordon, Hall [5]): <u>Let</u> P_x <u>and</u> P_y <u>be commuting projectors from</u> $W_2^{(2n,2m)}(\mathcal{S})$ <u>onto</u> $S(L_x,\pi_x,x_x)$ <u>and</u>

$S(L_y, \pi_y, z_y)$ respectively, determined by the interpolation schemes (26i,26ii). Then the Boolean sum projection $s_f \equiv (P_x \oplus P_y)[f]$ satisfies the following interpolation properties:

$$(i)\quad \begin{aligned} &(f-s_f)^{(k,0)}(x_i,y) = 0, \\ &0 \le k \le z_x-1, \\ &0 < i < N+1, \quad y \in [0,1] \\ &(f-s_f)^{(0,\ell)}(x,y_j) = 0, \\ &0 \le \ell \le z_y-1, \\ &0 < j < M+1, \quad x \in [0,1] \end{aligned}$$

$$(ii)\quad \begin{aligned} &(f-s_f)^{(k,0)}(0,y) \\ &\qquad = (f-s_f)^{(k,0)}(1,y) = 0 \\ &0 \le k \le n-1, \quad y \in [0,1] \\ &(f-s_f)^{(0,\ell)}(x,0) \\ &\qquad = (f-s_f)^{(0,\ell)}(x,1) = 0, \\ &0 \le \ell \le m-1, \quad x \in [0,1] \end{aligned} \tag{27}$$

where $f \in W_2^{(2n,2m)}(\mathcal{S})$. Moreover, if β_x and β_y are bounded as $h \to 0$, then

$$\begin{aligned} &||(f-P_x \oplus P_y[f])^{(k,\ell)}||_{L_2(\mathcal{S})} \\ &\qquad \le K_x K_y h_x^{2n-k} h_y^{2m-\ell} ||f||_{W_s(2n,2m)(\mathcal{S})} \\ &\qquad = O(h_x^{2n-k} h_y^{2m-\ell}) \end{aligned} \tag{28}$$

for $0 \le k \le 2n-1$ and $0 \le \ell \le 2m-1$ (cf. Figure 2).

Next, let $\bar{\pi}_x$: $0 = \bar{x}_0 < \bar{x}_1 < \ldots < \bar{x}_{\bar{N}+1} = 1$ and $\bar{\pi}_y$: $0 = \bar{y}_0 < \bar{y}_1 < \ldots < \bar{y}_{\bar{M}+1} = 1$ be partitions of [0,1] of mesh gauges $\bar{h}_x$ and $\bar{h}_y$ respectively, and such that $\bar{\pi}_x$ is a refinement of π_x and $\overline{\pi}_y$ is a refinement of π_y. Choose operators $\overline{L}_x$ and $\overline{L}_y$ and integers $\overline{z}_x \geq z_x$, $\overline{z}_y \geq z_y$. If $\overline{P}_x$ and $\overline{P}_y$ are the corresponding L-spline interpolation projectors, then the discrete $(L_x,z_x;L_y,z_y;\overline{L}_x,\overline{z}_x;\overline{L}_y,\overline{z}_y)$ blended interpolant S_f to f, is given by

$$S_f \equiv [\overline{P}_y P_x + \overline{P}_x P_y - P_x P_y][f], \tag{29}$$

and we have

THEOREM 2 (Cavendish, Gordon, Hall [5]): *The discrete* $(L_x,z_x;L_y,z_y;\overline{L}_x,\overline{z}_x;\overline{L}_y,\overline{z}_y)$ *blended interpolant* S_f *defined in* (29) *interpolates to* f *in the sense that*

$$\begin{gathered}(S_f-f)^{(k,\ell)}(s_i,\overline{y}_j) = 0\\ 0 \leq k \leq z_x-1,\ 0 < i < N+1\\ 0 \leq k \leq n-1,\ i = 0,N+1\\ 0 \leq \ell \leq \overline{z}_y-1,\ 0 < j < \overline{M}+1\\ 0 \leq \ell \leq \overline{m}-1,\ j = 0,\overline{M}+1\end{gathered} \tag{30}$$

and

$$\begin{gathered}(S_f-f)^{(k,\ell)}(\overline{x}_i,y_j) = 0\\ 0 \leq k \leq \overline{z}_x-1,\ 0 < i < \overline{N}+1\\ 0 \leq k \leq \overline{n}-1,\ i = 0,\overline{N}+1\\ 0 \leq \ell \leq z_y-1,\ 0 < j < M+1\\ 0 \leq \ell \leq m-1,\ j = 0,M+1\end{gathered} \tag{31}$$

Further, if the projectors $P_x,P_y,\overline{P}_x,\overline{P}_y$ *satisfy* $P_xP_y = P_yP_x$, $\overline{P}_xP_y = P_y\overline{P}_x$, *and* $P_x\overline{P}_y = \overline{P}_yP_x$ *and if* $f \in W_2^{(2p,2q)}(\mathcal{R})$, *where* $p = \max(n,\overline{n})$ *and* $q = \max(m,\overline{m})$, *then*

$$\|(f-S_f)^{(k,\ell)}\|_{L_2(\mathcal{R})} = O(h_x^{2n-k}h_y^{2m-\ell} + \overline{h}_x^{2\overline{n}-k} + \overline{h}_y^{2\overline{m}-\ell}) \tag{32}$$

as $h \equiv \max(h_x,h_y) \to 0$, $0 \leq k \leq 2\min(n,\overline{n}) - 1$ *and* $0 \leq \ell \leq 2\min(m,\overline{m}) - 1$.

Proof: As in Gordon [9, Lemma A1],

$$f-S_f = [I - (\overline{P}_y P_x + \overline{P}_x P_y - P_x P_y)][f]$$
$$= [(I-\overline{P}_x) + (I-\overline{P}_y) + (I-P_x \oplus P_y)$$
$$- (I-\overline{P}_x \oplus P_y) - (I-P_x \oplus \overline{P}_y)][f].$$

The desired result follows from (28) and (26iii). Q.E.D.

SOME REMARKS ON BEST SIMULTANEOUS APPROXIMATION

A.S.B. Holland and B.N. Sahney*

1. The problem of simultaneous Chebyshev approximation of two real-valued functions f_1 and f_2 defined on the interval [0,1] has been studied by C.B. Dunham [4] and by J.B. Diaz and H.W. McLaughlin [1,2,3]. The problem of best simultaneous approximation of two functions in abstract spaces has been discussed by Holland, Sahney etc. in [5,6], and a recent paper by Holland, Sahney and Tzimbalario [7] deals with the problem for more than two functions. Recently, Phillips and Sahney [8] have considered the problem of simultaneous approximation of two functions with respect to the L_1 and L_2 norm.

The purpose of this paper is to study the following two questions:

(i) What is the behaviour of the simultaneous approximation of n functions, in the L_2 norm?

(ii) What is the behaviour, in the L_p norm, of the simultaneous approximation of two functions which are L_p-integrable, given that p is an even natural number.

2. In this section we deal with the first question. Let us denote by S the non-empty set of real or complex-valued square integrable functions defined on an interval [a,b].

---

*The research in part, is supported by the National Research Council of Canada and the Scientific Affairs Division of N.A.T.O.

Definition. If there exists an element $s^* \in S$ such that

$$\inf_{s \in S} \left[\sum_{i=1}^{n} \|f_i - s\|_2^2 \right] = \sum_{i=1}^{n} \|f_i - s^*\|_2^2 , \qquad (2.1)$$

then we say that s^* is the best simultaneous approximation (b.s.a.) to the n functions $f_1, f_2, \ldots, f_n$ in the L_2 norm.

The following theorem is based on the above definition.

Theorem I. For s and f_i defined as above,

$$\inf_{s \in S} \sum_{i=1}^{n} \|f_i - s\|_2^2 = \inf_{s \in S} n \|\frac{1}{n} \sum_{i=1}^{n} f_i - s\|_2^2 + \frac{1}{n} \sum_{i<j} \|f_i - f_j\|_2^2 . \qquad (2.2)$$

Corollary. An element $s^* \in S$ is the b.s.a. to the n functions $f_1, f_2, \ldots, f_n$ (in an L_2 norm) if and only if it is a best approximation to the mean value of the functions, i.e. to

$$\frac{1}{n} \sum_{i=1}^{n} f_i .$$

Proof of Theorem I. The following identity can be easily verified,

$$\left(\sum_{i=1}^{n} a_i \right)^2 + \sum_{i<j} \left(a_i - a_j \right)^2 = n \sum_{j=1}^{n} a_j^2 . \qquad (2.3)$$

Let us choose $a_i = f_i - s$ where $s \in S$. Then we have

$$\left\{ \sum_{i=1}^{n} \left(f_i(x) - s(x) \right) \right\}^2 + \sum_{i<j} \left(f_i(x) - f_j(x) \right)^2 = n \sum_{j=1}^{n} \left(f_j(x) - s(x) \right)^2 \qquad (2.4)$$

Integrating each side from a to b and dividing by n we obtain,

$$n \| \frac{1}{n} \sum_{i=1}^{n} f_i - s \|_2^2 + \frac{1}{n} \sum_{i<j} \|f_i - f_j\|_2^2 = \sum_{j=1}^{n} \|f_j - s\|_2^2 . \qquad (2.5)$$

Taking the infimum over all $s \in S$, we obtain the theorem.

3. We now study the second question.

Let us assume that f_1, f_2 and all $s \in S$ are L_p integrable.

Definition. If there exists an element $s \in S$ such that

$$\inf_{s \in S}\left[\|f_1 - s\|_p^p + \|f_2 - s\|_p^p\right] = \|f_1 - s^*\|_p^p + \|f_2 - s^*\|_p^p \tag{3.1}$$

then we say that s^* is the b.s.a. to the functions f_1 and f_2 in the L_p norm.

We now prove the following theorem, in the case when p is an even natural number.

Theorem II. For s, f_1 and f_2 as above, and p an even natural number,

$$\inf_{s \in S}\left[\|f_1 - s\|_p^p + \|f_2 - s\|_p^p\right] = 2\inf_{s \in S}\left\{\sum_{k=0}^{p/2}\binom{p}{2k}\int_a^b\left[\frac{f_1(x)+f_2(x)}{2} - s(x)\right]^{p-2k}\left[\frac{f_1(x)-f_2(x)}{2}\right]^{2k}dx\right\} \tag{3.2}$$

This theorem essentially says the following: the b.s.a. to the two functions (in the L_p norm), is equivalent to the best approximation to the mean value of the functions in a certain sense, such that weight functions become involved as multipliers.

Proof. We first show the existence of the right hand side in the sense of the L_p norm.

Since

$$\int_a^b g(x)h(x)dx \le \left\{\int_a^b g^r(x)dx\right\}^{\frac{1}{r}}\left\{\int_a^b h^s(x)dx\right\}^{\frac{1}{s}} \tag{3.3}$$

where $\frac{1}{r} + \frac{1}{s} = 1$, and $g(x)$ and $h(x)$ are suitably integrable, we have for any $s \in S$,

$$2\left\{\sum_{k=0}^{p/2}\binom{p}{2k}\int_a^b\left[\frac{f_1(x)+f_2(x)}{2} - s(x)\right]^{p-2k}\left[\frac{f_1(x)-f_2(x)}{2}\right]^{2k}dx\right\}$$

$$\leq 2\left\{\sum_{k=0}^{p/2}\binom{p}{2k}\left[\int_a^b\left(\frac{f_1(x)+f_2(x)}{2} - s(x)\right)^p dx\right]^{\frac{p-2k}{p}}\right.$$

$$\left.\left[\int_a^b\left(\frac{f_1(x)-f_2(x)}{2}\right)^p dx\right]^{\frac{2k}{p}}\right\}$$

$$= 2\left\{\sum_{k=0}^{p/2}\binom{p}{2k}\left\|\frac{f_1+f_2}{2} - s\right\|_p^{p-2k}\left\|\frac{f_1-f_2}{2}\right\|_p^{2k}\right\}, \tag{3.4}$$

which implies the existence of the righthand side of (3.2).

By using the identity

$$(a+b)^p + (a-b)^p = 2\sum_{k=0}^{p/2}\binom{p}{2k}a^{p-2k}b^{2k} \tag{3.5}$$

for p even, and writing $a+b = f_1-s$, $a-b = f_2-s$ we have

$$(f_1-s)^p + (f_2-s)^p = 2\sum_{k=0}^{p/2}\binom{p}{2k}\left(\frac{f_1+f_2}{2} - s\right)^{p-2k}\left(\frac{f_1-f_2}{2}\right)^{2k}. \tag{3.6}$$

Integrating the last expression from a to b and taking the infimum over all $s \in S$, we have the desired result.

4. Error Bounds.

Since

$$2\sum_{k=0}^{p/2}\binom{p}{2k}\left\|\frac{f_1+f_2}{2} - s\right\|_p^{p-2k}\left\|\frac{f_1-f_2}{2}\right\|_p^{2k}$$

$$= \left[\left\|\frac{f_1+f_2}{2} - s\right\|_p + \left\|\frac{f_1-f_2}{2}\right\|_p\right]^p$$

$$+ \left[\left\|\frac{f_1+f_2}{2} - s\right\|_p - \left\|\frac{f_1-f_2}{2}\right\|_p\right]^p, \tag{4.1}$$

if s^* is the b.s.a. to f_1 and f_2 in the L_p norm and if s^{**} is the best approximation to $\frac{f_1+f_2}{2}$ in the L_p norm, then

$$\|f_1-s^*\|_p^p + \|f_2-s^*\|_p^p \le \left[\left\|\frac{f_1+f_2}{2} - s^{**}\right\|_p + \left\|\frac{f_1-f_2}{2}\right\|_p\right]^p$$

$$+ \left[\left\|\frac{f_1+f_2}{2} - s^{**}\right\|_p - \left\|\frac{f_1-f_2}{2}\right\|_p\right]^p. \tag{4.2}$$

5. The Hilbert space case.

Let H be a Hilbert space with the inner product $<\cdot,\cdot>$ and let $\|a\|^2 \equiv <a,a>$ be the norm induced by the inner product. If we define the b.s.a. to n elements $f_1, f_2, \ldots, f_n$ with respect to some set $S \subset H$, by an element $s^* \in S$ which satisfies

$$\inf_{s \in S} \sum_{i=1}^{n} \|f_i - s\|^2 = \sum_{i=1}^{n} \|f_i - s^*\|^2 \tag{5.1}$$

we can obtain a result similar to Theorem I.

Remark. It would be interesting to study the problem of b.s.a. with p real, positive and not even.

The authors wish to thank Dr. J. Tzimbalario for very helpful remarks.

References

1. J.B. Diaz and H.W. McLaughlin. Simultaneous Chebyshev approximation of a set of bounded complex-valued functions. Jour. Approx. Theory (1969) pp. 419-432.

2. J.B. Diaz and H.W. McLaughlin. Simultaneous approximation of a set of bounded real functions. Math. Comp. 23 (1969) pp. 583-594.

3. J.B. Diaz and H.W. McLaughlin. On simultaneous Chebyshev approximation and Chebyshev approximation with an additive weight function. Jour. Approx. Theory 6 (1972) pp. 68-71.

4. C.B. Dunham. Simultaneous Chebyshev approximation of functions on an interval. Proc. Amer. Math. Soc. 18 (1967) pp. 472-477.

5. D.S. Goel, A.S.B. Holland, C. Nasim and B.N. Sahney. On best simultaneous approximation in a normed linear space. To appear in Canadian Mathematical Bulletin.

6. D.S. Goel, A.S.B. Holland, C. Nasim and B.N. Sahney. Characterisation of an element of best L^p-simultaneous approximation. To appear in S.R. Ramanujan memorial volume.

7. A.S.B. Holland, B.N. Sahney and J. Tzimbalario. On best simultaneous approximation. To appear in Jour. Approx. Theory.

8. G.M. Phillips and B.N. Sahney. Best simultaneous approximation in the L_1 and L_2 norms. To appear, Proc. of Conf. on Theory of Approximation at University of Calgary, August 1975, Academic Press, N.Y.

THE NUMERICAL SOLUTION OF THE HILBERT PROBLEM

Y. Ikebe,[‡] T. Y. Li, and F. Stenger[†]

Abstract

In this paper we study the numerical solution of the homogeneous and nonhomogeneous Hilbert problems for regions which are sufficiently general for most practical applications. The numerical method proposed in this paper converges faster than any power of the number of function evaluations and remains stable near the singular points of the contour defining the region under consideration. As an example, both the homogeneous and the nonhomogeneous Hilbert problem for an L-shaped region are numerically solved.

1. Introduction and Summary

Let S^+ be a region bounded by a system of contours

$$(1.1) \qquad L = \bigcup_{j=0}^{p} L_j,$$

as shown in Fig. 1, and where the contour L_0 may be absent, although we shall assume that each L_j, $j = 0,1,\ldots,p$, has finite length.

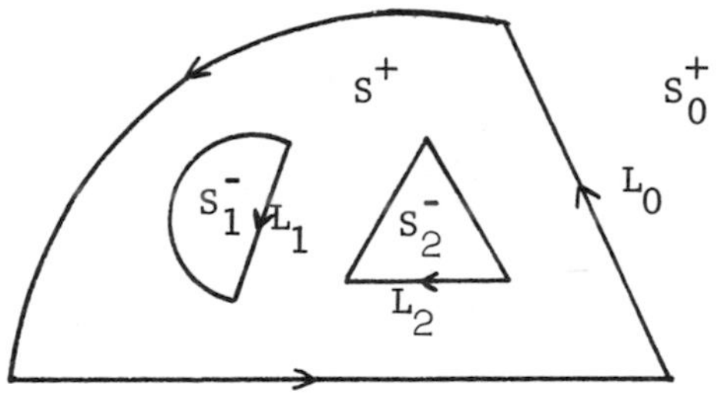

Figure 1

Let

$$(1.2) \qquad S^- = \bigcup_{j=0}^{p} S_j^-,$$

where S_j^- is the region bounded by the contour L_j (see Fig. 1).

In this paper we shall consider the numerical solution of

[‡]Professor Ikebe is presently a Visiting Professor at The University of Utah; his permanent address is The University of Texas at Austin.

[†]Supported by U. S. Army Research Grant DAHC-04-0175.

the homogeneous Hilbert problem [7, p. 86]

(1.3) $X^+(t) = G(t) \cdot X^-(t)$, t on L,

in which a sectionally analytic function X(z) [7, p. 35] is sought which satisfies (1.3), where $X^+(t)$ and $X^-(t)$ represent the limits from left or right of X(z) [7, p. 35] and where G(t) is a given nonvanishing function defined on L, satisfying the Hölder condition [7, p. 11]. We will also study the numerical solution of the nonhomogeneous Hilbert problem [7, p. 92]

(1.4) $\Phi^+(t) = G(t) \cdot \Phi^-(t) + g(t)$, t on L,

in which a sectionally analytic function $\Phi(z)$ is sought which satisfies (1.4), where $\Phi^+(t)$, $\Phi^-(t)$, and G(t) have the similar meaning as in the homogeneous case, and where g(t) represents a given function defined on L, satisfying the Hölder condition. We shall assume throughout this paper that G(t) has already been modified for the homogeneous problem (1.3) so that

(1.5) $$\int_{L_j} d[\log G(t)] = 0.$$

This corresponds to the case $\kappa = 0$ in [7, Chap. 5]. The exact solutions to (1.3) and (1.4) in this case are given by (2.1) and (3.1), respectively (see Sec. 2 and Sec. 3).

In Sec. 2 which follows, we shall describe a numerical procedure for finding an approximation $X_N(z)$ of the solution X(z) of the homogeneous problem (1.3). In Sec. 3 we shall describe a numerical procedure for solving the nonhomogeneous problem (1.4).

The method of the present paper is based on subdividing the contour L into a finite number of analytic arcs $C_1, C_2, \ldots, C_n$, such that log G(t) has an analytic extension to the interior of each arc C_j for the case of the homogeneous problem, and both log G(t) and g(t) have analytic extensions to the interior of each arc C_j for the case of the nonhomogeneous problem. We emphasize that we allow log G(t) or both of log G(t) and g(t) to have integrable singularities at the end points of any arc C_j, $j = 0,\ldots,n$.

In the above paragraph, an analytic arc C_j is assumed to be definable by a map $\psi_j: (-1,1) \to C_j$, where ψ_j is analytic and univalent in a region containing the open interval (-1,1).

In Sec. 4 we obtain a bound for the error of the approximations. For the approximation of the homogeneous problem (1.3), we prove that if we use 2N+1 points to approximate each integral

$$\int_{C_j} \frac{\log G(t)}{t - z}\, dt, \qquad j = 0,1,\ldots,n,$$

by means of the special quadrature formulas based on the trapezoidal rule (e.g. [12, Formula (1.10)]) then the difference between the exact solution X(z) and the approximate solution $X_N(z)$ is of order $O(e^{-\beta\sqrt{N}})$ as $N \to \infty$, where β is a positive constant. For the case of the nonhomogeneous problem (1.4) we require $(2N+1)^2$ points on each arc C_j, $j = 1,\ldots,n$, to compute an approximation $\Phi_N(z)$ of the exact solution $\Phi(z)$. In this case the error $\Phi_N(z) - \Phi(z)$ proves to be $O(Ne^{-\beta\sqrt{N}})$ as $N \to \infty$.

In Sec. 5, we state a few simple elementary conformal maps which are useful in practical applications. One of these is used in the example in Sec. 6.

In Sec. 6 we study a simple but nontrivial example to illustrate the method of the earlier sections for solving both the homogeneous and nonhomogeneous Hilbert problems (1.3) and (1.4). In this example the region S^+ is taken to be a L-shaped region and log G(t) and g(t) are allowed to vanish at end points.

Approximate methods for solving the Hilbert problems (1.3)

and (1.4) have previously been carried out by others, but only in the special cases where the contour L is the circle or the interval $(-\infty, +\infty)$. See [1, 2, 3, 4, 8, 9, 10, 11, 12]. It is expected that the assumptions of the present paper are especially suited to problems arising in applications, where the boundary functions $G(t)$ and $g(t)$ are not necessarily of class C^p but, rather, piecewise analytic. If so, then the rate of convergence to zero of the error for the approximate method discussed in this paper is faster than any power of the number of evaluation points. In practical applications this would often make the method of this paper more favorable than those methods which are based on the usual Gaussian or Newton-Cotes type integration formulas. It should also be pointed out that the method of this paper remains stable for the computation of $X(z)$ and $\Phi(z)$ where z is near the end points of arcs C_j where the singularities of $\log G(t)$ and of $g(t)$ may exist. We consider such an example in Sec. 6.

2. The Homogeneous Hilbert Problem

We subdivide the contour L (see Fig. 1, Sec. 1) into a finite number of arcs $C_1, \ldots, C_n$ such that the following conditions are satisfied on each arc C_j, $j = 1, \ldots, n$:

(a) C_j has end-points a_j and b_j, and except for a_j and b_j, C_j lies wholly in a simply connected domain D_j of the complex plane;

(b) the function $\gamma(t) \equiv \log G(t)$ is analytic in D_j;

(c) there exists an explicit function φ_j which conformally maps D_j into one and the same domain

$$D' = \{w = u + iv : |v| < \frac{\pi}{2}\},$$

such that

$$\varphi_j(a_j) = -\infty, \quad \varphi_j(b_j) = +\infty, \quad \varphi_j(C_j) = (-\infty, +\infty).$$

The solution of the homogeneous problem (1.3) is given by (see [7, Sec. 35])

$$X(z) = P(z)\cdot \exp \Gamma(z), \tag{2.1}$$

where P(z) is an arbitrary entire function of z and where for z not lying on L,

$$\Gamma(z) = \frac{1}{2\pi i} \int_L \frac{\gamma(t)}{t-z}\, dt$$

or

$$\Gamma(z) = \frac{1}{2\pi i} \sum_{j=1}^{n} \int_{C_j} \frac{\gamma(t)}{t-z}\, dt; \quad \gamma(t) = \log G(t), \tag{2.2}$$

since L is the union of $C_1, \ldots, C_n$.

We shall approximate each integral

$$I_j(z) = \int_{C_j} \frac{\gamma(t)}{t-z}\, dt \tag{2.3}$$

as follows.

(a) At the outset we decompose $\gamma(t)$ into a linear part and nonlinear part $\mu(t)$ in the sense that

$$\gamma(t) = \frac{t-b_j}{a_j-b_j}\, \gamma(a_j) + \frac{t-a_j}{b_j-a_j}\, \gamma(b_j) + \mu(t), \tag{2.4}$$

where the sum of the first two terms on the right-hand side represents the linear part (linear interpolation of $\gamma(t)$). Substitution of (2.4) into (2.3) yields

$$I_j(z) = \delta_j(z) + J_j(z), \tag{2.5}$$

where

$$\delta_j(z) = \gamma(b_j) - \gamma(a_j) + \frac{(z-a_j)\gamma(b_j) + (b_j-z)\gamma(a_j)}{b_j - a_j} \log\frac{b_j-z}{a_j-z}, \tag{2.6}$$

and

$$J_j(z) = \int_{C_j} \frac{\mu(t)}{t-z}\, dt. \tag{2.7}$$

(b) If z does not lie on D_j or else if $\mathrm{Im}|\varphi_j(z)| > 3\frac{\pi}{8}$, we simply approximate the integral (2.7) by means of the trapezoidal formula on $(-\infty,\infty)$,

$$(2.8) \qquad J_j(z) \cong h \sum_{k=-\infty}^{\infty} \frac{\mu(z_k^j(h))}{(z_k^j(h) - z)\varphi_j'(z_k^j(h))},$$

where $h > 0$ (e.g., 0.25, 0.5) and $z_k^j(h)$ is the inverse image of kh, k = 0, ± 1, ± 2, ..., under φ_j:

$$z_k^j(h) = \varphi_j^{-1}(kh).$$

Note that $z_k^j(h)$ lies on the contour C_j since kh is real.

(c) If z lies on D_j and $\mathrm{Im}\ |\varphi_j(z)| \leq 3\frac{\pi}{8}$, we approximate $J_j(z)$ by means of the formula

$$(2.9) \qquad J_j(z) \cong \mu(z)\log\frac{b_j - z}{a_j - z} + h \sum_{k=-\infty}^{\infty} \frac{\mu(z_k^j(h)) - \mu(z)}{(z_k^j(h) - z)\cdot\varphi_j'(z_k^j(h))},$$

where $z_k^j(h)$ has the same meaning as in (b).

We recommend care in evaluating the logarithm in (2.6) and (2.9), i.e.,

$$(2.10) \qquad \log\frac{b_j - z}{a_j - z} = \int_{a_j}^{b_j} \frac{dt}{t - z},$$

where the integral is taken along C_j. For example, if C_j is a line segment and if z is not on C_j, then the value of the logarithm (2.10) equals

$$\log\frac{|b_j - z|}{|a_j - z|} + i\cdot\theta,$$

where θ equals $\arg(b_j - z) - \arg(a_j - z)$ and $-\pi < \theta < \pi$ (see Fig. 2: if $z \to t$ on L, $\theta = \theta_+$ if $z \in S^+$; $\theta = \theta^-$ if $z \in S^-$).

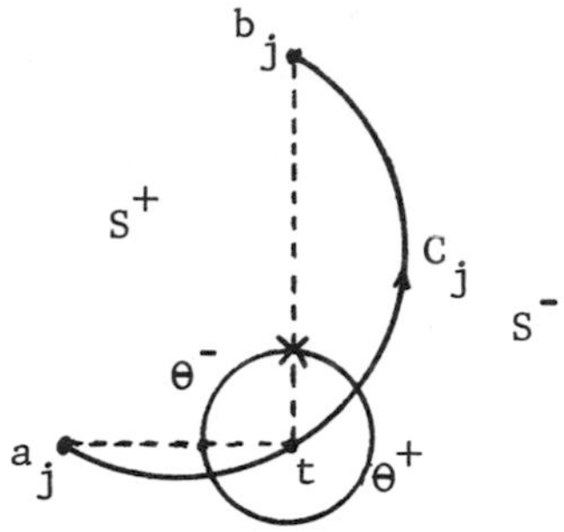

Figure 2

It would be convenient for the reader that we discuss at this time the method for computing the boundary values of X(z),

$$X^+(t) = \lim_{z \to t} X(z), \qquad t \text{ on } L,$$

where z approaches t along any curve in S^+, and

$$X^-(t) = \lim_{z \to t} X(z), \qquad t \text{ on } L,$$

where z approaches t along any curve in S^-. The boundary values $X^+(t)$ of X(z) are needed for the solution of the nonhomogeneous Hilbert problem (1.4), as $X^+(t)$ explicitly appears in the formula (3.1) for the solution of the nonhomogeneous problem (see the next section). Examination of the expressions in (2.3)-(2.9) shows that it is only necessary to consider the limit as z approaches a point t on C_j of the algorithm (2.10) which appears in (2.6) and (2.9), and the quotient

$$(2.11) \qquad \frac{\mu(z_k^j(h)) - \mu(z)}{z_k^j(h) - z}$$

which appears in (2.9), in order to compute an approximate value of $X^+(t)$ for any given point t on L ($t \neq$ end points).

We next consider the quotient (2.11) in the case where $z = z_k^j(h)$. Recall that the $z_k^j(h)$ lie on L. By the hypothesis (a) which was made earlier in this section, the function $\mu(t)$ has an analytic extension in a neighborhood of any point t on C_j which is not an end point of any C_j; we conclude that

$$\text{(2.12)} \qquad \lim_{z \to z_k^j(h)} \frac{\mu(z_k^j(h)) - \mu(z)}{z_k^j(h) - z} = \frac{d}{dz}\, \mu(z_k^j(h)).$$

As may be seen from the above discussions, we need, in general, two different types of subroutines for each arc C_j for the numerical solution of the nonhomogeneous problem (1.4). The subroutines of the first type are the ones which compute the integration of the type (2.3) with z not lying on C_j. The second ones are the ones which compute the same integration but with "z lying on C_j" in the sense explained in the above paragraphs.

3. The Nonhomogeneous Hilbert Problem

The solution $\Phi(z)$ of the nonhomogeneous Hilbert Problem (1.4) is given by

$$\text{(3.1)} \qquad \Phi(z) = \frac{X(z)}{2\pi i} \int_L \frac{g(t)dt}{X^+(t)(t-z)} + X(z)P(z),$$

where X(z) is the solution of the homogeneous Hilbert problem (1.3), P(z) is an arbitrary entire function, g(t) is a given continuous function on L, and

$$\text{(3.2)} \qquad X^+(t) = \lim_{z \to t} X(z), \qquad t \text{ on } L,$$

where z approaches t along a curve entirely lying in S^+.

We again assume that the contour L can be subdivided into a finite number of arcs $C_1, \ldots, C_n$ such that the conditions (a)-(c) in Section 2 are satisfied. In addition, we assume that the given function g(t) is analytic in each D_j, $j=1,\ldots,n$, where D_j is defined in Sec. 2. Under these assumptions, the solution (3.1) of the nonhomogeneous problem (1.4) is given by

$$\Phi(z) = \frac{X(z)}{2\pi i} \sum_{j=1}^{n} I_j'(z) + X(z)P(z), \tag{3.3}$$

where

$$I_j'(z) = \int_{C_j} \frac{g(t)dt}{X^+(t)(t-z)}, \qquad j = 1,\ldots,n. \tag{3.4}$$

The method of computing $X^+(t)$, where t lies on L, has already been discussed in the last section.

Each of the integrals $I_j'(z)$ can also be evaluated by the method of the last section. To this end, we apply the quadrature formulas (2.8) and (2.4) for approximating the integral (2.3)

$$I_j = \int_{C_j} \frac{\gamma(t)}{t-z}dt$$

to (3.4) with $\gamma(t)$ replaced by $g(t)/X^+(t)$. In this process one needs the values of $X^+(t)$ for each of those values of $t = z_k^j(h)$ on the contour so that the numerical evaluation of $I_j'(z)$ can be carried out by these same formulas.

4. Convergence

We first consider the homogeneous Hilbert problem (1.3). We shall assume that $\mu(t)$ (see Eq. (2.4)) satisfies the inequality

$$|\mu(t)| \le K_0 \cdot e^{-\alpha|\varphi_j(t)|} |\varphi_j'(t)| \tag{4.1}$$

for all t on C_j, j = 1,...,n, where K and α are positive constants independent of j. In addition, we assume that

$$\int_{\partial D_j} |\mu(t)dt| < \infty , \tag{4.2}$$

where ∂D_j denotes the boundary of the region D_j.

To illustrate the nature of the conditions (4.1) and (4.2), we consider the case where the arc C_j in question represents the interval [-1,1] and where the map $\varphi_j(t)$ is given by

$$\varphi_j(t) = \log \frac{1+t}{1-t} .$$

The function $\varphi_j(t)$ maps the unit disc in the t-plane onto $\mathcal{D}'$ such that $\varphi_j(-1) = -\infty$, $\varphi_j(+1) = +\infty$, as required by the hypothesis (c) in Sec. 2. In this case (4.1) is satisfied if $\mu(t)$ satisfies

$$|\mu(t)| \le K_0(1-t^2)^{\alpha-1}, \quad -1 < t < 1,$$

for some positive constants K_0 and α. If so, (4.2) is also satisfied if the region D_j containing (-1,1) is sufficiently small. Note that the function $(1-t^2)^{\alpha-1}$, $-1< t <1$, has integrable singularities at $t = \pm 1$ if $\alpha > 0$. In practice one always has $\alpha \ge 1$, since μ is continuous on L.

We now proceed as in [12, Sec. 1] and obtain

$$(4.3) \qquad \left| J_j(z) - h \sum_{k=-N}^{N} \frac{\mu(z_k^j(h))}{(z_k^j(h)-z)\varphi_j'(z_k^j(h))} \right| < K_1^j \cdot e^{-\beta\sqrt{N}} ,$$

where K_1^j is a constant and $\beta = 3\pi\, 2^{\frac{1}{2}}/4$, provided that $|\mathrm{Im}\, \varphi_j(z)| > 3\,\frac{\pi}{8}$. If $|\mathrm{Im}\, \varphi_j(z)| \le 3\,\frac{\pi}{8}$, then by taking $h = N^{\frac{1}{2}}$, we obtain

$$(4.4) \qquad \left| J_j(z) - \left\{ \mu(z)\log\left(\frac{b_j-z}{a_j-z}\right) + h \sum_{k=-N}^{N} \frac{\mu(z_k^j(h))-\mu(z)}{(z_k^j(h)-z)\varphi_j'(z_k^j(h))} \right\} \right|$$
$$\le K_2^j e^{-\beta\sqrt{N}} ,$$

where K_2^j is a constant and β is the same as in (4.3).

We, therefore, require a total of $n(2N+1)$ points in order to approximate X(z) to within an error of $K \cdot e^{-\beta\sqrt{N}}$, where $K = \max_j (K_1^j, K_2^j)$.

For the case of the nonhomogeneous problem (1.4) we require

$n(2N+1)$ points to approximate each of the function values $X^+(z_k^j(h))$ (where, as before, $h = N^{-\frac{1}{2}}$) to within an error of $K_1^j\, e^{-\beta\sqrt{N}}$, for $k = -N, -N+1, \ldots, N$, and $j = 1, \ldots, n$, or a total of $[n(2N+1)]^2$ function evaluations. In addition, we require $n(2N+1)$ points to approximate $X(z)$ for a given z not lying on the contour L. Consequently, a total of

$$[n(2N+1)]^2 + n(2N+1)$$

function evaluations is needed to approximate $\varphi(z)$. There is an error bounded by $K\cdot e^{-\beta\sqrt{N}}$ in each of the approximations of $X^+(z_k^j(h))$. Hence if $g(t)/X^+(t)$ is decomposed in a fashion analogous to the way $\gamma(t)$ was decomposed in Sec. 2, and if the corresponding function $\mu(t)$ (i.e., nonlinear part of $g(t)/X^+(t)$) satisfies (4.1) for each j, $j = 1, \ldots, n$, then the cumulative error in our approximation of $\Phi(z)$ is bounded by

$$n(2N+1)K\, e^{-\beta\sqrt{N}},$$

where K and β are constants defined earlier.

It should be pointed out that the preceding error analysis produces error bounds only. The actual magnitude of the errors mentioned may be considerably smaller than those predicted by the above error bounds. In fact, the step size h in (2.8) and (2.9) need not be chosen to be $N^{-\frac{1}{2}}$ for a prescribed N. In practice, the values of h and N may be determined during the actual computational process, given a prescribed accuracy for the numerical value of the type (2.3). We shall illustrate such a procedure in Sec. 6.

5. Some Useful Elementary Transformations

We remark at the outset that if the given function $\gamma(t) = \log G(t)$ for the homogeneous problem (1.3) is analytic in a region D containing a connected subset C of the contour L, where C has end-points a and b, then we may replace the possibly cumbersome C by a simpler analytic arc lying in D and connecting the end-points a and b, such as by a line segment or by an arc of a circle. A similar remark applies to the functions $\gamma(t) = \log G(t)$ and $g(t)$ in the nonhomogeneous problem (1.4). This is a consequence of the fact that $X^{\pm}$ and $\Phi^{\pm}$ are then analytic in D.

Many elementary transformations of analytic arcs are contained in Kober [6]. We mention some of the simplest.

The transformation

$$(5.1) \qquad w = \log \frac{1+z}{1-z} \Longleftrightarrow z = \tanh\,(w/z) = \frac{e^w-1}{e^w+1}$$

is a conformal map of the unit disc $\{z : |z| < 1\}$ of the z-plane onto the strip $D' = \{w : |\text{Im } w| < \pi/2\}$ of the w-plane. By this transformation the analytic arc $(-1,1)$ is transformed onto $(-\infty, +\infty)$ and the eye-shaped region R consisting of all z such that $|z+i\cdot(\rho^2-1)^{\frac{1}{2}}| < \rho$ and $|z-i\cdot(\rho^2-1)^{\frac{1}{2}}| < \rho$, where $\rho = \dfrac{1}{\sin(\alpha\pi/2)}$, is mapped conformally onto the strip $\{w: |\text{Im}\, w| < \alpha\pi/2\}$. See Fig. 3 below.

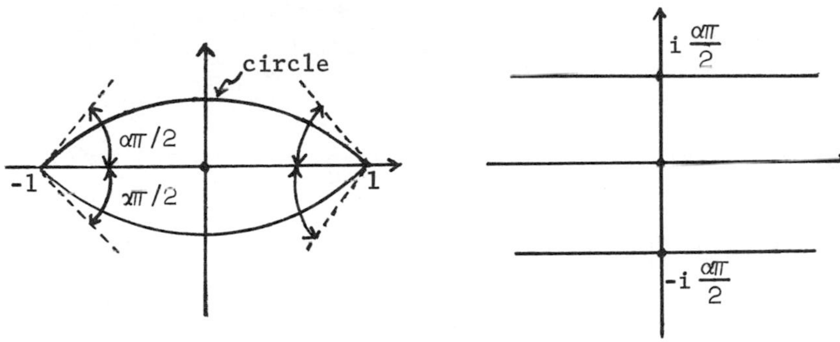

Figure 3

The transformation

$$w = \frac{z-b}{z-a} \cdot \frac{c-a}{c-b} ,$$

where a, b, c are distinct points on a circle C, transforms C onto the real axis $(-\infty,+\infty)$. The point a is mapped both onto $w = -\infty$ and $w = +\infty$. The circle C lies in a region R bounded by two circles that are tangent to C at a. The region R is mapped by the same transformation onto a parallel strip containing the x-axis.

A proper sub-arc of a circle having end-points a and b, $a \neq b$, and containing a point c distinct from a and b, may be mapped onto the segment $(-1,1)$ by means of the transformation

$$\frac{2z}{z+1} = \frac{w-c}{w-a} \cdot \frac{b-a}{b-c} .$$

Notice that this transformation maps circles passing through a and b onto circles passing through -1 and 1.

6. Example

We consider both the homogeneous and nonhomogeneous Hilbert problems for the L-shaped region described below, where the

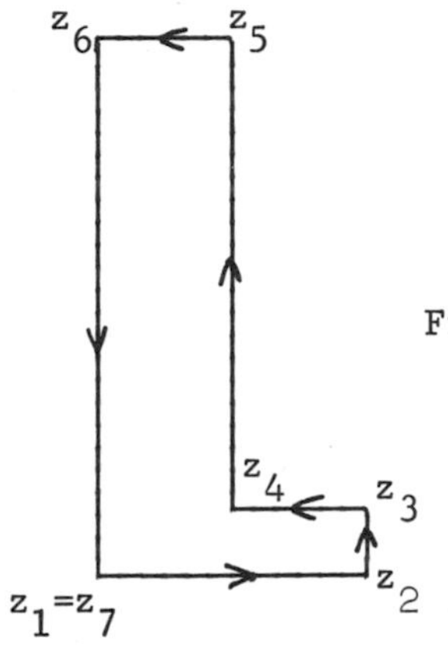

Figure 4

positively oriented contour L is determined by the points

$$(6.1)\qquad \begin{array}{lll} z_1 = 0 = z_7, & z_2 = 1, & z_3 = 1 + i\cdot 0.25 \\ z_4 = 0.5 + i\cdot 0.25, & z_5 = 0.5 + i\cdot 2, & z_6 = i\cdot 2 \end{array}$$

We subdivide L into 6 arcs $C_1,\ldots,C_6$ $(n = 6)$ in the most natural way:

$$(6.2)\qquad C_j = \text{the line segment from } z_j \text{ to } z_{j+1},\ j = 1,\ldots,6.$$

In this example we shall choose G(t) and g(t) in (1.3) and (1.4) so that these C_j will play the same rôle as those C_j which are defined in Sec. 2.

In order to compute X(z), the solution of the homogeneous problem (1.3), it is necessary to compute $\Gamma(z)$ defined by (2.2) and the integrals (2.3) where $\gamma(t) = \log G(t)$, as before. The $I_j(z)$ are, in turn, computed from (2.5). Consequently, we must compute the integrals (2.7) where $\mu(t)$ is the nonlinear part of $\gamma(t)$ defined by (2.4).

The region D_j of analyticity of $\gamma(t)$ which is to contain the arc C_j, and the associated map φ_j which maps D_j onto the strip $|\text{Im } w| < \frac{\pi}{2}$ depend on $\gamma(t)$. In this section we choose $\gamma(t)$ (and g(t) for the nonhomogeneous case) in such a way that D_j may be taken to be the interior of the circle which passes through two end-points of C_j and whose diameter equals the length of C_j (see Eqs. (6.13) and (6.14) below). In the case φ_j may be taken to be the composition map

$$(6.3)\qquad D_j \to \text{unit disk } |z| < 1 \to \text{strip } |\text{Im } w| < \frac{\pi}{2},$$

where the first map is realized via the linear parametrization of C_j

$$(6.4)\qquad t = \frac{1}{2}(z_{i+1} - z_i)s + \frac{1}{2}(z_{i+1} + z_i), \quad -1 \leq s \leq 1,$$

and the second map is given by (5.1). Hence $\varphi_j(t) = w$,

$$(6.5)\qquad \varphi_j'(t) = \frac{(e^w + 1)^2}{2e^w} \cdot \frac{2}{z_{j+1} - z_j},$$

and

$$(6.6)\qquad z_k^j(h) = \varphi_j^{-1}(hk) = \frac{z_{j+1}-z_j}{2} \frac{e^{kh}-1}{e^{kh}+1} + \frac{z_{j+1}+z_j}{2}.$$

Therefore, the formulas (2.8) and (2.9) for the approximate computation of $J_j(z)$ now read, respectively,

$$(6.7)\qquad J_j(z) \cong \frac{z_{j+1}-z_j}{2} \cdot h \cdot \sum_{k=-\infty}^{\infty} \frac{2e^{kh}}{(e^{kh}+1)^2} \frac{\mu(z_k^j(h))}{z_k^j(h)-z},$$

and

$$(6.8)\qquad J_j(z) \cong \mu(z) \log \frac{z_{j+1}-z}{z_j-z} + \frac{z_{j+1}-z_j}{2} \cdot h \cdot \sum_{k=-\infty}^{\infty} \cdot \frac{2e^{kh}}{(e^{kh}+1)^2} \frac{\mu(z_k^j(h)) - \mu(z)}{z_k^j(h)-z}.$$

Note that the integral $J_j(z)$ itself is transformed by (6.4) to

$$(6.9)\qquad J_j(z) = \int_{-1}^{1} \frac{\mu(t(s))}{s + \frac{z_{j+1}+z_j-2z}{z_{j+1}-z_j}} ds = \mu(z) \log \frac{z_{j+1}-z}{z_j-z} +$$

$$+ \int_{-1}^{1} \frac{\mu(t(s)) - \mu(z)}{s + \frac{z_{j+1}+z_j-2z}{z_{j+1}-z_j}} \, ds \ ,$$

where the integral is taken along the line segment from -1 to +1. We could, of course, obtain (6.7) and (6.8) from (6.9) by applying the trapezoidal formula

$$\int_{-1}^{1} f(z)dz \cong h \, . \sum_{k=-\infty}^{\infty} \frac{2e^{kh}}{(e^{kh}+1)^2} f\left(\frac{e^{kh}-1}{e^{kh}+1}\right)$$

$$= h\left[\frac{1}{2}f(0) + \sum_{k=1}^{\infty} \frac{2e^{kh}}{(e^{kh}+1)^2}\left\{f\,\frac{e^{kh}-1}{e^{kh}+1} + f\left(-\frac{e^{kh}-1}{e^{kh}+1}\right)\right\}\right] . \tag{6.10}$$

In this regard, we again refer the reader to [12].

For the numerical solution of the nonhomogeneous problem (1.4) we need approximate values of

$$X^+(z_k^j(h)) = \lim_{z\to z_k^j(h)} X(z),$$

$k = 0, \pm 1, \pm 2, \ldots$, $j = 1,\ldots,6$. For the computation of $X^+(z_{k_0}^j(h))$, where k_0 is any prescribed integer, we proceed exactly in the same way as we compute $X(z)$ when z is not on L, except that we modify (6.8) for the jth segment C_j on which $z_{k_0}^j(h)$ lies, to the form

$$J_j(z_{k_0}^j(h)) \cong \mu(z_{k_0}^j(h))\left\{\log\frac{|z_{j+1}-z_{k_0}^j(h)|}{|z_j-z_{k_0}^j(h)|} + i\pi\right\} \tag{6.11}$$

$$+ \frac{z_{j+1}-z_j}{2} \cdot h \sum_{k=-\infty}^{\infty}{}' \frac{2e^{kh}}{(e^{kh}+1)^2} \frac{\mu(z_k^j(h))-\mu(z_{k_0}^j(h))}{z_k^j(h) - z_{k_0}^j(h)} \ ,$$

where the prime (') on the summation sign Σ indicates that the $k_0{}^{th}$ term which has the form 0/0 should be replaced by

$\frac{d\mu}{dz}$ at $z = z^j_{k_0}(h)$ (see Eq. (2.12)).

For the actual numerical computations we have used real arithmetic. Thus, the expressions in (6.7), (6.8), and (6.11) were computed by separating real and imaginary parts.

The formulas (6.7), (6.8), and (6.11) involve infinite series of the form $\sum_{k=-\infty}^{\infty}$, which are approximated in the actual numerical computation by the finite sum of the form $\sum_{k=-N}^{N}$. We now discuss this truncation error. From the nature of the weight $2e^{kh}/(e^{kh}+1)^2$ in the trapezoidal formula (cf. (6.10)) and from the assumption on the integrand, the following error estimate may be obtained by proceeding as in [12]. We consider the trapezoidal rule (6.10). Given $\epsilon > 0$. Choose $h > 0$ and an N so large that

$$\left| \sum_{-(N+1)}^{N+1} - \sum_{-N}^{N} \right| < \frac{\epsilon}{4} .$$

Then, approximate the right-hand side of (6.10) by $h \cdot \sum_{-N}^{N}$. Call this sum T(h). Then compute T(h/2) with the same ϵ. If

$$|T(h) - T(h/2)| < \frac{\epsilon}{2} ,$$

then

$$|T(h/2) - \int_{-1}^{1} f(z)dz| \le \epsilon. \tag{6.12}$$

This is a practical convergence criterion. Our computer program is arranged in such a way that the above estimate is effectively used as a stopping criterion for the convergence test. Actually, ϵ is taken to be a prescribed accuracy factor (say, 10^{-4}) times the approximate integral, so that the convergence test is done in terms of the relative accuracy.

We now describe the functions $\gamma(t)$ and $g(t)$ which we use in our example:

$$\gamma(t) = \log G(t) = \begin{cases} (t-z_3)\sqrt{t-z_4}\ , \text{ on } C_3\left(\sqrt{t-z_4} \ge 0\right) \\ (e^{t-z_3}-1)(t-z_4), \text{ on } C_j,\ j \ne 3, \end{cases} \tag{6.13}$$

and

$$(6.14) \qquad g(t) = \begin{cases} \sqrt{t} \geq 0, & \text{on } C_1 \\ t \quad , & \text{on } C_j,\ j \neq 1. \end{cases}$$

Note that $\gamma(t)$ is analytic on C_3 except at the left end-point z_4 and that $g(t)$ is analytic on C_1 except at the left end-point z_1. It is clear that $\gamma(t)$ and $g(t)$ satisfy the condition on analyticity stated earlier in this section (see the paragraph containing Eq. (6.3)).

For definiteness, we take $P(z) = 1$ in (2.1) so that the solution of the homogeneous problem (1.3) is given by

$$(6.15) \qquad X(z) = \exp \Gamma(z),$$

where $\Gamma(z)$ is given by (2.2). For the nonhomogeneous problem, we let $P(z) = 0$ in (3.1) so that the solution of (1.4) is given by

$$(6.16) \qquad \Phi(z) = \frac{X(z)}{2\pi i} \int_L \frac{g(t)dt}{X^+(t)(t-z)} .$$

The computations were done on the UNIVAC 1108 system at The University of Utah with the double-precision real arithmetic which provides approximately 18 decimal significant digits with overflow/underflow limit approximately equal to $10^{\pm 308}$. The computer programs were written for the L-shaped region stated at the beginning of this section with $\gamma(t)$ and $g(t)$ as formal parameters. The programs were thoroughly checked by taking $\gamma(t)$ and $g(t)$ to be simple entire functions such as polynomials or exponentials, in which cases the exact solutions of (1.3) and (1.4) are easily computed from Cauchy's integral theorem.

Some of our numerical results follow (Tables I and II). To avoid excessive amount of tabulation we state the values of $\Gamma(z)$ and $\Phi(z)$ only at a number of points z in "critical" areas, i.e., in a neighborhood of the corner $z = z_1$ where $g(t)$ is not analytic and a neighborhood of $z = z_4$ where $\gamma(t)$ is not analytic. In fact, we choose 3 points on the line segment joining $z_1 = 0$ and $z_4 = 0.5 + i\cdot 0.25$:

$$z_4 \cdot \left(\frac{255}{256}\right), \quad \frac{z_4}{2} \quad \text{and} \quad \frac{z_4}{256} .$$

The first point is near z_4 and the last point is near z_1. The mid-point $z_4/2$ of the segment is chosen for the sake of comparison.

Also included in Tables I and II are the cases where $\gamma(t)$ and $g(t)$ are entire functions:

$$\gamma(t) = \left(e^{t-z_3}-1\right)(t-z_4), \tag{6.17}$$

and

$$g(t) = t. \tag{6.18}$$

In this case, the exact solutions are given by

$$\Gamma(z) = \gamma(z) = \left(e^{z-z_3}-1\right)(z-z_4), \tag{6.19}$$

and by

$$\Phi(z) = g(z) = z. \tag{6.20}$$

The inclusion of this analytic case is again for the sake of comparison.

We now give a brief exposition on Tables I and II. The tables are self-explanatory except perhaps for the meaning of h and N. For definiteness, consider Table I, for example. The value $h = 1.0$ refers to the step size in the approximate integration formulas (6.7), (6.8), and (6.11). For the meaning of N refer to the earlier paragraph containing (6.12). The fact $N = 16$ means that the integration on each segment C_j, $j = 1,\ldots,6$, requires 16 points on some segment (in fact, on C_3) but no more than 16 points on any segment C_j, in order to obtain all the correct significant figures which are obtainable from the approximate integration with $h = 1.0$.

We pass to Table II. Take, for example, Case 1, Part (a). The fact $\epsilon = 10^{-4}$ indicates that the boundary values $X^+(t)\cdot$ $(t = z_k^j(h))$ are computed correct to the relative accuracy 10^{-4}. The number of the boundary points on a given segment C_j $(j = 1,\ldots,6)$ to be used for the approximate integration of (3.4) is given by $2N+1$ $(= 33)$ points. Thus, in Case 1, Part (a), 6×33 $(= 198)$ boundary values $X^+(z_k^j(h))$, $j = 1,\ldots,6$, $k = -N, -N+1,\ldots,N$, are computed with the indicated value of h $(= 1.0)$ before the integrals (3.4) are computed.

TABLE I. NUMERICAL RESULTS FOR HOMOGENEOUS HILBERT PROBLEMS

$\epsilon = 10^{-4}$	$\gamma(t)$ given by Eq. (6.17)	$\gamma(t)$ given by Eq. (6.13)
z (Compare Fig. 4)	$\Gamma(z)$	$\Gamma(z)$
$\frac{255}{256} z_4$	Exact: (7.7022 95378 + i 3.8655 80288)(-4) h = 1, N = 12: (7.702 + i 3.86)(-4) h = 0.5, N = 24: (7.7022 9537 - i 3.8655 802)(-4)	h = 1, N = 16: (-.12814 - i 4.7743 4)(-2) h = 0.5, N = 32: (-.12814 9 - i 4.7743 472)(-2)
$\frac{1}{2} z_4$	Exact: (1.2546 82303 20 + i .81137 91954 24)(-1) h = 1, N = 12: (1.2546 82 + i .8113)(-1) h = 0.5, N = 24: (1.2546 82303 + i .81137 91954)(-1)	h = 1, N = 16: (1.2224 565 + i .71102)(-1) h = 0.5, N = 32: (1.2224 56571 + i .71102 7703)(-1)
$\frac{1}{256} z_4$	Exact: (2.9750 79618 106 + i 2.0530 96654 95)(-1) h = 1, N = 12: (2.9750 79 + i 2.0530 9)(-1) h = 0.5, N = 24: (2.9750 79618 10 + i 2.0530 96654 9)(-1)	h = 1, N = 16: (2.9524 67 + i 1.9935)(-1) h = 0.5, N = 32: (2.9524 67357 + i 1.9935 79928)(-1)

Note: $7.702\ (-4) = 7.702 \times 10^{-4}$.

TABLE II. NUMERICAL RESULTS FOR NONHOMOGENEOUS HILBERT PROBLEMS

$\epsilon = 10^{-4}$	$\gamma(t)$ given by Eq. (6.17); $g(t)$ by Eq. (6.18)	$\gamma(t)$ given by Eq. (6.13); $g(t)$ by Eq. (6.14)
z (Compare Fig. 4)	$\Phi(z)$	$\Phi(z)$
$\frac{255}{256} z_4$	Exact solution = $g(z)$: (4.9804 6875 + i 2.4902 34375)(-1) h = 1, N = 12: (4.9803 + i 2.4903)(-1) h = 0.5, N = 28: (4.9804 70 + i 2.4902 32)(-1)	h = 1, N = 16: (5.6660 + i 2.4968)(-1) h = 0.5, N = 28: (5.6657 + i 2.4966)(-1)
$\frac{1}{2} z_4$	Exact solution = $g(z)$: (2.5 + i 1.25)(-1) h = 1, N = 12: (2.4998 + i 1.2500 7)(-1) h = 0.5, N = 28: (2.5000 02 + i 1.2499 98)(-1)	h = 1, N = 16: (3.4396 + i .8408)(-1) h = 0.5, N = 28: (3.4398 + i .8407)(-1)
$\frac{1}{256} z_4$	Exact solution = $g(z)$: (1.9531 25 + i .97656 25)(-3) h = 1, N = 12: (1.93 + 1.98)(-3) h = 0.5, N = 28: (1.9534 + i .9764)(-3)	h = 1, N = 12: (.2949 - i 1.7195)(-1) h = 0.5, N = 28: (.2951 - i 1.7195)(-1)

References

[1] Baxter, R., A norm inequality of a "finite section" Weiner-Hopf equation, Ill. J. Math. 7 (1963), pp.97-103.

[2] Dombrouskia, I. N., Approximate solution of singular integral equations (Russian), Ural. Gos. Univ. Math. Zap. 4, tetrad 2 (1963), pp.38-45.

[3] Gabdulhaev, B. G., Approximate solution of singular integral equations by mechanical quadratures, Soviet Math. Dokl. 9 (1968), pp.329-332.

[4] ______, A certain general quadrature process and its application to the approximate solution of singular integral equations, Soviet Math. Dokl. 9 (1968), pp.386-389.

[5] Gakhov, F. D., Boundary Value Problems, Pergamon Press Ltd., London (1966).

[6] Kober, H., A Dictionary of Conformal Representations, Dover, New York (1957).

[7] Muskhelishvili, N. I., Singular Integral Equations, Nordhoff, Holland (1963).

[8] Orth, D. L., An algorithm for discrete Wiener-Hopf equations, Princeton University preprint (1969).

[9] Stenger, F., The approximate solution of Wiener-Hopf equations, J. Math. Anal. Appl. 37 (1972), pp.687-724.

[10] ______, The approximate solution of convolution-type integral equations, SIAM J. Math. Anal. 4 (1973), pp.536-555.

[11] ______, An algorithm for solving Wiener-Hopf integral equations, Comm. ACM 16 (1973), pp.708-710.

[12] ______, Integration formulas based on the trapezoidal formula, J. Inst. Maths. Applics. 12 (1973), pp.103-114.

[13] ______, Constructive proofs for approximation by inner functions, J. Approximation Theory 4 (1971), pp.372-386.

Acknowledgement. The authors express their appreciation to Mrs. Dorothy Baker of the Center for Numerical Analysis, The University of Texas at Austin, for her excellent typing of this manuscript.

ON ADAPTIVE PIECEWISE POLYNOMIAL APPROXIMATION

JOHN R. RICE

1. ADAPTIVE APPROXIMATION

A crucial element in spline and piecewise polynomial approximation is the optimal (or at least good) placement of the knots. An adaptive algorithm is one that attempts to achieve good knot placement by adapting the computation to the specific function being approximated. There are many ways to do this and a particular one ADAPT, Rice [1976b] is described in the next section. The objectives of ADAPT are to be highly reliable and flexible, to be fast and to produce smooth approximations. The objective of this paper is to report on studies made to determine the properties of ADAPT and to verify that it meets these objectives.

The degree of convergence possible for piecewise polynomial approximation was determined recently by Brudnyi [1974], Burchard [1976] and Peetre [1976]. Their results are rather complicated so we quote an earlier result Burchard [1975]. To state this, let $S_{k,r}(x)$ be a spline (or piecewise polynomial) of order r (or degree r-1) with k knots. Then we have:

<u>Suppose</u> f(x) <u>is such that</u>

$$\int_0^1 [f^{(r)}(x)]^{\frac{1}{r+1/p}} dx < \infty .$$

<u>Then the order of the best L_p - approximation, $S^*_{k,r}(x)$ to f(x), satisfies</u>

$$\text{dist}(f, S^*_{k,r}) = \mathcal{O}\left(\frac{1}{k^r}\right) .$$

Note that the above integral exists whenever f(x) is piecewise smooth with r continuous derivatives and has a finite number of algebraic singularities (i.e. behaves locally like x^α for $\alpha > -1/p$, see Rice, [1969]). The remarkable fact about this result is that the degree of approximation for such unsmooth functions is the same as for functions which simply have r continuous derivatives and no singularities. This is, of course, provided that the knots are chosen suitably well.

A class of adaptive approximation methods based on local approximation operators (e.g. local Hermite interpolation) has been analyzed by Rice, [1976a] and he proves that the approximations obtained this way give the optimal degree of convergence indicated above.

We may summarize the results of this study as follows: Smooth Functions. The use of high polynomial degree is most efficient for functions like e^x. For something that oscillates a few times (say, sin x on [-10, 10]) a few pieces are essential to obtain a good approximation. The rates of convergence predicted by the theory are seen immediately in these experiments.

Rough Functions and Singularities.

A. Too few knots or too low a degree are both bad for efficiency. As the accuracy requirement increases one should increase the degree slowly and the number of knots will increase somewhat more rapidly.

B. The rates of convergence measured in these experiments are approximately those predicted by the theory and they are seen immediately.

C. Smoothness requirements (e.g. ninth degree piecewise polynomials with 0, 1, 2, ... continuous derivatives) have a small but definitely observable effect on efficiency. For a given smoothness requirement it seems that one should take the polynomial degree slightly higher than the minimum necessary.

D. Very strong singularities have little adverse effect as long as f(x) remains bounded. Unbounded singularities such as $1/\sqrt{x}$ cause extensive computations even in the L_1-norm.

E. There appears to be little systematic difference in the L_1, L_2 and L_∞ norms as far as computational effort is concerned. There are, of course, substantial differences in the nature of the approximations obtained. Other norms were tried (e.g. $L_{.2}$, $L_{.5}$, $L_{1.5}$, L_4, L_{40}) and no advantages and occasional disadvantages were observed.

Error Distribution. There is essentially no difference between the fixed and proportional error distributions for extremely smooth functions. For rougher and more oscillating functions the fixed error distribution is significantly more efficient. The proportional error distribution tends to make the approximation overly accurate in regions of rapid oscillation. This behavior is suggested, but not implied, by the theoretical results in Rice, [1976a].

Estimation of Derivatives by Finite Differences. This is satisfactory provided a proper balance is maintained between the required approximation accuracy, the errors from round-off and the discretization errors. The standard approaches for estimating these effects seem to serve adequately.

Special Schemes for Singularities. ADAPT is very good at handling singularities in the following sense. A number of special schemes were tried in the neighborhood of known singularities. None of these improved the performance of ADAPT for singularities involving infinite values of some derivative. If jump discontinuities are known along with the correct left and right values of the appropriate derivative then this information can be (and may be) used to significantly improve the efficiency of ADAPT via the breakpoint facility.

The third part of this paper gives detailed summaries of some of the experiments performed to determine the actual behavior of ADAPT.

2. AN ADAPTIVE SMOOTH APPROXIMATION METALGORITHM

The experimental results in this paper are based on a particular algorithm (computer program) to be published elsewhere [Rice, 1976b]. The description given here of the algorithm is at a somewhat abstract level (compared to Fortran) and there are a multitude of concrete realizations of this abstract description. We have previously introduced the term metalgorithm for classes of algorithms all of which have the same description at some abstract level.

The key ingredient in this metalgorithm is stack of subintervals of the interval of approximation. The top subinterval is the leftmost of those on the stack and one attempts to obtain an acceptable approximation on this subinterval by smooth interpolation. That is, one determines the polynomial of specified degree which interpolates f(x) in this subinterval and the interpolation is osculatory at each end, matching the number of derivatives required by the smoothness condition. This, of course, places a constraint on the combinations of smoothness and polynomial degree. The maximum smoothness is that of the Hermite piecewise polynomials, i.e. smoothness is at most half the (degree-1). It is our experience and conjecture that all stack based algorithms are unstable if more smoothness is imposed. Analytical results show this in certain limited cases.

The error on the subinterval is estimated (by some numerical quadrature rule) and if it is small enough then

(a) The subinterval is discarded from the stack,

(b) The approximation for this subinterval is saved,

(c) The right end-point of the subinterval becomes a knot.

If the error is too large, then the subinterval is divided into two halves, they are placed on the interval stack to replace the previous one and the process is repeated. The metalgorithm also provides for break points where the smoothness condition is specified to be changed. The location, derivative to be broken and the left and right values of this derivative (including the zero derivative as function value) are specified at a break point. The presence of break points modifies how intervals are subdivided. Break points are not of great theoretical interest but there are applications where they are required.

A final point is how to judge whether the error is small enough on a particular subinterval. There are two basic type of error distributions: proportional and fixed. In the proportional error distribution the requested accuracy of approximation is distributed according to the length of the subinterval. Thus if an accuracy of .001 is required on (0, 1) then .0001 is required on (.6, .7). In the fixed error distribution the requested accuracy of approximation is used on all subintervals regardless of length. The attraction of the proportional error distribution is that it guarantees that the requested accuracy is achieved (provided the numerical quadrature estimates are not too bad - they are made conservatively), whereas the other one gives an overall error (except in the L_∞ norm) of the requested accuracy times the final number of subintervals required. The computer program used has a mode where it tries to estimate the number of subintervals and uses that to achieve a requested accuracy while still using a fixed error distribution. This scheme is highly reliable (> .99 in our experience) but not foolproof.

If f(x) has singularities then the degree of convergence results are much better for the fixed error distribution and, as is seen from experimental results below, this advantage shows up dramatically in real computations involving such functions.

This general description of the metalgorithm gives the background to understand the following more formal and more detailed description of the metalgorithm.

F	.	FUNCTION TO FIT	CHARF	.	LIMIT ON LENGTH OF SUBINTERVALS OF (A,B)
A,B	.	INTERVAL ENDPTS			
ACCUR	.	ACCURACY DESIRED	EDIST	.	TYPE OF ERROR CONTROL
DEGREE	.	POLYNOMIAL DEGREE			

```
SMOOTH  .  NO. CONT. DERIVS     NBREAK . NUMBER OF SPECIFIED
NORM    .  MEAS. OF L-P ERROR            BREAK POINT IN FIT.
             P IN (0, INFINITY)          HAS RELATED ARGUMENTS
                                         GIVING DETAILS.

   CALL ADAPT - TO DO THE APPROXIMATION

SUBPROGRAM ADAPT

   CALL SETUP - CHECK INPUT AND INITIALIZE THINGS

   *** LOOP OVER PROCESSING INTERVALS ***

   CALL TAKE   - AN INTERVAL OFF THE STACK
   CALL COMPUT - AN APPROX ON THIS INTERVAL
   CALL CHECK  - FOR DISCARDING OR DIVIDING INTERVAL
   CALL PUT    - NEW INTERVALS ON STACK, UPDATE ALGORITHM
                 STATUS
   CALL TERMIN - TEST FOR FINISH, PRINT INTERMEDIATE OUTPUT
   IF NOT FINISHED - REPEAT LOOP

END ADAPT

SUBPROGRAM SETUP
   SET LIMITS ON COMPUTATION PARAMETERS
   CHECK ALL INPUT DATA
   INITIALIZE VARIABLES AND INTERVAL STACK
END SETUP

SUBPROGRAM TAKE
   CHECK FOR BREAK POINT IN TOP INTERVAL
      IF SO - ADJUST XKNOTS TO MAKE IT A PARTITION POINT
      ELSE  - DO NOTHING
END TAKE

SUBPROGRAM PUT

   CHECK FOR DISCARDING INTERVAL
      IF SO - UPDATE ERROR ESTIMATE
              ADJUST STACK
              CALL PTRANS - TO OBTAIN COEFS FOR THIS INTERVAL
              UPDATE XKNOTS AND COEFS

      ELSE  - SUBDIVIDE INTERVAL, PLACE 2 NEW ONES ON STACK
              CHECK FOR EXCEEDING MAX STACK SIZE OR GETTING
                 AN INTERVAL WHICH IS TOO SHORT.  SUCH SHORT
                 INTERVALS ARE DISCARDED WITHOUT REGARD TO
```

```
                 ERROR CONTROL POLICY AND WITH MESSAGE
END PUT

   SUBPROGRAM PTRANS - OF PUT
      CHANGES POLYNOMIAL REPRESENTATION FROM NEWTON DIVIDED
      DIFFERENCE FORM TO POWER FORM WITH ORIGIN SHIFTED TO
      XKNOT VALUE ON LEFT OF INTERVAL. USE SYNTHETIC DIVISION
   END PTRANS

SUBPROGRAM COMPUT
   OBTAIN      - VALUES OF F AND DERIVATIVES. MAKE ADJUSTMENTS
                 IF A BREAK POINT IS INVOLVED
   CALL NEWTON - FOR DIVIDED DIFFERENCES OF INTERPOLATING
                 POLYNOMIAL FOR THIS INTERVAL
   CALL ERRINT - TO ESTIMATE LOCAL ERROR IN L-P NORM
END COMPUT

   SUBPROGRAM NEWTON - OF COMPUT
      BUILD UP TRUE DIVIDED DIFFERENCE TABLE WITH MULTIPLE
      POINTS AT THE INTERVAL ENDS PLUS INTERPOLATION POINTS
   END NEWTON

   SUBPROGRAM ERRINT - OF COMPUT
      USES 4-POINT GAUSS QUADRATURE TO ESTIMATE ERROR NORM
      ON INTERVAL.  SPECIAL COMPUTATION FOR MAX-NORM.
   END ERRINT

SUBPROGRAM CHECK
   USE ERROR DISTRIBUTION TYPE AND CHARF TO DECIDE ON DISCARD
END CHECK

SUBPROGRAM TERMIN
   TEST - FOR TERMINATION    EMPTY STACK - NORMAL
                             EXCEEDED XKNOTS LIMIT - ABNORMAL
END TERMIN
```

3. THE BEHAVIOR OF THE ALGORITHM ADAPT

This section consists primarily of graphical summaries that illustrate the behavior of ADAPT and support the conclusions of the first section. These are only selected values from a large set of computational experiments. The combinatorial nature of the situation is formidable; there are a good number, at least 10, of independent variables in the algorithm input and function characteristics. All combinations of even a few values for each variable implies a prohibitively expensive computational effort. All told about 30 different functions have been used for ADAPT and about 2500 approximations computed.

The results reported here involved only seven functions; e^x, $\sqrt{|\sin x|}$, a broken line, and the four functions shown in Figure 1.

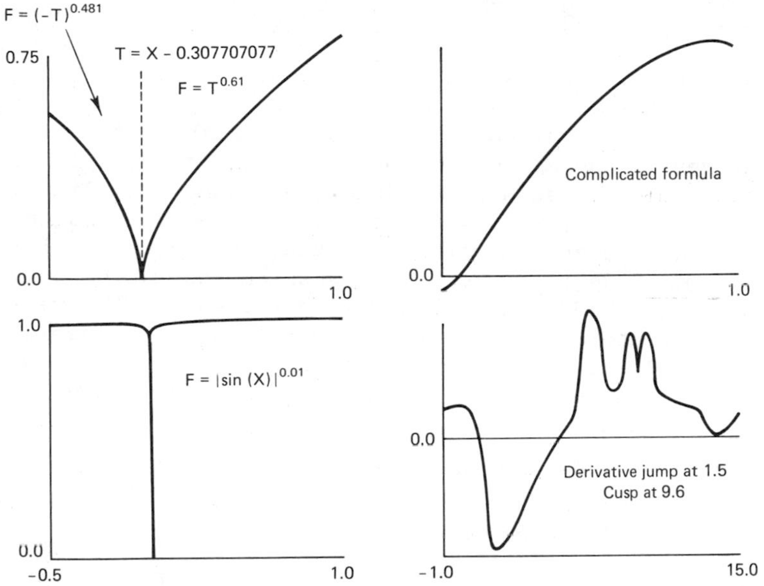

Figure 1. (left) Two examples of functions used to study the algorithm properties. Both have infinite first derivative at one point. (right) Second two examples of functions used. The third one is smooth and analytic but has a complicated formula. The fourth has a jump in the 1.5 derivative at 1.5 plus a cusp and is otherwise quite variable but smooth in nature.

The actual formula for the third one (labeled Complicated Formula) is

$$\sin(x^2 - 3.1e^{x/(1+x^2)}) \quad (x + \ln(2 + x^2/(1 + x))),$$

and the formulas for its derivatives expand very rapidly with the order. The fourth function is referred to as the complicated function and its actual formula is

$$e^{-\sqrt{|x + 2|}} + 2 - 4x^2 e^{-|x - 1.5|} + 1/(1+150(x-7)^6)$$
$$+ 6/(1 + 5(x-7)^4) + 10/(1 + 4(x-9.6)^4) \; |x-9.6|^{0.6}$$
$$- 2e^{-|x-14|^{2.5}} + \sin(4x)/(.5 + |x-16|^5) \; e^{\max(0,\, x-15)} .$$

3.1 <u>Approximation of the Entire Function e^x</u>. Figure 2 shows the effect of changing the smoothness requirement as everything else remains the same. All computer times in this paper are for the CDC 6500 owned by the Purdue University Computing Center. One sees that the smoothness requirement has little effect here. The variable SMOOTH is the number of continuous derivatives of the approximation throughout this paper. This effect can also be seen later (though not shown explicitly) in Figures 5, 6, 8, 9, 10, 11 and 12. Figure 3 shows the dramatic improvement in efficiency that occurs when raising the polynomial degree. For low requested accuracies only two knots (the end points -2 and 2) are used and the variations in computer time is due to the different polynomial degrees. All the approximations of Figure 3 are only continuous. Figure 4 is a graph of the actual least squares error versus the number of knots plotted on log-log paper. The straight lines indicate that the error is of the form k^{-r} where k is the number of knots. The measured values of these slopes are tabulated below.

Degree	5	5	7	7	9	9	9
SMOOTH	0	2	0	2	0	2	4
Measured Slope	6.0	6.0	7.5	7.5	12	12	12

There are more entries here than in Figure 4 because some of the lines are on top of one another. The theoretically predicted values are 6, 8 and 10, respectively. The discrepancies here may well be due to the inaccuracies in measuring the slopes with a ruler.

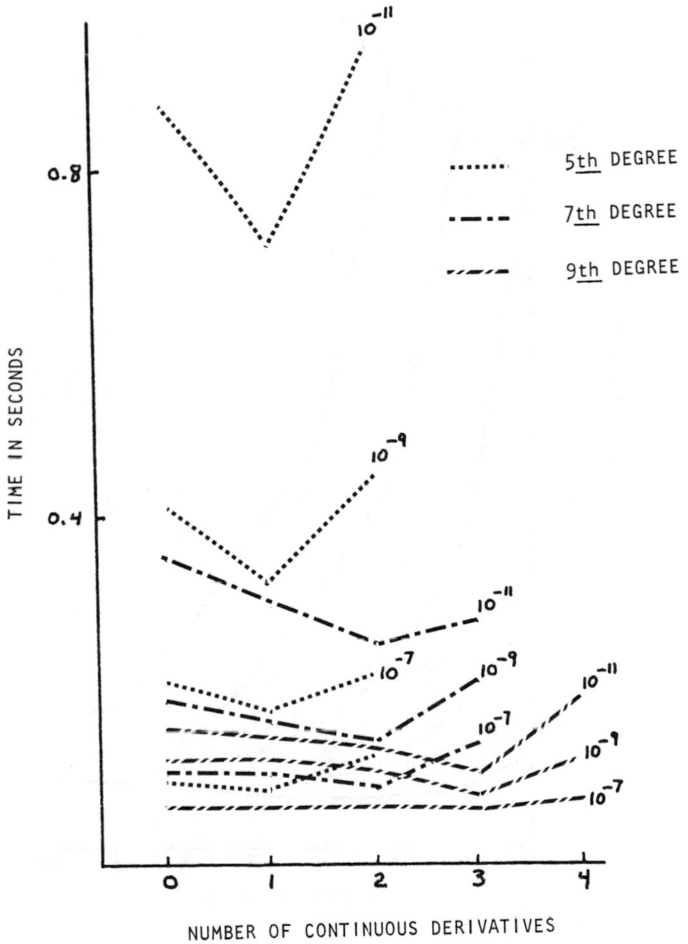

Figure 2. Plot of computer time versus smoothness for various requested accuracies and polynomial degrees in approximating e^x on [-2,2] by least squares.

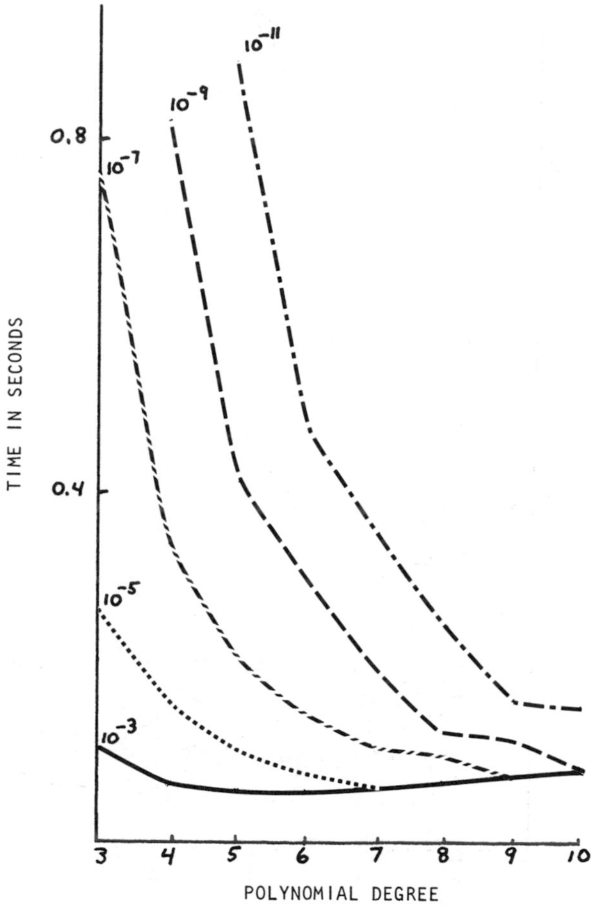

Figure 3. Plot of computer time versus polynomial degree for least squares approximation of e^x on [-2,2] for various requested accuracies.

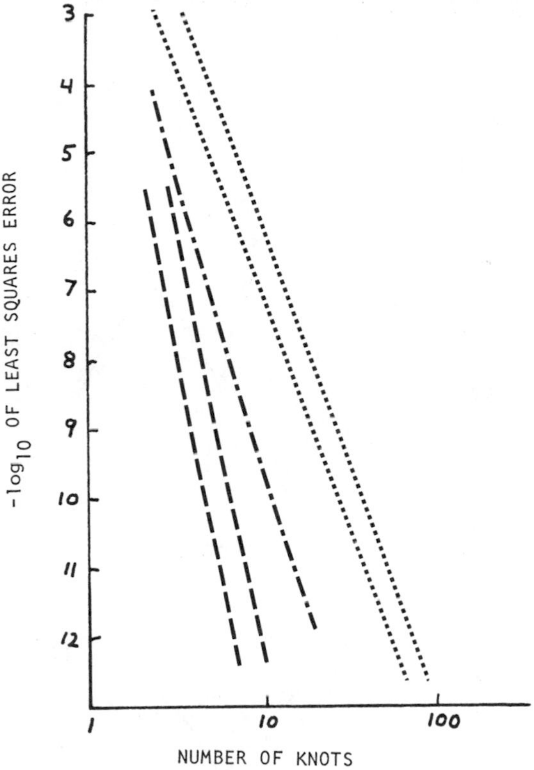

Figure 4. Plot to show experimentally measured degree of convergence for approximation of e^x. Dotted lines with slope about 6.0 are degree 5 with 0 (left) and 2 (right) continuous derivatives. Dot-dash line with slope about 7.5 is degree 7 with 0 and 2 continuous derivatives. Dashed line of slope about 12 is degree 9 with 0 (left) and 4 (right) continuous derivatives.

3.2 Approximation of Functions with Complications and Singularities. Figures 5 and 6 illustrate the effect of increasing the polynomial degree in approximating the fourth and first functions of Figure 1, respectively. There are two points to notice. First, for a given requested accuracy there is a polynomial degree beyond which it is pointless to go. This optimal degree is fairly low, but it does increase gradually as the requested accuracy increases. Note also that there is a much more severe penalty in efficiency in using too low a polynomial degree than one too high. Note also that requiring a continuous derivative rather than merely continuity does not have a major effect on the computation time.

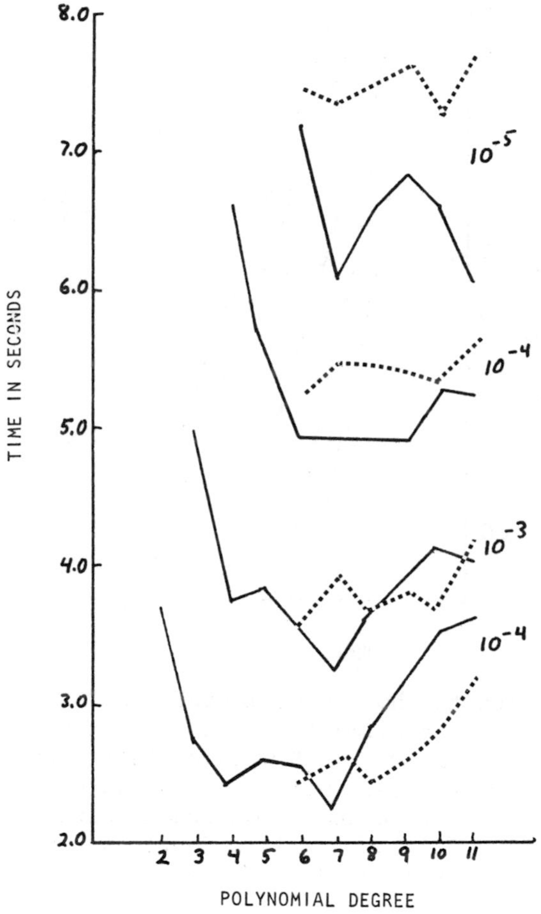

Figure 5. Plot of computer time versus polynomial degree for Tchebycheff approximation to the fourth (complex) function of Figure 1. Various requested accuracies are shown for C^0 (solid lines) and C^1 (dotted lines) continuity.

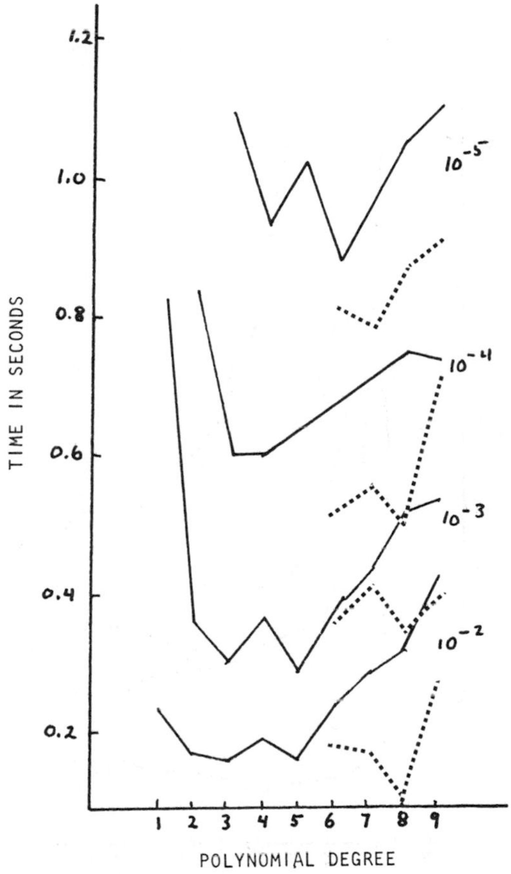

Figure 6. Plot of computer time versus polynomial degree for Tchebycheff approximation to the second function of Figure 1 (with singularity at x = .307707077). Various requested accuracies are shown for C^0 (solid lines) and C^1 (dotted lines) continuity.

Figure 7 presents another view of this data which also shows the existence of an optimal range of polynomial degrees. The vertical scale here is the number of knots rather than computer time. This difference in scale does not change the gross nature of the graphs in this paper, but it does tend to lower the right ends of the curves shown in Figure 7 compared to a graph using computer time.

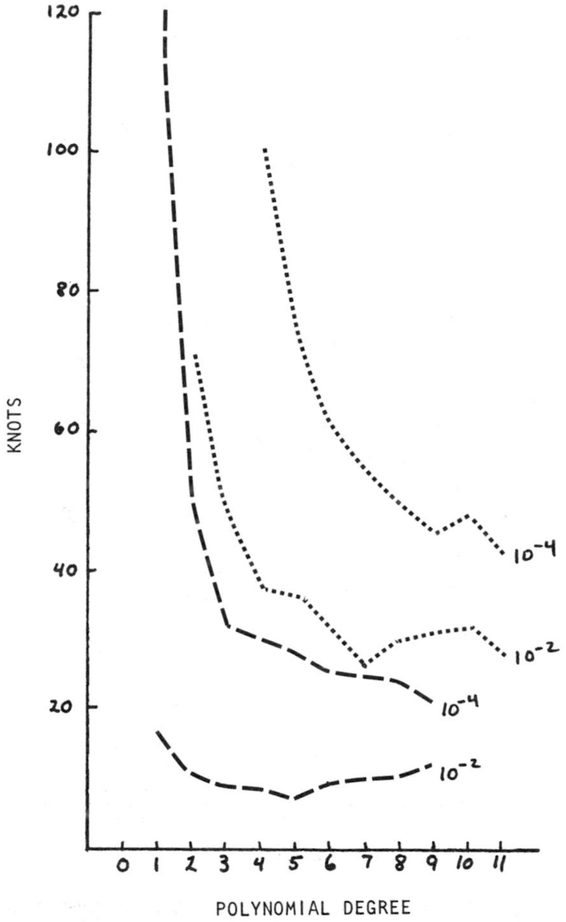

Figure 7. Plot of knots versus polynomial degree for Tchebycheff approximation to the second (dashed) and fourth (dotted) functions of Figure 1. The approximations are only continuous and the requested accuracies are as indicated.

Similar conclusions are reached from Figures 8, 9 and 10 which involve the functions $\sqrt{|\sin x|}$ and $\sqrt[100]{|\sin x|}$. Figures 8 and 9 are from identical approximations and only the vertical scale is changed from number of knots to computer time. One can compare Figures 8 and 10 to see the small effect resulting from a drastic strengthening of the singularity in the function being approximated.

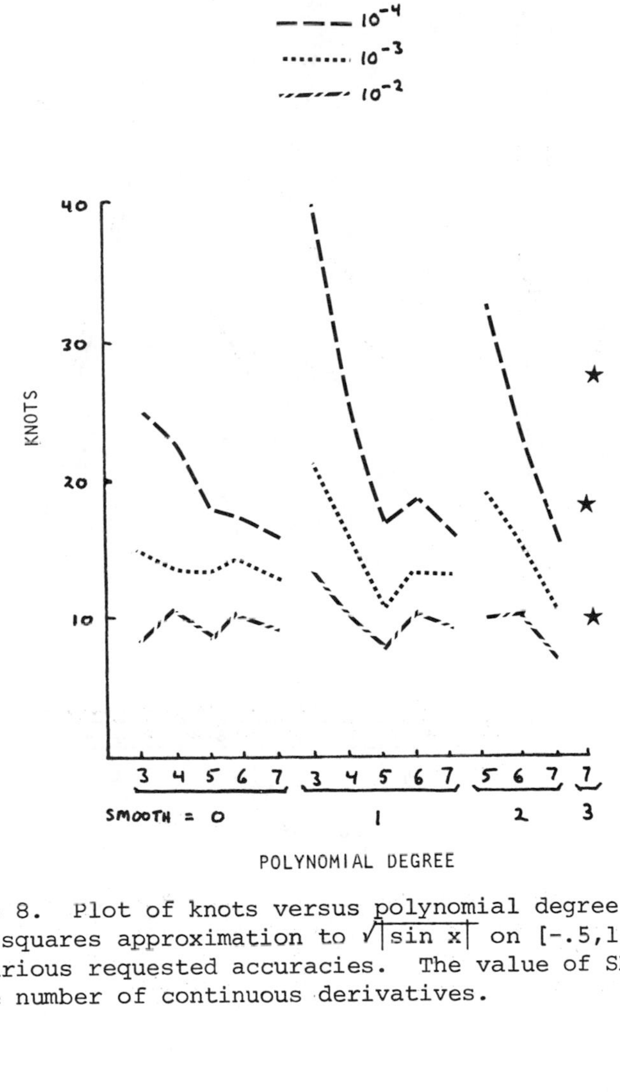

Figure 8. Plot of knots versus polynomial degree for least squares approximation to $\sqrt{|\sin x|}$ on [-.5,1.0] for various requested accuracies. The value of SMOOTH is the number of continuous derivatives.

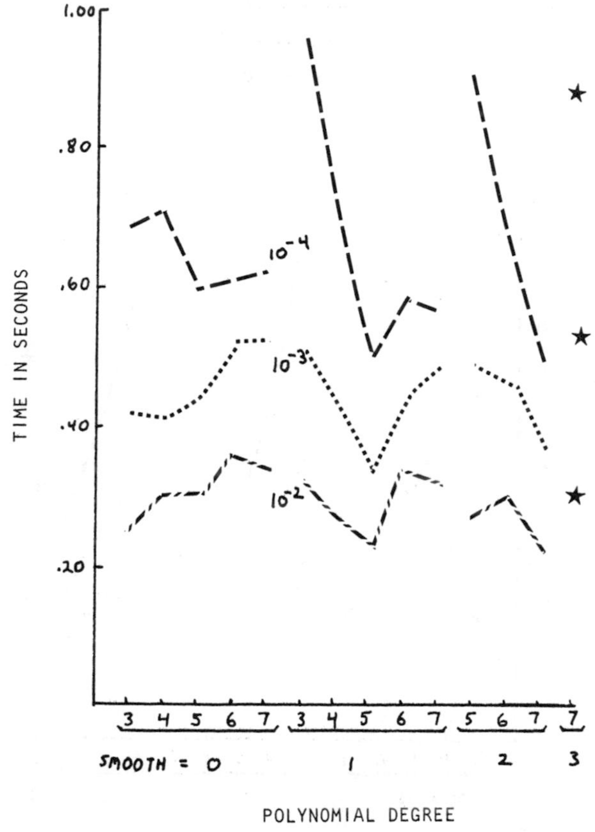

Figure 9. Plot of computer time versus polynomial degree for least squares approximation to $\sqrt{|\sin x|}$ on [-.5,1.0] for various requested accuracies. The number of continuous derivatives is SMOOTH.

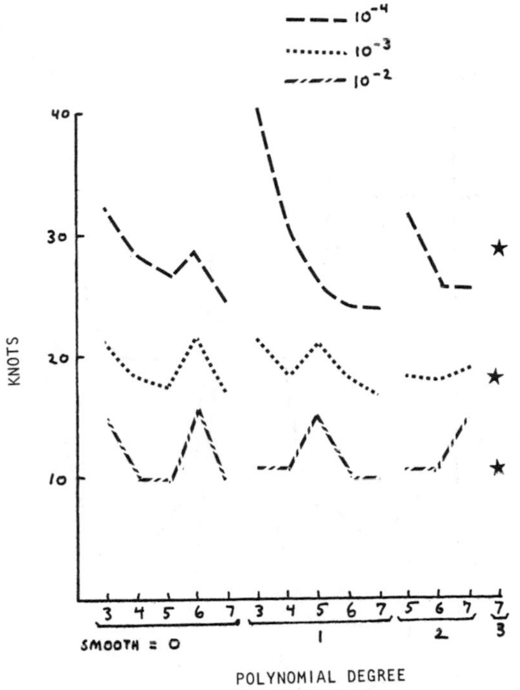

Figure 10. Plot of knots versus polynomial degree for least squares approximation to $^{100}\sqrt{|\sin x|}$ on [-.5, 1.0] for various requested accuracies. The number of continuous derivatives is SMOOTH.

Figures 11, 12 and 13 illustrate the effect of using various norms of approximation for functions with singularities. Again one can see that the relationship between computer time and number of knots is very strong. The computations for Tchebycheff approximation are somewhat more efficient because special code is used to compute the error estimates in this case. These estimates form a substantial part of the computation and for $p < \infty$ they involve the Fortran **p operation (even if p = 1 or 2).

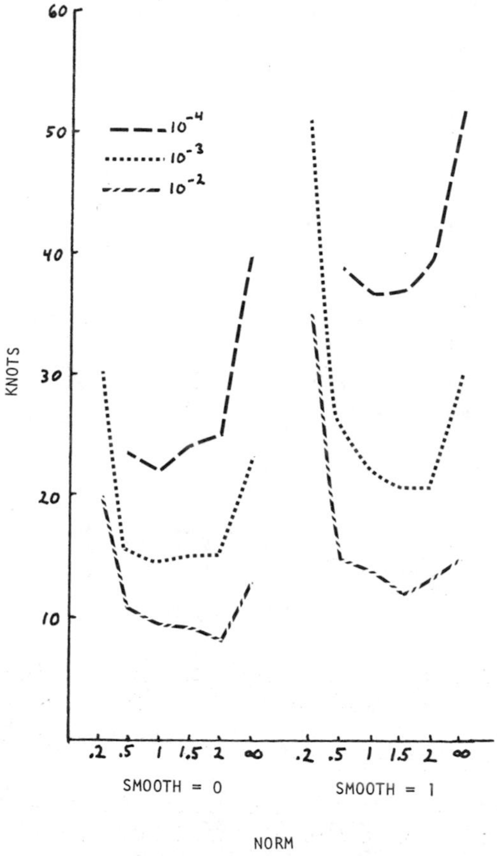

Figure 11. Plot of knots versus norm for third degree approximation to $\sqrt{|\sin x|}$ on [-.5,1.0] for various requested accuracies. SMOOTH is the number of continuous derivatives.

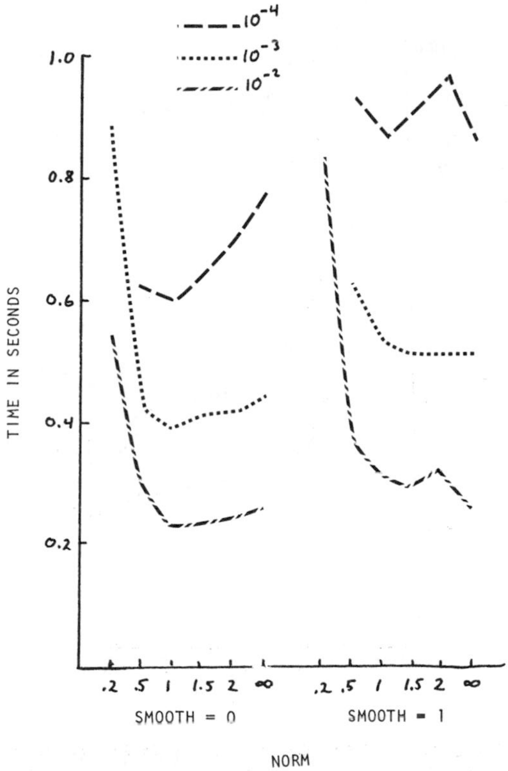

Figure 12. Plot of computer time versus norm for third degree approximation to $\sqrt{|\sin x|}$ on [-.5,1.0] for various requested accuracies. SMOOTH is the number of continuous derivatives.

One should not judge the L_p norms for $p < 1$ by their performance here. ADAPT does produce approximations whose L_p error is within the requested accuracy but it is doubtful that the knots are anywhere near optimal. The logic of the error distribution (either proportional or fixed) does not apply to these non-convex norms and an examination of the approximations produced suggests that the knots are poorly placed in these cases.

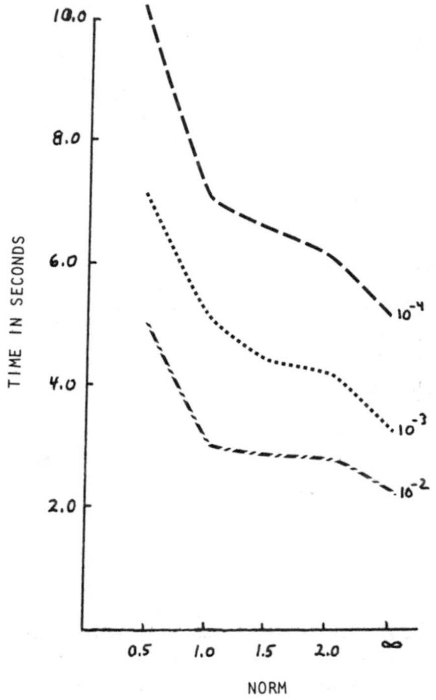

Figure 13. Plot of Computer time versus norm for C^1, fifth degree approximation to the fourth function of Figure 1 for various accuracies.

Figure 14 illustrates the effect of three types of error distribution: proportional, approximate fixed and fixed which are indexed by 0, 1 and 2, respectively. All of the functions involved here have singularities and the proportional scheme performs much less efficiently than the other two. Note that the final accuracy achieved for the fixed distribution is lower than that of the other two. The extra computation of the proportional scheme achieves excessively high accuracy at the singularities and the overall least squares error is little, if any, better than for the approximate fixed scheme.

The approximate fixed error distribution attempts to achieve the requested accuracy by dynamically estimating the final number of knots and modifying the fixed accuracy requirement appropriately. Such a scheme cannot, of course, be completely reliable and a little ingenuity suffices to construct examples where it fails by arbitrarily large amounts.

Nevertheless, a check of the approximations made with this scheme in this set of computations reveals a reliability of about .99 and in no case did a dramatic failure occur.

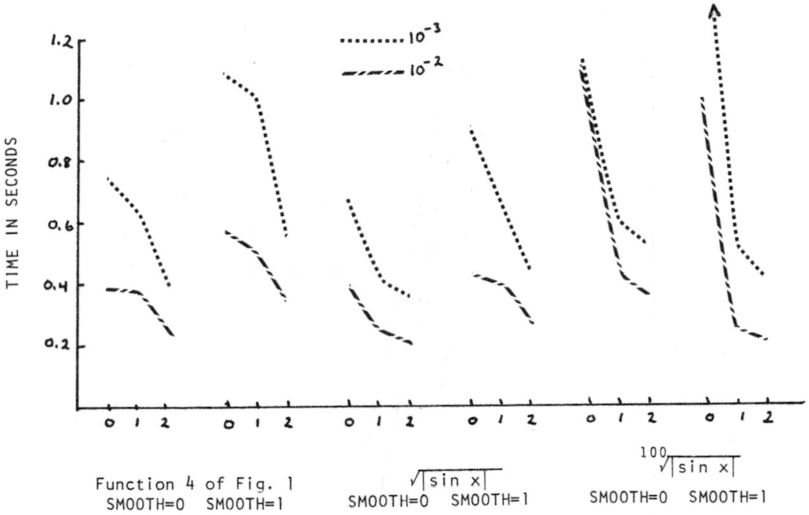

Figure 14. Study of the effect of the error distribution type: (0 = proportional, 1 = approx. fixed, 2 = fixed). The approximations are least squares by C^0 and C^1 piecewise cubic polynomials. The vertical scale is to be multiplied by 10 for the complicated function.

Experiments not illustrated here show that these three error distribution methods are approximately equivalent for very smooth and well behaved functions. This partly explains the different behavior of the results for the complicated functions. While it has two singularities, it also has large pieces where it is smooth but rather variable. On these two pieces the three schemes perform about the same and since this constitutes a large part of the computation, there is less overall difference between the schemes.

Figures 15 and 16 show the experimentally measured rates of convergence for least squares approximations to $\sqrt{|\sin x|}$ and $\sqrt[100]{|\sin x|}$. They agree reasonably well with theoretical expectations except for the case of degree five with SMOOTH = 2. This occurs for both functions and the difference between the observed slope of 3.5 and the expected slope of 6 is too large to be a measurement error. We have found no explanation for this anomaly beyond saying (somewhat unsatisfactorily) that it is an unusual interaction between the particular characteristics of the algorithm and the function.

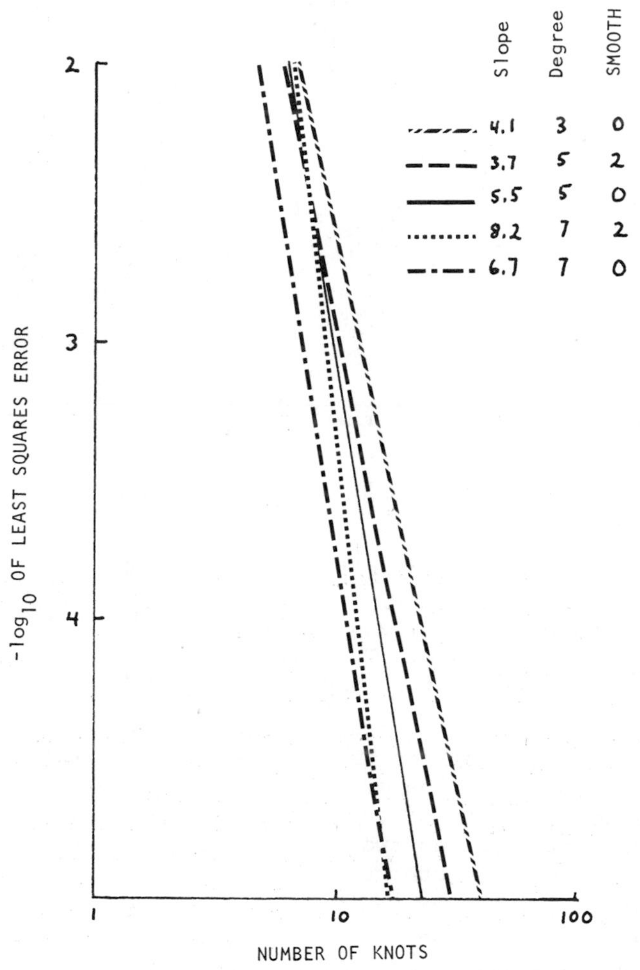

Figure 15. Plot to show experimentally determined degree of convergence for least squares approximation to $\sqrt{|\sin x|}$ on [-.5,1.0] by piecewise polynomials of various degrees with SMOOTH continuous derivatives.

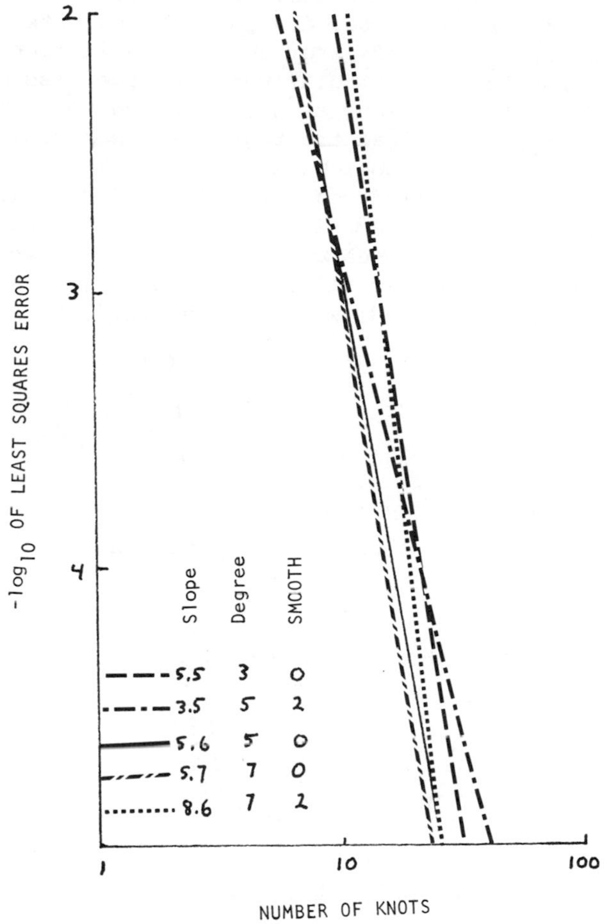

Figure 16. Plot to show experimentally determined degree of convergence for least squares approximation to $100\sqrt{|\sin x|}$ on [-.5,1.0] by piecewise polynomials of various degrees with SMOOTH continuous derivatives.

3.3 <u>The Use of Finite Difference Estimates of Derivatives</u>. The algorithm ADAPT uses Hermite interpolation and hence smooth approximations require that one have the appropriate derivatives of f(x). These are sometimes difficult or impossible to obtain and thus one would attempt to estimate them using finite difference formulas. Figure 17 shows the results of this approach for the third function of Figure 1. The simplest symmetric difference formulas were used to estimate the first, second and third derivatives. There is a standard procedure in numerical analysis to estimate the effects of machine round-off error and discretization (or truncation) error for a particular value of the discretization step H. We found that as long as H was chosen so that these errors are somewhat less than the requested approximation accuracy, then everything goes well. If this is not the case, then this effectively introduced a pseudo-random oscillation in the derivatives which dramatically degrades the efficiency of the computation. The cases of degree = 7 with two and three continuous derivatives illustrate this well. Note that these computations where made with a long word length machine

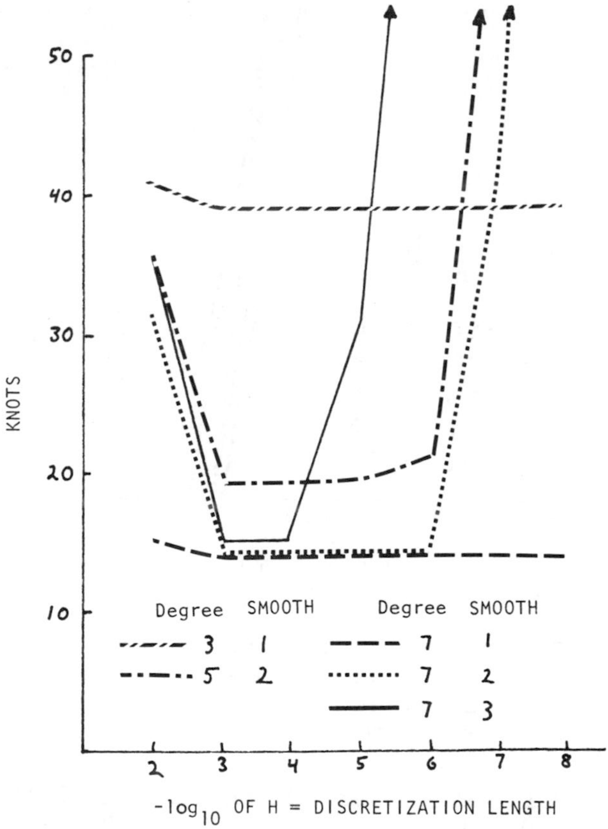

Figure 17. Effect of using numerically estimated derivatives for least squares approximation to the third function (complicated formula) of Figure 1. Piecewise polynomials of various degrees with SMOOTH continuous derivatives are used.

(14 decimal digits). Machines with much shorter word lengths would require higher order difference formulas than used here (especially for second and third derivatives) if high accuracy approximations are desired.

3.4 Special Strategies for Singularities. It seems plausible that if one knows the location of a singularity then one can make use of this information to improve the efficiency of the computation. We attempted two different types of special strategies. One, labeled as 2 in the graphs, uses the location of the singularity. One then tries to modify the knot selection near the singularity so as to improve the convergence. None of the several modifications that were tried proved to be very successful and finally this feature was removed from the algorithm.

The other, labeled 3, 4 and 5 in the graphs, uses not only the location of the singularity but also knowledge of which derivative is discontinuous and the left and right limits of that derivative. A knot is then placed at the singularity, the smoothness of the approximation lowered (if must be) and these limiting values imposed on the approximation.

Figures 18 and 19 show the results for two cases, Figure 18 for the first function of Figure 1 and Figure 19 for a broken line with one break. Strategy 3 corresponds to use of a break point in the derivative with the singularity (the first in these two cases) and the use of the correct limit values. For the infinite derivative example, the "correct" value was chosen to be the order of 10^{10}. Strategy 4 is the same except that approximate limit values are used. For the infinite example these were about 10^{3} and they were about 15% in error for the broken line. Strategy 5 uses a break in the derivative one order lower than that with the singularity. The limit values then are used to reimpose continuity for this derivative. This allows the limit values of the derivative with the singularity to be determined by the algorithm. Strategy 1 is ADAPT without any additional information.

Figure 18 shows that strategies 2, 3 and 4 are always worse than 1 or 5. Strategy 5 is erratic in nature, but sometimes better than Strategy 1. One may safely conclude that it is not safe to try to estimate the "correct" value of infinity for infinite singularities.

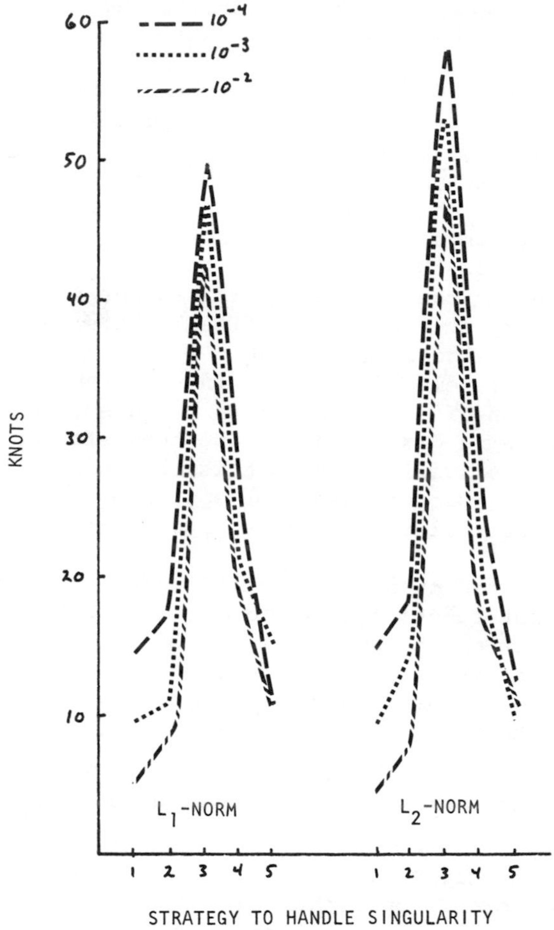

Figure 18. Evaluation of five strategies to handle the singularity in the first function of Figure 1. Various requested accuracies are used for C^1 fifth degree piecewise polynomials in the L_1 and L_2 norms. See text for description of strategies.

Figure 19 shows that significant improvements in efficiency are possible for jump discontinuities. Strategies 3 and 5 both exactly reproduce the broken line with the minimum number (3) of knots. Other examples have confirmed the effectiveness of the break point strategy to handle jump discontinuities well.

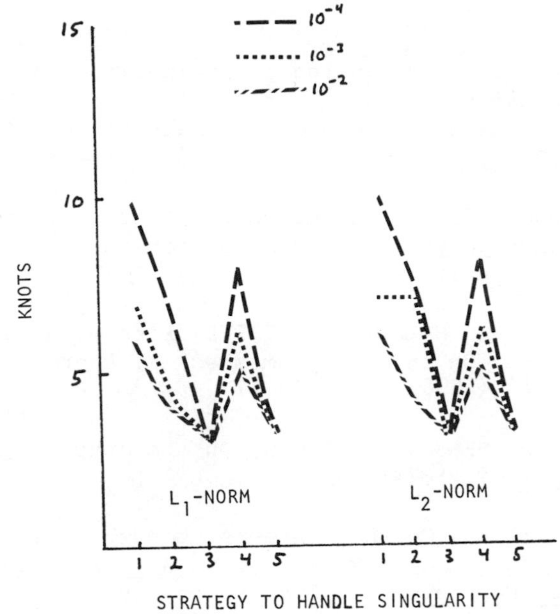

Figure 19. Evaluation of five strategies to handle the singularity in a broken line. Various requested accuracies are used for C^1 fifth degree piecewise polynomials in the L_1 and L_2 norms. See text for description of strategies and function.

4. ACKNOWLEDGEMENTS

This work was supported in part by the National Science Foundation under grant GP 32960X. I also thank the Purdue University Computing Center for their support of this work with additional funds for computing.

5. REFERENCES

BRUDNYI, Ju.A., [1974] Spline approximation and functions of bounded variation, Doklady Akad. Nauk SSR, 215 (English translation: Soviet Mathematics Doklady), 15 (1974) pp. 518-521.

BURCHARD, H.G., [1976] On the degree of convergence of piecewise polynomial approximation on optimal meshes II, to appear.

__________ and HALE, D.F., [1975] Piecewise polynomial approximation on optimal meshes, J. Approximation Theory 14 (1975) pp. 128-147.

PEETRE, J. and BERGH, J., [1976] On the spaces V_p $(0 < p \leq \infty)$, to appear.

RICE, J.R., [1969] On the degree of convergence of nonlinear spline approximation, in Approximation with Special Emphasis on Spline Functions (I.J. Schoenberg, ed.), Academic Press, pp. 349-365.

__________, [1976a] Adaptive Approximation, J. Approximation Theory, to appear.

__________, [1976b] ADAPT - Adaptive smooth curve fitting, to appear.

Extremals and Zeros in Markov Systems are Monotone Functions of One Endpoint

ROY L. STREIT

Abstract. The synthesis of optimum field patterns for discrete linear antenna arrays leads naturally to the study of the behavior of generalized Chebyshev polynomials as a function of one endpoint of the interval of definition. Of particular importance in this application is the variation of the zeros and the extreme points of the generalized Chebyshev polynomials as the left-hand endpoint of the interval is shifted to the right. All the zeros and all the extreme points of the classical Chebyshev polynomials defined on the intervals $[t, b]$ are strictly increasing functions of t. In Haar Systems, this property does not hold, but in Markov Systems with unit element it does. An apparently new extremal property in Haar Systems is proved and then used to show that in every Haar System the first zero must be an increasing function of t. If, in addition, the Haar System has a unit element, then the first zero must be strictly increasing.

INTRODUCTION AND MOTIVATION

Visualize any number M of fixed points in the plane, all collinear and spaced symmetrically about the center of the shortest line segment containing all the points. Take the origin to be the center of this smallest line segment. If each of these points is taken as the position of a sensor of an antenna array, then the directional response of this array is directly proportional to the absolute value of

$$P(u) = \sum_{k=1}^{N} a_k \cos \xi_k u \quad , \; 0 \le u \le \pi \quad , \tag{1}$$

where $N = \left[\frac{M+1}{2}\right]$, ξ_k is a constant multiple of the distance of the k-th point to the right of the origin, the variable u can be regarded as an angle measured from a normal to the line of points, and the coefficients a_k are any real constants. The coefficients a_k are the design parameters and are chosen to enhance the directionality of the array.

Usually, the direction perpendicular to the line of points is the desirable direction and all other directions are of less interest. Thus, coefficients a_k are chosen so that $P(u)$ has its largest magnitude near or at the point $u = 0$, while keeping $P(u)$ as small as possible in magnitude elsewhere. With this objective in mind, the u domain is split into two parts: the "mainlobe" region $[0, u_0]$ and the "sidelobe" region $[u_0 , \pi]$. The coefficients a_k are defined to be optimal if and only if the ratio of L_∞ norms

$$\| P(u) \|_{[0,\, u_0]} \div \| P(u) \|_{[u_0,\, \pi]} \tag{2}$$

is the largest possible for any coefficient set. Theorem 2 of this paper gives conditions under which this ratio is maximized by minimizing the denominator independently of the numerator. That is, if the set of cosines in (1) forms a Haar System, then the optimal coefficients are proportional to the coefficients of the generalized Chebyshev polynomial for the interval $[u_0, \pi]$. With these optimal coefficients, the peak point of a sidelobe is just an extreme point of a uniform function approximation.

The ratio (2) is therefore maximized as a function of the parameter u_0 dividing the mainlobe and the sidelobe regions. In engineering applications, it is important to know how this ratio changes with u_0 and how this change shows up in the actual field pattern (1). Theorem 3 gives necessary and sufficient conditions for the maximum value of the ratio (2) to be a strictly increasing function of u_0 , while Theorem 1 gives conditions under which all the zeros and all the sidelobes of the optimal field pattern shift strictly to the right with increasing u_0 . Finally, since the first zero of $P(u)$ is sometimes used as a measure of the mainlobe region $[0, u_0]$, Theorem 4 gives weaker conditions under which this measure is also strictly increasing with increasing u_0 .

Pokrovskii [1] studied this problem in detail, but for equispaced sensors only. In the equispaced case, a transformation of variables reduces the problem to a study of ordinary polynomials. An explicit solution to the maximization of the ratio (2) for equispaced sensors was given by Dolph [2]. A generalization of Dolph's method to symmetrically spaced sensors can be found in Streit [3], [4].

THE MATHEMATICS

Let $T_n(u) \equiv \cos [n \cos^{-1} u]$ be the Chebyshev polynomial of degree n on the closed interval $[-1, 1]$. Then the Chebyshev polynomial on the closed interval $[t, b]$ is just

$$T_n\left(\frac{2}{b-t}x - \frac{b+t}{b-t}\right) \quad , \; t \leq x \leq b \; ,$$

which has the n zeros (counting from left to right)

$$z_k(t) = \frac{t}{2}\left[1 - \cos\left(n - k + \frac{1}{2}\right)\frac{\pi}{n}\right]$$

$$+ \frac{b}{2}\left[1 + \cos\left(n - k + \frac{1}{2}\right)\frac{\pi}{n}\right] \quad ,$$

$$k = 1 \, , \, . \, . \, . \, , \, n \, ,$$

and the $(n + 1)$ extremal points

$$x_k(t) = \frac{t}{2}\left[1 - \cos\,(n - k + 1)\,\frac{\pi}{n}\right]$$

$$+ \frac{b}{2}\left[1 + \cos(n - k + 1)\,\frac{\pi}{n}\right] \quad ,$$

$$k = 1 \, , \, . \, . \, . \, , \, n + 1 \; .$$

By inspection, except for $x_{n+1}(t)$, all the zeros and all the extremals are strictly increasing continuous functions of the left-hand endpoint t. Theorem 1 extends this property to more general spaces.

Let $C[a, b]$ be the linear space of all real valued and continuous functions on the closed finite interval $[a, b]$. Meinardus [5] defines the finite or infinite sequence of linear subspaces V_n , $n = 0 , 1, 2, . . .$ of $C[a, b]$ to be a Haar System provided it has the three properties:

i. $V_n \subset V_{n+1}$

ii. The dimension of V_n is $n + 1$

iii. V_n satisfies the Haar condition, i.e., $f \in V_n$ and $f \not\equiv 0$ implies that f has at most n zeros in $[a, b]$.

Define $V_n(t)$ to be that linear subspace of $C[t, b]$ obtained by restricting every function in V_n to the interval $[t, b] \subset [a, b]$. Thus, $V_n(t)$ forms a Haar System for each $t \in [a, b)$. For $f \in C[a, b]$, define

$$\| f \|_{\mathcal{S}} = \max_{x \in \mathcal{S}} | f(x) | \quad , \mathcal{S} = \overline{\mathcal{S}} \subset [a, b] ,$$

and

$$\rho_n(f; t) = \min_{g \in V_n(t)} \| f - g \|_{[t, b]} \quad .$$

A Haar System with unit element is a Haar System for which V_o contains the constant functions. Zielke [6] proves the following lemma in a more general form.

<u>Lemma 1</u>. If V_n is a Haar System with unit element on the closed interval $[a, b]$, and if $f \in V_n \backslash V_{n-1}$, then there exist at most $(n + 1)$ points $x_k \in [a, b]$ such that f is strictly monotone on each interval $[x_k, x_{k+1}]$, $k = 1, 2, \ldots, n + 1$.

Define $S_0^{(t)}(x) \equiv 1$ for all x and t . The following lemma, due to Meinardus [7], [5], is an immediate consequence of de la Vallée Poussin's Theorem and Lemma 1.

<u>Lemma 2.</u> For a Haar System V_n with unit element, the functions $S_n^{(t)}$ satisfying the conditions

a. $S_n^{(t)} \in V_n(t)$,

b. $\| S_n^{(t)} \|_{[t,b]} = 1$,

c. $\rho_{n-1}(S_n^{(t)} ; t) = 1$,

d. $S_n^{(t)}(b) = 1$

have the following properties:

1. $S_n^{(t)}$ is uniquely defined for each $n \geq 0$ and $t \in [a, b)$.
2. $S^{(t)}$ possesses precisely n+1 extremal points $x_k(t)$ $(k = 1,2,\ldots, n + 1)$ in the interval $[t, b]$. The points t and b are extremal points, and arranging the points in increasing order

$$t = x_1(t) < x_2(t) < \ldots < x_n(t) < x_{n+1}(t) = b \quad ,$$

we have the alternating property

$$S_n^{(t)}\left(x_k(t)\right) + S_n^{(t)}\left(x_{k+1}(t)\right) = 0 \ (k = 1,2,\ldots,n).$$

3. $S_n^{(t)}$ is strictly monotone (increasing or decreasing) in $x_k(t) \leq x \leq x_{k+1}(t)$ $(k = 1, 2,\ldots,n)$.

Lemma 1 does not imply that V_n' is a Haar System if V_n contains only continuously differentiable functions. Let $V_0 = \{1\}$ and $V_1 = \{1, h\}$ where h is any continuously differentiable strictly increasing function on $[a, b]$ that has, say, 15 inflection points in (a, b). Then V_0, V_1 is a continuously differentiable Haar System, but every function in V_1' has 15 zeros in (a, b).

The system V_n is defined here to be a Markov System with unit element on the closed interval [a, b] if V_n is a Haar System of functions continuous on [a, b], having continuous derivative on the open interval (a, b), and with the property that the spaces V_n' spanned by the derivatives of functions in V_n form a Haar System on (a, b). Therefore, $f \in V_n'$ and $f \not\equiv 0$ implies that f has at most n - 1 zeros in (a, b). Let $V_n'(t)$ be the restriction of V_n' to (t, b). We now prove an extension of Lemma 2 to Markov Systems with unit element.

<u>Theorem 1</u>. Let V_n be a Markov System with unit element on the closed interval [a, b]. Then the functions $S_n^{(t)}$ satisfying conditions a. through d. of Lemma 2 also have the following additional properties:

4. Each function $x_k(t)$ $(k = 1, \ldots, n)$ is a continuous and strictly monotonically increasing function of t.
5. The zeros $z_k(t)$ $(k = 1, \ldots, n)$ of $S_n^{(t)}$, arranged in the increasing order

   $$t < z_1(t) < z_2(t) < \ldots < z_{n-1}(t) < z_n(t) < b,$$

   are all continuous and strictly monotonically increasing functions of t.
6. $S_n^{(t)}(x)$ is strictly monotone (increasing or decreasing) in the interval $a \leq x \leq x_2(t)$.

<u>Proof</u>. To prove property 6, note that Rolle's Theorem implies that $\left[S_n^{(t)}(x)\right]'$ has exactly n-1 zeros in the interval $[x_2(t), z_n(t))$, and so must have none in the interval $[a, x_2(t))$, because any function in V_n' has at most n-1 zeros in [a, b]. Thus, $S_n^{(t)}$ must be monotone in the interval $[a, x_2(t)]$. The continuity of $x_k(t)$ and $z_k(t)$ $(k = 1, 2, \ldots, n)$ follows from a remark in Meinardus

[5, p. 85]. Now, fix $t \in [a, b)$. By continuity, there exists $\delta > 0$ such that for each $0 < \varepsilon < \delta$, the sets $I_k(\varepsilon)$ defined by

$$I_k(\varepsilon) = \begin{cases} [x_k(t),\ x_k(t+\varepsilon)], & \text{if } x_k(t) < x_k(t+\varepsilon) \\ \{x_k(t)\}, & \text{if } x_k(t) = x_k(t+\varepsilon) \\ [x_k(t+\varepsilon),\ x_k(t)], & \text{if } x_k(t) > x_k(t+\varepsilon) \end{cases}$$

$$(k = 1, 2, \ldots, n+1)$$

are pairwise disjoint. Put

$$S(x;\varepsilon) = S_n^{(t)}(x) - S_n^{(t+\varepsilon)}(x) .$$

Because of property 6, $S(x;\varepsilon)$ has no zero in $I_1(\varepsilon)$ and precisely one zero in each of the remaining intervals. Thus the intervals $I_2(\varepsilon), \ldots, I_{n+1}(\varepsilon)$ contain all the zeros of $S(x;\varepsilon)$. Because of property 2,

$$x_1(t) = t < t + \varepsilon = x_1(t+\varepsilon) .$$

Let $j < n + 1$ be the smallest integer such that there exists $\hat{\varepsilon} < \delta$, $\hat{\varepsilon} > 0$, and $x_j(t+\hat{\varepsilon}) \leq x_j(t)$. If strict inequality holds, then the interval $(x_{j-1}(t+\hat{\varepsilon}), x_j(t+\hat{\varepsilon}))$ is disjoint from the intervals $I_2(\hat{\varepsilon}), \ldots, I_{n+1}(\hat{\varepsilon})$ but must contain a zero of $S(x;\hat{\varepsilon})$. On the other hand, if $x_j(t+\hat{\varepsilon}) = x_j(t)$, then $S'(x;\hat{\varepsilon})$ has a zero at $x_j(t)$. But Rolle's Theorem already gives $S'(x;\hat{\varepsilon})$ at least n-1 zeros, one between each consecutive pair of zeros of $S(x;\hat{\varepsilon})$. Therefore, $S'(x;\hat{\varepsilon})$ contradicts the hypothesis on $V_n'(t)$. Hence, $j = n + 1$ and property 4 follows. Finally, property 5 is an immediate consequence of property 4, for otherwise the function $S(x;\hat{\varepsilon})$ will have too many zeros. This completes the proof.

The next object is to weaken the hypotheses of Theorem 1, specifically, by dropping the unit element and the differentiability conditions. One method of achieving this end leads to Theorem 2. First, define $\{h_0 , h_1 , \dots , h_n\}$ to be a basis for the Haar System V_n if $\{h_0 , h_1 , \dots , h_k\}$ is a basis for V_k , $k = 0, 1, \dots , n$. Also, the functions $S_n^{(t)}$ are defined above only for Haar Systems with unit element. For the general Haar System, define $S_n^{(t)} \equiv c(t)\cdot(h_n + g^*)$, where $g^* \in V_{n-1}(t)$ and the constants $c(t)$ are uniquely defined by

$$[c(t)]^{-1} \equiv \| h_n + g^* \|_{[t,b]} = \min_{g\in V_{n-1}(t)} \| h_n + g \|_{[t,b]}$$

<u>Theorem 2</u>. Let V_n be a Haar System on the closed interval $[a, b]$. Let $\{h_0, h_1, \dots , h_n\}$ be a basis for V_n . For fixed r and s satisfying $a \le r \le s \le t < b$, define

$$M_n(t) = \max_{g\in V_{n-1}} \left\{ \frac{\| h_n(x) + g\|_{[r,s]}}{\| h_n(x) + g\|_{[t,b]}} \right\} .$$

Then

a. the ratio of norms is maximized by the best approximation to h_n on $[t, b]$, namely $g^* \in V_{n-1}$,

b. $M_n(t) = \| S_n^{(t)} \|_{[r,s]}$,

c. $M_n(t)$ is a continuous increasing function of t . If, in addition, V_n has a unit element, then also

b′. $M_n(t) = | S_n^{(t)}(r) |$.

Proof. Suppose $h \in V_n$, $h \not\equiv S_n^{(t)}$, and

$$\frac{\| h \|_{[r,s]}}{\| h \|_{[t,b]}} > \frac{\| S_n^{(t)} \|_{[r,s]}}{\| S_n^{(t)} \|_{[t,b]}} . \tag{3}$$

There exists a constant $k > 0$ such that

$$\| kh \|_{[r,s]} = \| S_n^{(t)} \|_{[r,s]} , \tag{4}$$

so that,because of (3) ,

$$\| kh \|_{[t,b]} < \| S_n^{(t)} \|_{[t,b]} .$$

The alternating properties of $S_n^{(t)}$ (de la Vallée Poussin's Theorem) imply that both of the functions

$$S_n^{(t)}(x) \pm kh(x) \in V_n(t) \tag{5}$$

have n zeros in the open interval (t,b). But because of (4), one of these functions also has a zero in the interval $[r, s]$. Thus, one of the functions (5) violates the Haar condition on V_n . This contradiction establishes that either $h \equiv S_n^{(t)}$ or that (3) is an equality. Either way, $S_n^{(t)}$ maximizes the ratio of norms and part (a) is established. Part(b) follows easily from part (a). The continuity of $M_n(t)$ follows from part (b). To prove that $M_n(t)$ is increasing, suppose there exist $\beta > \alpha \geq s$ such that $M_n(\beta) < M_n(\alpha)$. Then for some constant $k \neq 0$,

$$\| kS_n^{(\beta)} \|_{[r,s]} = \| S_n^{(\alpha)} \|_{[r,s]}$$

and

$$\| kS_n^{(\beta)} \|_{[\beta,b]} > \| S_n^{(\alpha)} \|_{[\alpha, b]} .$$

Hence, both of the functions

$$kS_n^{(\beta)}(x) \pm S_n^{(\alpha)}(x)$$

have n zeros on the interval $[\beta, b]$. But one of them also has a zero in the interval $[r,s]$, contrary to the assumption that V_n is a Haar System. Thus, part (c) is established. If V_n has a unit element, then Lemma 1 guarantees that $S_n^{(t)}$ is monotone in $[r,s]$, so that (b') must be true. This completes the proof.

<u>Theorem 3</u>. Under the assumptions of Theorem 2, for all $\beta > \alpha \geq s$,

$$M_n(\alpha) = M_n(\beta) \tag{6}$$

if and only if

$$S_n^{(\alpha)}(x) = S_n^{(\beta)}(x) \tag{7}$$

for each $x \in [a,b]$.

Proof. If (7) holds, then (6) follows from Theorem 2. Suppose then that (6) holds, but that (7) does not. Let $x_1(t)$ denote the $(n+1)$st extremal point of $S_n^{(t)}$ counted from the right-hand endpoint to the left. If $h = x_1(\alpha) > \alpha$, then $S_n^{(\alpha)} \equiv S_n^{(t)}$ for all $t \in [\alpha, h]$, so that (7) holds if $\beta \leq h$. Hence without loss of generality, we may take $x_1(\alpha) = \alpha < \beta = x_1(\beta)$. Also, because $x_1(t)$ is continuous and $M_n(t)$ is monotone, it may be assumed without loss of generality that $x_1(\beta)$ is strictly less than the first zero of $S_n^{(\alpha)}$ and that $S_n^{(\beta)}(\beta)$ is of the same sign as $S_n^{(\alpha)}(\beta)$. Define the functions $T_n^{(t)}(x) \equiv S_n^{(t)}(x)\cdot c(t)$. (Thus, the coefficient of h_n is always unity.) Because $T_n^{(\beta)} \not\equiv T_n^{(\alpha)}$, the function $T(x) \equiv T_n^{(\alpha)}(x) - T_n^{(\beta)}(x)$ lies in V_{n-1} and cannot be identically zero. Because $[\beta, b]$ is a subset of $[\alpha, b]$ and $T(x) \not\equiv 0$,

$$\| T_n^{(\beta)} \|_{[\beta,\, b]} < \| T_n^{(\alpha)} \|_{[\alpha,\, b]}, \tag{8}$$

and because of (6),

$$\| T_n^{(\beta)} \|_{[r,s]} < \| T_n^{(\alpha)} \|_{[r,s]} . \tag{9}$$

Now if $|T_n^{(\beta)}(\beta)| \leq |T_n^{(\alpha)}(\beta)|$, then $T(x)$ has n zeros in $[\beta, b]$ because of de la Vallee Poussin's Theorem and (8), contradicting $T(x) \in V_{n-1}$ and $T(x) \not\equiv 0$. Therefore, $|T_n^{(\beta)}(\beta)| > |T_n^{(\alpha)}(\beta)|$ and $T(x)$ has only $n-1$ zeros in $[\beta, b]$. However, $T(x)$ has another zero in $[r, \beta)$ because of the Intermediate Value Theorem, (9), and the fact that $T_n^{(\beta)}(\beta)$ and $T_n^{(\alpha)}(\beta)$ are of the same sign. Thus, $T(x) \in V_{n-1}$ and $T(x) \not\equiv 0$ is a contradiction. This completes the proof.

Corollary. Under the assumptions of Theorem 2, if V_n has a unit element, then also

c'. $M_n(t)$ is strictly increasing in t .

Proof. $S_n^{(\alpha)} \equiv S_n^{(\beta)}$ cannot occur for $\alpha \neq \beta$ because α and β must be extremal points of $S_n^{(\alpha)}$ and $S_n^{(\beta)}$, respectively.

The next theorem is the weakened form of Theorem 1 that was sought earlier.

Theorem 4. Let V_n be any Haar System, and let $\{h_0, h_1, \ldots, h_n\}$ be a basis for the system. Let $z_1(t)$ be the smallest zero in the interval $[t, b]$ of $S_n^{(t)}$. Then $z_1(t)$ is a monotonically increasing function of t . Furthermore, if V_n has a unit element, then $z_1(t)$ is strictly monotone.

Proof. Consider first the case for V_n without a unit element. Suppose there exists $\beta > \alpha \geq a$ such that $z_1(\beta) < z_1(\alpha)$. Therefore, $S_n^{(\alpha)} \not\equiv S_n^{(\beta)}$ and by Theorem 3,

$$\frac{\| S_n^{(\beta)} \|_{\{a\}}}{\| S_n^{(\beta)} \|_{[\beta,b]}} > \frac{\| S_n^{(\alpha)} \|_{\{a\}}}{\| S_n^{(\alpha)} \|_{[\alpha,b]}} .$$

Thus, there is a constant $k \neq 0$ such that

$$\| kS_n^{(\beta)} \|_{\{a\}} = \| S_n^{(\alpha)} \|_{\{a\}} \tag{10}$$

and

$$\| kS_n^{(\beta)} \|_{[\beta,b]} < \| S_n^{(\alpha)} \|_{[\alpha,b]} .$$

Because $z_1(\beta) < z_1(\alpha)$, de la Vallée Poussin's Theorem guarantees that both the functions

$$S_n^{(\alpha)} \pm kS_n^{(\beta)} \in V_n \tag{11}$$

have n zeros on $[\beta, b]$, while (10) guarantees that one of them has a zero at $x = a$. Therefore, one of the functions (11) violates the Haar condition of V_n . In the case where V_n has a unit element, $\beta > \alpha$ guarantees by Lemma 2 that $S_n^{(\alpha)} \neq S_n^{(\beta)}$. The supposition that $z_1(\beta) \leq z_1(\alpha)$ leads to a contradiction in the same manner as before. This completes the proof.

Finally, we note that Theorem 2 fails if V_n does not satisfy the Haar condition.

Example. The linear space spanned by the functions $\{1, \sin x\}$ on the interval $[0, \pi]$ is not Haar. For $[r, s] = [0, \pi/2]$,

$$\frac{\| \sin x + a \|_{[0, \pi/2]}}{\| \sin x + a \|_{[\pi/2, \pi]}} = 1$$

for all constants a, including the constant of best approximation to $\sin x$ on the internal $[\pi/2, \pi]$, namely, $a = 1/2$. When the interval $[r, s]$ is replaced by the point $x = \pi/6$, then

$$\frac{\| \sin x \|_{\{\pi/6\}}}{\| \sin x \|_{[\pi/2,\pi]}} = \frac{1}{2} > 0 = \frac{\| \sin x - \frac{1}{2} \|_{\{\pi/6\}}}{\| \sin x - \frac{1}{2} \|_{[\pi/2,\pi]}} .$$

Thus, the results of Theorem 2 do not hold and $M_n(t)$ is not achieved by minimizing the denominator.

Acknowledgment. I would like to thank the referee for suggesting a helpful change in notation, and for pointing out the articles by Zielke and Pokrovskii.

REFERENCES

[1] V. L. Pokrovskii, "On a class of polynomials with extremal properties," AMS Translations, vol. 19, 1962, pp 199-219.

[2] C. L. Dolph, "A current distribution of broadside arrays which optimizes the relationship between beam width and side-lobe level," Proc. IRE Waves Elections, vol. 34, June 1946, pp 335-348.

[3] R. Streit, "Sufficient conditions for the existence of optimum beam patterns for unequally spaced linear arrays," IEEE Trans. Antennas Propag., vol. AP-23, no. 1, Jan 1975. pp 112-115.

[4] R. Streit, "Optimized symmetric discrete line arrays," IEEE Trans. Antennas Propag., (to appear November, 1975).

[5] G. Meinardus, Approximation of Functions: Theory and Numerical Methods, Springer-Verlag, New York, 1967.

[6] R. Zielke, "Alternation properties of Tchebyshev-Systems and the existence of adjoined functions," J. Approximation Theory, vol. 10, 1974, pp 172-184.

[7] G. Meinardus, "Über Tschebyscheffsche Approximationen," Arch. Rat. Mech. Anal., vol. 9, 1962, pp 329-351.

Author Index

Aronszajn, N., 46

Bauer, F. L., 87
Bennet, C., 146
Bjorck, A., 87
Blatter, J., 3
Brosowski, B., 1, 3
Brudnyi, Ju. A., 359
Burchard, H. G., 359
Businger, P., 87

Calderon, A. P., 146, 147
Carleson, L., 78, 85
Carlson, F., 251
Cavaretta, A. S., 257, 271, 272
Cavendish, J. C., 330
Cheney, E. W., 157
Chipman, D. M., 196
Cody, W. J., 111
Curry, H. B., 120

Davis, C., 3, 4
Davis, P. J., 44, 47
de Boor, C., 146
de Branges, L., 59
DeLeeuw, K., 82
DeVore, R., 227
Diaz, J. B., 213, 332
Dolph, C. L., 389
Douglas, J. Jr., 124
Dunham, C. B., 213, 332
Dupont, J., 124

Fisher, S. D., 106
Fitzgerald, C. H., 4

Golomb, M., 46
Golub, G. H., 87
Goncar, A. A., 238
Gordon, W. J., 308, 309, 328, 330
Gronwall, T. H., 162, 163, 165, 167

Hall, C. A., 328, 330
Handscomb, D. C., 44
Hanson, R. H., 87
Hobby, C. R., 5
Holland, A. S. B., 218, 332
Holmes, R., 106

Jacobs, S., 78
Jerome, J. W., 106, 107, 126
Johnson, R. S., 269

Kammerer, W. J., 4
Kaniel, S., 95

Karlin, S., 4
Kilgore, T. A., 1, 4, 5
Kober, H., 348
Krasnoselskii, M. A., 102
Kuelbs, J., 48, 57

Larkin, F. M., 45, 48, 49, 57
Laurent, P. J., 283, 285
Lawson, C. L., 87
Ling, W. H., 213, 214
Lorentz, G. G., 226, 227
Luke, Y. L., 183, 186, 191, 192
Luxemburg, W. A. J., 102

Mangasarian, O. L., 109
McLaughlin, H. W., 213, 332
Meinardus, G., 1, 111, 112, 242, 249, 393
Morris, P., 3
Myers, D., 230

Newman, D. J., 111, 112, 115, 116, 130, 228

Osborne, E. E., 87

Passow, E., 227, 228, 229, 231
Paszkowski, S., 221
Peetre, J., 359
Phillips, G. M., 332
Pokrovskii, V. L., 389

Raymon, L., 227, 228, 229, 231
Reddy, A. R., 112
Rice, J. R., 5
Rockafellar, R. T., 104, 108, 109
Rosenbaum, P. D., 66
Rosman, B. H., 66
Roulier, J. A., 112, 226, 227, 228
Rudin, W., 82
Rutickii, Y. B., 102

Sahney, B. N., 332
Sard, A., 44
Schoenberg, I. J., 120, 131, 133, 139, 257, 259, 265, 269
Schonhage, A., 111
Schumaker, L. L., 4, 109, 126
Sharma, A., 268
Shisha, O., 225
Singer, I., 3
Sippel, W., 64, 65, 66, 69, 71, 72
Studden, W. J., 4
Swartz, B. K., 327

Taylor, G. D., 64, 67, 112, 171
Tepper, D., 85
Titchmarsh, E. C., 182
Tzimbalario, J., 218

Vacendish, J. C., 328
Varga, R. S., 111, 112, 242, 249, 327
Videnskii, V. S., 4
Vlasov, L. P., 2

Wahlbin, L., 124
Walsh, J. L., 238
Weinberger, H. F., 46
Wilkinson, J. H., 87
Williamson, J., 48, 57
Wimp, J., 183, 186
Wolibner, W., 223
Wulbert, D. E., 3

Zeller, K. L., 226, 227
Ziegler, Z., 265
Zielke, R., 391
Zienkiewicz, O. C., 316

Subject Index

A

algorithm, 304, 365, 383
constrained, 298
metalgorithm, 361
Neville-Aitken, 200
Newton, 200
Remez, 278, 284, 286, 307
alternation, 291
theory, 64
analytic extension, 238, 339, 345
approximation
adaptive, 359
and interpolation, 221
best uniform, 76, 85
by moments, 7
comonotone, 220, 228
convex, 278
ε-interpolator, 278, 282
H_∞, 76
least squares, 95, 366, 373
monotone, 278, 294
Padé, 243, 246, 256
piecewise polynomial, 359, 380, 386
polygonal, 317
product, 24
rational, 40, 119, 253, 255
restricted range, 171, 278
simultaneous, 213, 217, 221, 332
spline, 145, 146, 386
Tchebycheff, 40, 218, 298
trigonometric, 9
weighted polynomial, 9, 23,

B

B-spline, 120, 132
Banach space, 104, 147
separable, 106
basis
spline, 152
Besov space, 154
best approximation, 5, 26, 279, 301, 333
polynomial, 25
operator, 25
simultaneous, 213, 217, 332
uniform, 76
unique, 300
blending functions, 308, 311
Boolean sum, 313, 327

C

cardinal spline, 257, 265
closure
  L_p,10
comonotone approximation, 220, 228
computation
  recursive, 199
computer
  process control, 325
condition
  Haar, 278, 302
  Hölder, 359
  Lipschitz, 29
convergence
  angular, 238
  geometric, 111, 114, 238, 249, 256
  over convergence, 243, 255
convex
  coconvex, 229
  functions, 10, 102
  hull, 205, 283
  non-convex norms, 377
cubic spline, 101, 312

D

decomposition, 9, 17
degree
  of convergence, 369
  of freedom, 315
  of monotone approximation, 225
differences, 311
  finite, 320, 360
  divided, 364
dual, 77, 123

E

eigenvalue, 52
Euler spline, 133, 139

F

finite element method, 308
fixed point theorems, 1
Fourier
  expansion, 86
  projection, 161
functionals
  atomic, 176
  convex, 110
  K-functionals, 9
  quadratic, 43
  tame, 49
functions
  additive weight, 218
  Bessel, 250
  blending, 308
  cardinal, 59
  Christoffel, 22
  convex, 10, 102
  copositive, 232
  hypergeometric, 250
  interpolation, 144
  Lebesgue, 4, 158
  Stekloff, 149
  Young, 109

G

gamma spline, 268, 270
geometric convergence, 111, 114, 238, 249, 256

H

Haar
  condition, 278, 302
  space, 8
  subspace, 64
  system, 387, 397
Hilbert
  space 6, 43, 87, 88, 336
  problem 339, 341, 345

I

inequality, 306
  Hölder, 77
  Minkowski, 16
integral equations, 358
interpolation, 303, 361
  and approximation, 221
  bicubic spline, 312
  cardinal, 145
  constrained, 235
  function, 144
  Hermite, 210, 298, 381
  Lagrange, 4, 167
  Newton-Hermite, 203
  oscillatory, 307
  piecewise monotone, 220
  piecewise polynomial, 325
  polynomial, 114
  projection, 155
  Riesz-Thorin, 16
  transfinite, 308, 311, 316, 322, 326

L

L-spline, 259, 314, 326
L_1
  approximation, 5, 337
  norm, 5, 384
  weighted approximation, 21
L_2, 384
  approximation, 337
  least squares problem, 98
L_p
  best approximation, 337
  closure, 10
least squares approximation, 95, 366, 373
Lebesgue function, 4, 158
linear operators
  contractive, 1

M

majorant, 9, 13
map, 318
  conformal, 160
  continuous, 25
  minimax, 157
  onto, 159
  proximity, 6
matrix
  pyramid, 288
modulus of continuity, 9, 12
moment
  approximation, 7
  problem, 7
  space, 7
monotone mapping operator, 7

N

network, 308
norm
  L_∞, 267
  non-convex, 377
  supremum, 9
  t-norm, 262
  Tchebicheff, 111

O

operator, 201, 318
  approximation, 359
  idempotent, 157
  monotone, 110
  product approximation, 25, 40
  self adjoint, 47
  translation, 10
optimal, 54
  control, 109
Orlicz space, 101, 147

P

Padé approximation, 243, 246, 256
perfect splines, 132, 257, 268, 270
piecewise approximation,
  polynomial, 359, 380, 386
  polygonal, 317
polynomial
  interpolating, 114
  orthonormal, 21
  piecewise, 359, 380, 386
  Tchebicheff, 155, 180, 183, 221
  weighted approximation, 23
polytopes, 6
problem
  eigenvalue, 52
  Hilbert, 339
  initial value, 7
  least squares, 98
  moment, 7
  quadrature, 43
  restricted range, 64
projection, 151, 323
  discrete Fourier, 161
  Fourier, 161
  idempotent, 308
  Lagrange, 155, 167
  metric, 6
  orthogonal, 47, 328

Q

quadrature, 43, 340, 343, 352

R

rational approximation, 40, 119, 253, 255
refinement, 98
Remez algorithm, 278, 284, 286, 307
RRAS, 171, 174
  non-RRAS, 175

S

SAIN approximation, 179
separable
  Banach space, 106
  Hilbert space, 88
Sets
  A-variant, 1
  alternation, 31
  compact, 1
  convex, 1
  sun, 2
  Tchebicheff, 2
simplex, 3, 142
simultaneous approximation, 213, 217, 221, 332
Sobolev space, 7, 104, 148, 319
Space
  Besov, 154
  Banach, 104, 147
  Hilbert 6, 43, 87, 88, 336
  Orlicz, 101, 147
  Sobolev, 7, 104, 148, 319
spline, 7, 101, 268, 326, 386
  approximation, 146
  B-spline, 120
  basis, 152
  bicubic, 325
  cardinal, 257, 265
  cubic, 101, 312
  Euler, 133, 139
  extremal, 265
  gamma, 268, 270
  interpolation, 105, 312
  L-spline, 259, 314, 326
  local, 145
  monospline, 257, 276
  perfect, 257
  perfect B-spline, 132
  t-perfect, 258, 260, 268, 270, 272

1-perfect, 274
support, 128

T

Tchebicheff
approximation, 40, 337, 370, 375
nodes, 168
norm, 111, 156
polynomial, 155, 180, 183, 221, 387, 389
set, 2, 8, 261
theorem
alternation, 31
fixed point, 1
Hahn-Banach, 77
strong uniqueness, 301
unicity, 33
transfinite interpolation, 308

V

variance
minimum, 45
varisolvent families, 74

W

weighted approximation, 9, 23

A 6
B 7
C 8
D 9
E 0
F 1
G 2
H 3
I 4
J 5